AF597851

METHODS IN MOLECULAR BIOLOGY

For further volumes:
http://www.springer.com/series/7651

The BAM Complex

Methods and Protocols

Edited by

Susan K. Buchanan

Laboratory of Molecular Biology, National Institute of Diabetes and Digestive and Kidney Diseases, National Institutes of Health, Bethesda, MD, USA

Nicholas Noinaj

Department of Biological Sciences, Markey Center for Structural Biology, Purdue University, West Lafayette, IN, USA

Editors
Susan K. Buchanan
Laboratory of Molecular Biology
National Institute of Diabetes
and Digestive and Kidney Diseases
National Institutes of Health
Bethesda, MD, USA

Nicholas Noinaj
Department of Biological Sciences
Markey Center for Structural Biology
Purdue University, West Lafayette
IN, USA

ISSN 1064-3745 ISSN 1940-6029 (electronic)
Methods in Molecular Biology
ISBN 978-1-4939-2870-5 ISBN 978-1-4939-2871-2 (eBook)
DOI 10.1007/978-1-4939-2871-2

Library of Congress Control Number: 2015949089

Springer New York Heidelberg Dordrecht London

Printed on acid-free paper

Humana Press is a brand of Springer
Springer Science+Business Media LLC New York is part of Springer Science+Business Media (www.springer.com)

Preface

Cells are encapsulated by a single lipid bilayer called a membrane that forms the boundary separating the inside of the cell from the outside. The membrane serves many essential functions for the cell including nutrient import, signaling, motility, adhesion, endocytosis, and replication. These functions are accomplished by a large family of proteins called membrane proteins that are either partially or fully integrated into the membrane. Fully integrated membrane proteins are embedded into the membrane by hydrophobic domains that contain either an α-helical fold or a β-barrel fold. While α-helical membrane proteins can be found in nearly all known membranes in nature, β-barrel membrane proteins can only be found within the outermost membranes of mitochondria, chloroplasts, and Gram-negative bacteria, all of which are unique in that they contain two concentric membranes (inner and outer) and are related by their endosymbiotic lineage. The mechanism for how these β-barrel membrane proteins are folded and inserted into the outer membrane remains unknown. However, within the past 10 years, significant advancements have been made to understand this process, particularly in Gram-negative bacteria where genetic analyses, mutagenesis studies, biochemical assays, in vitro assays, and structural biology techniques have all contributed.

Early work identified a multicomponent complex that we now refer to as the β-barrel assembly machinery (BAM) complex, which is required in Gram-negative bacteria to integrate newly synthesized β-barrel membrane proteins into the outer membrane. From the initial identification of the BAM complex and its individual components to the recent structural characterization of all individual proteins, much has been learned about the role the BAM complex plays in the biogenesis of β-barrel membrane proteins. In this volume of the *Methods in Molecular Biology* series, we have assembled a collection of experimental protocols for common techniques and strategies used to study the biogenesis of β-barrel membrane proteins in Gram-negative bacteria. This volume contains step-by-step methods based on the protocols that were used during the research efforts performed in determining what is currently known about the regulation and function of the BAM complex, the roles played by each of the individual components, the expression and purification of the components, crystallization and structure determination of the components, and how the individual Bam components may assemble into a functional complex. Given that several studies have reported the folding of β-barrel membrane proteins from Gram-negative bacteria in mitochondria and vice versa, one chapter focuses on methods used to study the evolutionarily conserved system that exists in mitochondria.

The methods and protocols here will appeal to a wide variety of scientists in academia, government, and industry including microbiologists, biochemists, bacteriologists, structural biologists, and those looking to target the BAM complex for therapeutic discovery and development. It is our hope that this volume will serve as an invaluable reference for those interested in studying the BAM complex and how it functions at the outer membrane, as well as for those who may want to apply the protocols communicated here to other interesting biological systems.

Last but certainly not least, this volume would not have been possible without the contributions from the authors, to whom we are truly indebted.

Bethesda, MD, USA *Susan K. Buchanan*
West Lafayette, IN, USA *Nicholas Noinaj*

Contents

Contributors

ARVIND ANAND • *Department of Medicine, University of Connecticut Health, Farmington, CT, USA*
SURAAJ AULAKH • *Department of Molecular Biology and Biochemistry, Simon Fraser University, Burnaby, BC, Canada*
INGO B. AUTENRIETH • *Interfaculty Institute of Microbiology and Infection Medicine, University of Tübingen, Tübingen, Germany*
MATTHEW J. BELOUSOFF • *Department of Microbiology, Monash University, Melbourne, VIC, Australia*
HARRIS D. BERNSTEIN • *Genetics and Biochemistry Branch, National Institute of Diabetes and Digestive and Kidney Diseases, National Institutes of Health, Bethesda, MD, USA*
GUSTAVO BODELÓN • *Department of Microbial Biotechnology, Centro Nacional de Biotecnología, Consejo Superior de Investigaciones Científicas, Madrid, Spain*
MARTINE P. BOS • *Department of Molecular Microbiology and Institute of Biomembranes, Utrecht University, Utrecht, The Netherlands; Department of Medical Microbiology and Infection Control, VU University Medical Center, Amsterdam, The Netherlands*
RIA TOMMASSEN-VAN BOXTEL • *Department of Molecular Microbiology and Institute of Biomembranes, Utrecht University, Utrecht, The Netherlands*
SUSAN K. BUCHANAN • *Laboratory of Molecular Biology, National Institute of Diabetes and Digestive and Kidney Diseases, National Institutes of Health, Bethesda, MD, USA*
PAMELA ARDEN DOERNER • *Department of Chemistry and Biochemistry, University of Colorado, Boulder, Boulder, CO, USA*
ENGUO FAN • *Institute of Biochemistry and Molecular Biology, ZBMZ, University of Freiburg, Freiburg, Germany*
LUIS ÁNGEL FERNÁNDEZ • *Department of Microbial Biotechnology, Centro Nacional de Biotecnología, Consejo Superior de Investigaciones Científicas, Madrid, Spain*
KAREN G. FLEMING • *T.C. Jenkins Department of Biophysics, Johns Hopkins University, Baltimore, MD, USA*
DENNIS GESSMANN • *T.C. Jenkins Department of Biophysics, Johns Hopkins University, Baltimore, MD, USA*
FABIAN GRUSS • *Biozentrum, University of Basel, Basel, Switzerland*
CHRISTOPHER S. HAYES • *Department of Molecular, Cellular and Developmental Biology, University of California, Santa Barbara, CA, USA; Biomolecular Science and Engineering Program,University of California, Santa Barbara, CA, USA*
SEBASTIAN HILLER • *Biozentrum, University of Basel, Basel, Switzerland*
YIHUA HUANG • *National Laboratory of Biomacromolecules, National Center of Protein Science-Beijing, Institute of Biophysics, Chinese Academy of Sciences, Beijing, China*
MARK JEEVES • *School of Cancer Sciences, University of Birmingham, Edgbaston, Birmingham, UK*
KELLY H. KIM • *Department of Molecular Biology and Biochemistry, Simon Fraser University, Burnaby, BC, Canada*

TIMOTHY J. KNOWLES • *School of Cancer Sciences, University of Birmingham, Edgbaston, Birmingham, UK*
ADAM J. KUSZAK • *Laboratory of Molecular Biology, National Institute of Diabetes and Digestive and Kidney Diseases, National Institutes of Health, Bethesda, MD, USA*
JACK C. LEO • *Department of Biosciences, University of Oslo, Oslo, Norway*
DENISSE L. LEYTON • *Research School of Biology, Australian National University, Canberra, ACT, Australia*
DIRK LINKE • *Department of Biosciences, University of Oslo, Oslo, Norway; Max Planck Institute for Developmental Biology, Tübingen, Germany*
TREVOR LITHGOW • *Department of Microbiology, Monash University, Melbourne, VIC, Australia*
DAVID A. LOW • *Department of Molecular, Cellular and Developmental Biology, University of California, Santa Barbara, CA, USA; Biomolecular Science and Engineering Program, University of California, Santa Barbara, CA, USA*
AMIT LUTHRA • *Department of Medicine, University of Connecticut Health, Farmington, CT, USA*
TIMM MAIER • *Biozentrum, University of Basel, Basel, Switzerland*
ELVIRA MARÍN • *Department of Microbial Biotechnology, Centro Nacional de Biotecnología, Consejo Superior de Investigaciones Científicas, Madrid, Spain*
MATTHIAS MÜLLER • *Institute of Biochemistry and Molecular Biology, ZBMZ, University of Freiburg, Freiburg, Germany*
DONGCHUN NI • *National Laboratory of Biomacromolecules, National Center of Protein Science-Beijing, Institute of Biophysics, Chinese Academy of Sciences, Beijing, China*
NICHOLAS NOINAJ • *Department of Biological Sciences, Markey Center for Structural Biology, Purdue University, West Lafayette, IN, USA*
DERRICK NORELL • *Institute of Biochemistry and Molecular Biology, ZBMZ, University of Freiburg, Freiburg, Germany; Faculty of Biology, University of Freiburg, Freiburg, Germany*
PHILIPP OBERHETTINGER • *Interfaculty Institute of Microbiology and Infection Medicine, University of Tübingen, Tübingen, Germany*
MARK PAETZEL • *Department of Molecular Biology and Biochemistry, Simon Fraser University, Burnaby, BC, Canada*
NAGARAJAN PARAMASIVAM • *Department 1, Max Planck Institute for Developmental Biology, Tübingen, Germany; Computational Oncology, Theoretical Bioinformatics, DKFZ, Heidelberg, Germany*
ASHLEE M. PLUMMER • *T.C. Jenkins Department of Biophysics, Johns Hopkins University, Baltimore, MD, USA*
JUSTIN D. RADOLF • *Department of Medicine, University of Connecticut Health, Farmington, CT, USA; Department of Pediatrics, University of Connecticut Health, Farmington, CT, USA; Department of Genetics and Genomics Sciences, University of Connecticut Health, Farmington, CT, USA; Department of Immunology, University of Connecticut Health, Farmington, CT, USA; Department of Molecular Biology and Biophysics, University of Connecticut Health, Farmington, CT, USA*
DORON RAPAPORT • *Interfaculty Institute of Biochemistry, University of Tübingen, Tübingen, Germany*
DANTE P. RICCI • *Department of Developmental Biology, Stanford University School of Medicine, Stanford, CA, USA*

Giselle Roman-Hernandez • *Genetics and Biochemistry Branch, National Institute of Diabetes and Digestive and Kidney Diseases, National Institutes of Health, Bethesda, MD, USA*
Zachary C. Ruhe • *Department of Molecular, Cellular and Developmental Biology, University of California, Santa Barbara, CA, USA*
Marcelo Carlos Sousa • *Department of Chemistry and Biochemistry, University of Colorado, Boulder, CO, USA*
Pooja Sridhar • *School of Cancer Sciences, University of Birmingham, Edgbaston, Birmingham, UK*
Jan Tommassen • *Department of Molecular Microbiology and Institute of Biomembranes, Utrecht University, Utrecht, The Netherlands*
Thomas Ulrich • *Interfaculty Institute of Biochemistry, University of Tübingen, Tübingen, Germany*
Chaille T. Webb • *Department of Microbiology, Monash University, Melbourne, VIC, Australia*
Kornelius Zeth • *Laboratory of Biomolecular Research, Paul Scherrer Institute, Villigen, Switzerland; Ikerbasque, Basque Foundation of Science, Bilbao, Spain*

Chapter 1

The β-Barrel Assembly Machinery Complex

Denisse L. Leyton, Matthew J. Belousoff, and Trevor Lithgow

Abstract

The outer membranes of gram-negative bacteria contain integral membrane proteins, most of which are of β-barrel structure, and critical for bacterial survival. These β-barrel proteins rely on the *β*-barrel *a*ssembly *m*achinery (BAM) complex for their integration into the outer membrane as folded species. The central and essential subunit of the BAM complex, BamA, is a β-barrel protein conserved in all gram-negative bacteria and also found in eukaryotic organelles derived from bacterial endosymbionts. In *Escherichia coli*, BamA docks with four peripheral lipoproteins, BamB, BamC, BamD and BamE, partner subunits that add to the function of the BAM complex in outer membrane protein biogenesis. By way of introduction to this volume, we provide an overview of the work that has illuminated the mechanism by which the BAM complex drives β-barrel assembly. The protocols and methodologies associated with these studies as well as the challenges encountered and their elegant solutions are discussed in subsequent chapters.

Key words Outer membrane, β-barrel, Outer membrane β-barrel proteins (OMPs), Periplasmic chaperones, β-barrel assembly machinery (BAM), Omp85, BamA

1 The Cell Envelope

The defining feature of gram-negative bacteria is the double membrane cell envelope comprising lipid bilayers that enclose two aqueous compartments called the cytoplasm and the periplasm (Fig. 1a). Both of these membranes are enriched in phosphatidylethanolamine (PE) and contain phosphatidylglycerol (PG) as well as a small amount of cardiolipin (CL) [1] (Fig. 1b), yet the two are distinguished: whereas the inner membrane is a phospholipid bilayer, the outer membrane is asymmetrical and is composed of an inner leaflet of phospholipid and an outer leaflet of phospholipid and lipopolysaccharide (LPS) (Fig. 1c). The elaborate chemical moieties found in the LPS component of the outer membrane decorate the outer surface of the cell and reduce the fluidity of the outer membrane, a property that is largely responsible for its low permeability [2] and one that protects the cell from the surrounding environment.

Susan K. Buchanan and Nicholas Noinaj (eds.), *The BAM Complex: Methods and Protocols*, Methods in Molecular Biology, vol. 1329, DOI 10.1007/978-1-4939-2871-2_1, © Springer Science+Business Media New York 2015

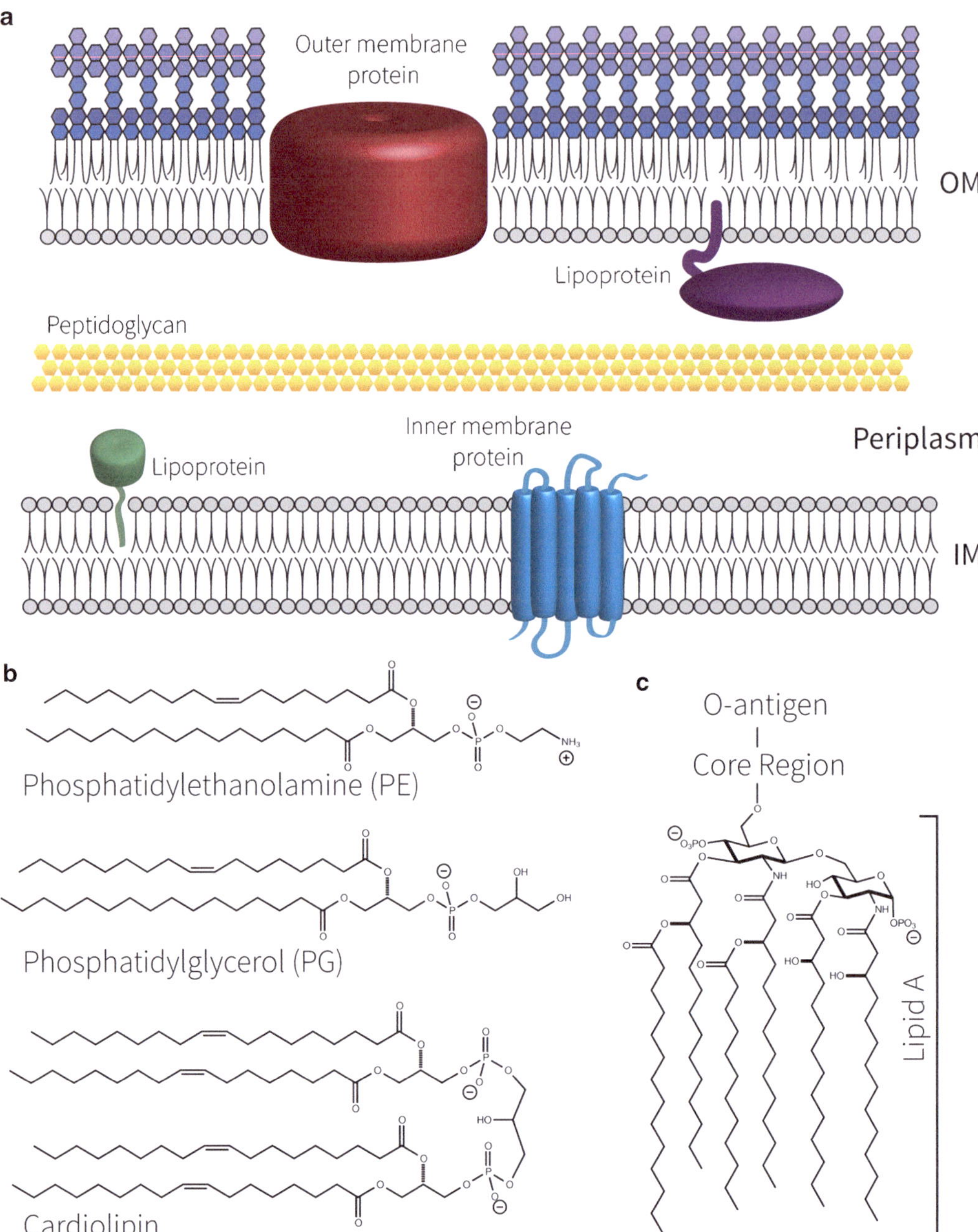

Fig. 1 The double membrane cell envelope of gram-negative bacteria. (**a**) The outer membrane (OM) and inner membrane (IM) are separated by an aqueous periplasm containing a thin peptidoglycan layer. Integral proteins of the IM are primarily α-helical while integral proteins of the OM predominantly form a β-barrel structure. Peripheral lipoproteins are found associated with both membranes. Structures of (**b**) phosphatidylethanolamine (PE), phosphatidylglycerol (PG), cardiolipin (CL), and (**c**) lipopolysaccharide (LPS)

2 Outer Membrane β-Barrel Proteins (OMPs)

Proteins associated with outer membranes are of two general types and include peripheral lipoproteins that are securely anchored to the inner phospholipid leaflet of the outer membrane through covalently affixed lipid moieties [3] and integral outer membrane β-barrel proteins (OMPs) which span the lipid bilayer. Although unified by many common structural features (detailed below), OMPs carry out a variety of functions that underpin bacterial survival and competitive fitness in a given environment. For example, OMPs allow the uptake of nutrients and ions, the efflux of antibiotics and metabolic waste, adhesion to host cells, the translocation of effector molecules such as bacterial toxins and hydrolytic enzymes into the extracellular milieu, outer membrane biogenesis as well as maintenance of its integrity and barrier function [4].

3 The β-Barrel Scaffold

While there are some important exceptions [5, 6], integral proteins of the outer membrane predominantly form a cylindrical structure known as a β-barrel, which is composed of β-strands that associate laterally in an antiparallel fashion to form an extensive hydrogen bond network. This intricate network of hydrogen bonds within the outer membrane confers substantial stability on the OMPs [7]. For example, many β-barrel proteins, once folded, are largely resistant to denaturation in SDS with folded and unfolded forms separable according to temperature conditions in SDS-PAGE [8]. This useful property is known as "heat modifiability" and can be used in in vitro studies to follow the formation of native protein.

Protein crystallography has shown how β-barrel proteins use between 8 and 26 antiparallel β-strands, typically connected by tight turns on the periplasmic side of the membrane and longer surface-exposed loops that are often highly mobile [9], to span the bacterial outer membrane (Fig. 2). These β-strands are comprised of alternating hydrophobic (extending outwards, into the membrane) and hydrophilic (extending into the lumen of the structure) residues [7]. In these structures, the β-strands have a tilt between 20 and 45° relative to the membrane normal and are stabilized in this position through the interactions of aromatic residues with the membrane interfacial region, thereby defining the positioning of the barrel within the plane of the outer membrane [10]. Notably, this common β-barrel scaffold can form multimers and can also contain periplasmic or extracellular domains that vary in structure and complexity, and that sometimes help to further stabilize the β-barrel structure [11, 12].

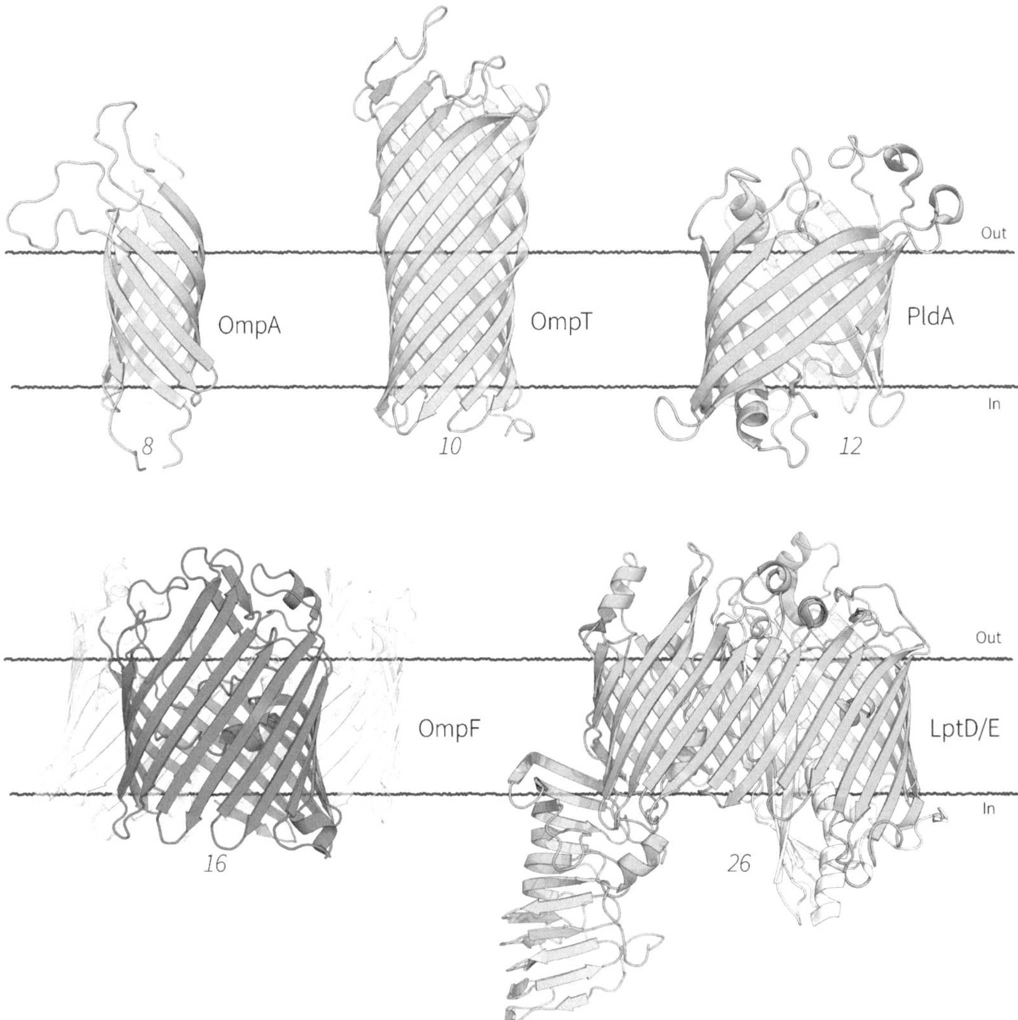

Fig. 2 Representative structures of different outer membrane β-barrel proteins (OMPs), illustrating the common β-barrel scaffold. Numbers indicate the number of β-strands. OmpA (PDB 1G90), OmpT (PDB 1I78), PldA (PDB 1QD5), OmpF (PDB 2ZFG), and LptD/E (PDB 4Q35)

4 OMP Biogenesis: Jumping a Topological Hurdle

OMPs have intrinsic features, including targeting sequences and motifs that are recognized by cellular machinery to facilitate their navigation through the cell envelope [13–17]. OMPs are synthesized in the cytoplasm with N-terminal signal sequences to target them to the inner membrane SecYEG translocon for energy-driven export into the periplasm. Molecular chaperones bind nascent OMPs while they are still in complex with, or as they leave, the SecYEG translocon [17]. Chaperone-bound OMPs then transit across the periplasm to the inner surface of the outer membrane, where they are recognized by the β-barrel *a*ssembly *m*achinery (BAM) complex. This recognition is mediated by binding of BamA, the central and essential subunit of the BAM complex, to a signature sequence termed the "β-signal" that is often, but not always, found in the C-terminal β-strand of bacterial OMPs [13, 15]. A highly conserved C-terminal phenylalanine residue within the β-signal is critical for the assembly of OMPs in vivo [18–20] and was recently shown to be required for optimal BamA-catalyzed folding of OMPs into liposomes comprised of naturally occurring *E. coli* lipids [21]. Intriguingly, while there are small differences between the β-signal in diverse substrate OMPs from diverse gram-negative bacteria [22], high species specificity in the functioning of BamA has been observed [15, 23], which is suggestive of a species-specific OMP recognition signal.

5 Periplasmic Chaperones Matter to OMP Biogenesis

Newly synthesized OMPs penetrate the periplasm in an unfolded conformation and are therefore in danger of off-pathway aggregation in this aqueous compartment. Periplasmic chaperones are thought to maintain nascent OMPs in a translocation-competent conformation as they transit through the periplasm to reach the outer membrane. Two parallel, partially redundant pathways for chaperone activity in the periplasm have been proposed where SurA functions in one pathway, and Skp and DegP in the other [24]. This concept is based on genetic data, which demonstrated that the loss of either pathway is tolerated because the other pathway is still functional, while the loss of both pathways simultaneously results in a lethal phenotype because both pathways are compromised. It has been suggested that SurA is the primary chaperone responsible for the biogenesis of most OMPs, with Skp and DegP functioning to rescue OMPs that have deviated from the SurA pathway [25].

The individual deletion or depletion of *surA* and *skp* results in a diminution of OMPs in the outer membrane and in the accumulation of their misfolded forms in the periplasm [25–27], reflecting the inability to correctly target OMPs to the outer membrane in their absence. The available evidence suggests that SurA and Skp can both interact with OMPs as they leave the SecYEG translocon [26, 28, 29] to deliver them to the BAM complex in the outer membrane. Certainly, interactions between SurA and BamA [25, 30], and Skp and BamA [31] have been observed. In addition, SurA has been shown to enhance BAM complex-dependent folding of OmpT in vitro [32, 33]. These biochemical studies have been substantiated by structural data showing SurA bound to a peptide mimicking the OMP β-signal [34] and a Skp-OmpA complex where the OmpA β-barrel is found in an unfolded state within the

Skp cavity, while the soluble part of OmpA that localizes in the periplasm remains outside of the cavity in a folded conformation [35]. However, detailed studies using OmpX showed that while this β-barrel can bind to Skp or to SurA, in structurally similar states of unfoldedness, the interaction with Skp is not determined by the β-signal region, but by the chemically denatured state of OmpX [36]. While the exact role of SurA and Skp are still under discussion, the chaperone function of DegP is even less clearly defined and controversial with some studies suggesting that it primarily functions as a protease to restore homeostasis by sequestering and then degrading misfolded OMPs in periplasm [37, 38].

6 The BAM Complex: Assisted Folding and Membrane Insertion of OMPs

The assembly of OMPs into bacterial outer membranes requires the BAM complex, a protein complex functionally conserved in all bacteria with outer membranes and in organelles of eukaryotes derived from endosymbiotic bacteria [39]. The core subunit of the BAM complex is BamA, an integral β-barrel protein originally identified in *Neisseria meningitidis* that is essential for β-barrel assembly [40, 41].

6.1 The Omp85 Superfamily: Structural and Functional Insights

Proteins of the Omp85/FhaC superfamily are defined by a "D15 domain" also known as a "bacterial surface antigen domain," which forms a β-barrel in the outer membrane [39]. It has recently become apparent that this superfamily is comprised of ten subfamilies, which are defined by distinct domain architectures linked to conserved Omp85 β-barrel domains [42]. While diverse domain architectures abound, one seemingly common feature is an N-terminal periplasmic region comprised of one or more *p*olypeptide *t*ransport *a*ssociated (POTRA) domains that serve as interfaces for protein–protein interactions [39, 42]. Structural and functional information is available for the BamA, FhaC, and the TamA subfamilies, all of which use POTRA domains to function in protein transport (Fig. 3) [43–45].

6.1.1 Insight from Crystal Structures

FhaC serves as a dedicated pore for the translocation of its soluble partner protein "filamentous hemagglutinin" (FHA) across the outer membrane [46]. However, in the FhaC crystal structure this pore is blocked by a large extracellular loop L6, which is folded back into a hairpin that spans the barrel lumen, with its tip extending into the periplasm [43]. As a result, opening of the pore is required for translocation through FhaC and, indeed, FHA-mediated conformational changes of L6 during secretion have been observed [43, 47]. A similar substrate-induced conformational change of L6 in BamA has been proposed to activate the BAM complex for a round of β-barrel assembly [48]. Interestingly, the tip of L6 contains a motif, VRGY/F, that is highly conserved

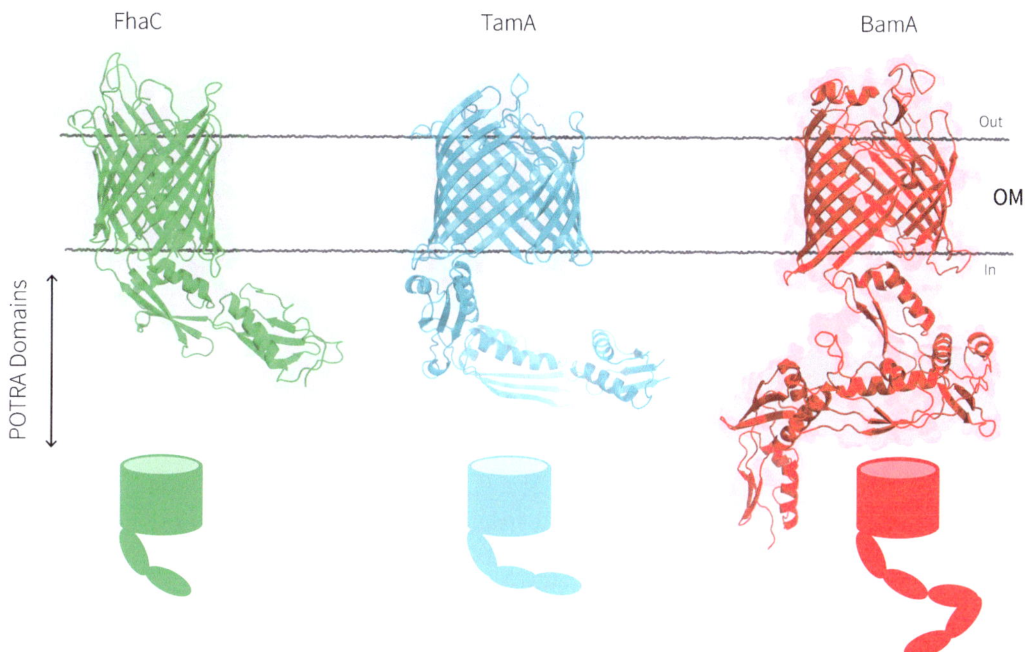

Fig. 3 Crystal structures of FhaC (PDB 2QDZ), TamA (PDB 4C00) and BamA (PDB 4K3B), illustrating key structural features of the Omp85 superfamily

among Omp85 family members [49]. While mutations in this motif have no effect on substrate recognition, they almost completely abolish FHA translocation and affect FhaC stability [49]. Mutation of residues within the VRGY/F motif in L6 of BamA produces a conditional lethal phenotype, concomitant with drastically reduced BamA levels due to its aberrant folding [50]. However, in both the TamA and BamA crystal structures, L6 only partially inserts into the barrel pore where residues within the VRGY/F motif interact with the interior of the barrel [44, 45, 51]. This interaction is important for proper folding of BamA and its stabilization [45, 51]. Notably, disulfide cross-linking of cysteine residues introduced into L6 (the residue immediately before the conserved VRGY/F motif) and into the inner wall along β-strand 12 were shown to support bacterial growth [52]. These results demonstrate that L6 remains in a mostly fixed conformation and that BamA function does not require a large conformational change of L6. The conformational differences in L6 between FhaC and BamA may be reflective of their related, yet distinct functions where BamA inserts proteins into, rather than across, the outer membrane.

6.1.2 What Is This Lateral Gate and Exit Pore?

Recent structural data and molecular dynamics simulations showed that the hydrophobic belt of BamA is greatly reduced in width along the last β-strand compared to the opposite side of the barrel.

This exciting observation results in a marked decrease in lipid order and membrane thickness [45]. Similarly, β-strands 1–4 and 16 of the *E. coli* BamA β-barrel are markedly shorter than the other β-strands of the barrel [51]: all of these observations are building a picture of a membrane protein that sits uncomfortably in the outer membrane. Perturbation of the outer membrane was predicted through lateral opening of the barrel at the interface between the first and last β-strands, which are only weakly associated in all BamA structures. By engineering paired cysteine mutants between β-strands 1 and 16, it was shown that prevention of lateral opening at this interface renders BamA non-functional [52]. These studies also pointed toward an exit pore identified just above the lateral opening site, which may allow exit of soluble loops or domains of substrate OMPs during assembly [51, 52]. The authors proposed a mechanism whereby lateral opening of BamA facilitates binding of the C-terminal β-strand of a substrate OMP to the first β-strand of BamA to create a hybrid barrel where the BamA β-strands serve as a structural template for nucleation of OMP folding/insertion via β-augmentation, while soluble loops or domains are transported through the substrate exit pore at the barrel surface [52]. A final fission event would disassociate the hybrid barrel, releasing the folded OMP laterally into the outer membrane and allowing BamA to return to its original state. Weak contacts between β-strands 1 and 16 were also evident in the crystal structure of TamA [44], the core component of the *t*ranslocation and *a*ssembly *m*odule (TAM) that is also comprised of the inner membrane protein, TamB [53]. The hybrid model fits well with the function of the TAM, which mediates the assembly of a subset of OMPs into the bacterial outer membrane [53].

Our knowledge of the bacterial outer membrane as a physical environment that kinetically retards the intrinsic folding ability of OMPs is slowly, but surely, advancing. A lipid-imposed activation barrier to OMP folding appears to function kinetically to partition OMPs away from the inner membrane and into complexes with periplasmic chaperones that will escort them to the correct biological membrane [21]. Importantly, this study showed that BamA enabled OMPs to overcome the kinetic barrier (purportedly) by creating local bilayer defects that result in a markedly thinner outer membrane with decreased lipid packing and lower lateral pressure. Thus, while BamA specifically recognizes substrate OMPs via their β-signal [15, 21–23], the extent to which BamA directs OMP folding/insertion as suggested by the hybrid model versus accentuating local membrane defects to accelerate the intrinsic folding of bacterial OMPs remains to be determined. What we do know is that these proposed perturbations are independent of POTRA domains 1–4, indicating that the mechanistic abilities of BamA are located in the membrane [21].

6.1.3 POTRA Domains in the Periplasm

NMR studies and crystallization of the isolated POTRA domains of BamA, TamA and FhaC revealed that, despite their low sequence identity, each domain consists of approximately 75 residues that are organized as a three-stranded β-sheet overlaid by two antiparallel α-helices [44, 54–57]. The POTRA domain is fundamental: in the case of BamA, the interaction interfaces between POTRA 1–2 and POTRA 3–5 maintain them as two rigid bodies bridged by a flexible linker. Consistent with this, SAXS analyses on the isolated POTRA domains of BamA revealed conformational flexibility between POTRA 2 and 3 that gives rise to multiple, yet highly stable conformations that may have functional implications during the OMP assembly pathway [55, 56, 57]. Certainly, the full-length crystal structures of BamA captured the POTRA domains in two distinct conformations relative to the β-barrel with the "open" and "closed" conformations hypothesized to allow and occlude access to the lumen of the barrel from the periplasm, respectively [45]. The authors suggested that such conformations could reflect a gating system for regulated entry of substrate OMPs into the β-barrel lumen.

A comprehensive comparative analysis of the Omp85 superfamily demonstrated that POTRA sequences from the Omp85 protein subfamilies display striking specialization with sequence identity that is sometimes so dissimilar that they conform to distinct Pfam profiles [42]. Several functions have been ascribed to the POTRA domains of BamA. Initial substrate binding, as well as subsequent folding, by β-augmentation might be mediated by the BamA POTRA domains [54, 55, 58]. In addition, the POTRA domains of BamA have been shown to serve intrinsically in the assembly of BamA into the outer membrane [54, 59] and, as detailed below, are essential for the docking of lipoprotein partner proteins to form a functional BAM complex.

6.2 Protein Partner Subunits

Through evolution BamA has acquired distinct combinations of membrane protein partner subunits, giving rise to various forms of the BAM complex in distinct bacterial lineages [39, 60–62]. In *E. coli*, BamA associates tightly with BamB, BamC, BamD and BamE, peripheral lipoprotein partner subunits anchored to the inner leaflet of the outer membrane [63–66], where the POTRA domains of BamA serve as the docking point for one copy of each partner protein [32]. Investigations into the architecture of the BAM complex in the outer membrane revealed that it is built of BamA:B and BamC:D:E modules that can be disassociated using non-ionic detergents [67], and functionally reconstituted into liposomes from purified BamA:B and BamC:D:E sub-complexes [32]. As detailed below, it now appears that dynamic interplay of these modules is important for connectivity and function of the BAM complex.

6.2.1 The BamA:B Module

Mutation of POTRA 5 of BamA separates the BAM complex into BamA:B and BamC:D:E modules [68] where BamA and BamB interact directly in a manner that is independent of the other protein partner subunits [33, 54]. This connection can be disrupted through removal of any POTRA domain except POTRA 1, and through point mutations or small insertions into POTRA 3 [54], where an exposed β-strand of this domain may interact with BamB via β-augmentation [69, 70]. POTRA domains 3-5 are essential for cell viability in *E. coli* [54] and while BamB is not essential, its loss results in a severe defect in OMP assembly [71] and in a severely compromised permeability barrier [72]. Together these data suggest that a major, but not exclusive role for POTRA domains 3-4 is to organize the BamA:B module.

BamB adopts an eight-bladed β-propeller fold that forms a ring-like structure (Fig. 4) [69, 70, 73, 74], which provides a continuous surface that has been proposed to mediate the simultaneous binding of BamA via conserved regions and diverse substrate OMPs in a sequence-independent manner [62, 73]. Support for the participation of BamB in the direct recruitment of OMP precursors is provided by several in vivo and in vitro studies. For example, in vivo photo-cross-linking experiments were able to demonstrate a direct interaction between a substrate OMP jammed in the BAM complex and BamB [75], and a BamB-dependent increase of OmpT maturation has been observed in vitro [32].

Further observation of the effects of the individual components of the BAM complex on β-barrel assembly in vitro showed that a membrane containing only BamB can catalyze BamA, but not

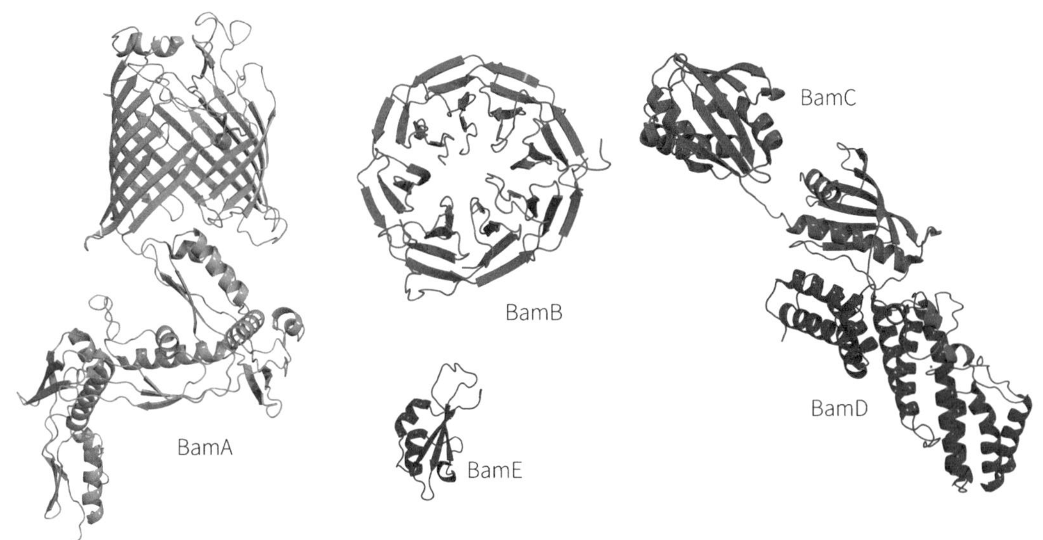

Fig. 4 Crystal structures of BamA (PDB 4K3B), BamB (PDB 3Q7M), BamC bound to BamD (PDB 3TGO), and BamE (PDB 2KXX), illustrating the components of the *E. coli* BAM complex

OmpA assembly, but this substrate was assembled if the BAM complex was preassembled in the membrane [76]. Given that BamA did not assemble efficiently into empty liposomes under the conditions tested [76], these data show that BamB is required for proper BamA assembly in vitro (this has now also been shown in vivo [77]) and suggests that if BamB is a substrate-capturing component, that this function may relate to the assembly of certain OMP substrates or alternatively, is mediated in combination with other components of the BAM complex.

6.2.2 The BamC:D:E Module

BamD makes a direct contact with BamA via POTRA 5 where C-terminal truncations of BamD prevent the association of BamA with BamC and BamE, suggesting that BamD serves as a scaffold for these two lipoproteins [65, 68, 78]. It is not surprising that POTRA 5 is required for viability since this domain is responsible for scaffolding BamD, which is an essential member of the BAM complex and the only essential lipoprotein; its absence results in stalled OMP assembly and cell death [65]. That *bamA* and *bamD* are essential genes implies that BamA and BamD work in concert throughout the OMP assembly process or are responsible for completing separate, yet equally important steps. For example, there is evidence to suggest that BamD mediates BamA folding [76, 77] and then regulates its activity once correctly assembled in the outer membrane [48, 68].

BamD is composed primarily of α-helices that form five tetratricopeptide-repeat (TPR) motifs (Fig. 4) [79–82], sequences that are often found in proteins with substrate-binding functions [62, 83]. BamD can be cross-linked to substrate OMPs in vivo [75, 84] and also with synthetic peptides harboring the OMP β-signal [79]. Structural and mutational analyses suggest that a pocket formed by the N-terminal half of BamD (TPR 1-3) mediates substrate binding [65, 79]. However, a crystal structure of BamD bound to BamC revealed that the highly conserved disordered N-terminus of BamC binds to the same region of BamD, thereby prompting the suggestion that BamC may regulate the substrate binding activity of BamD [80]. The remainder of BamC is comprised of two compact helix-grip domains (Fig. 4) [79, 80], which are exposed at the bacterial cell surface [67], a remarkable feature of yet unknown significance. In fact, the role of BamC in OMP biogenesis remains elusive, with its loss causing only a slight defect in β-barrel assembly [64]. However, a recent study has shown that the BAM complex is destabilized in cells lacking BamC such that increased amounts of the BamA:B module are detectable in the outer membrane [67]. These studies suggest that BamC functions to partially stabilize interactions of BamD and BamE to the BamA:B module.

BamE consists of an α-α-β-β-β topology where a three-stranded anti-parallel β-sheet is found packed against a pair of α-helices

(Fig. 4) [79, 80, 85]. NMR spectroscopy in combination with mutational analyses revealed that BamE binds preferentially to PG lipids where the lipid-binding region partially overlaps with that of the BamD binding surface [85]. It has been proposed that this dual protein- and lipid-anchoring mechanism of BamE may promote OMP insertion into the outer membrane. Similar to BamC, loss of BamE compromises the stability of the BAM complex, yet causes only mild OMP assembly defects [78]. Interestingly, while BamE is not required for BamA assembly [76], it modulates the conformation of BamA in the outer membrane, albeit indirectly through interactions with BamD [48, 86]. Thus taken together, the available evidence suggest that the BamC:D:E module acts both directly and indirectly to modulate the structural conformation of BamA in order to make the BAM complex as a whole more efficient in receiving and assembling substrate OMPs into the outer membrane.

7 Summary

The observation that OMPs can fold into their native states in vitro demonstrates that (1) the narrative instructing a protein on how to fold is provided by its amino acid sequence, and (2) that this process can occur in the absence of a chemical energy source. Fittingly, there is no obvious energy source in the periplasm to drive OMP assembly. However, this observation raises the important question of why OMPs require the BAM complex for their insertion into the outer membrane as folded β-barrel species and consequently, also of how the BAM complex mediates this assembly. The last decade has seen transformative breakthroughs in the field of OMP biogenesis that have provided significant insights into these fundamental processes, however, many questions remain. What is the precise architectural arrangement of the subunits of the BAM complex and how do they coordinate recognition, folding and membrane insertion of nascent OMPs? What is the exact mechanism of OMP folding/insertion? What is the specific biochemical function of each individual BAM complex subunit? Does the BAM complex coordinate together with periplasmic chaperones for OMP assembly? How does the BAM complex accommodate such a diverse range of OMP substrates? Do the BAM complex and the TAM work together or separately to assemble more "difficult" OMPs, such as those that oligomerize or those with elaborate extracellular domains? Do other, as yet unidentified factors, serve as modular appendages of the BAM complex to coordinate their assembly? A more thorough understanding of the mechanism by which the BAM complex catalyzes the final steps of OMP assembly will require a multidisciplinary approach, and productive interplay between the in vivo and in vitro systems discussed in subsequent chapters.

Acknowledgements

We thank Chaille Webb, Victoria Hewitt, and Christopher Stubenrauch for constructive comments on the manuscript. We acknowledge support from the Australian Research Council (ARC) for research funding through the Super Science Fellowship grant FS110200015 (to T.L.) and NHMRC Program Grant 606788 (to T.L.). D.L.L. is an ARC Super Science Fellow, M.J.B. is an NHMRC Biomedical Research Fellow and T.L. is an ARC Australian Laureate Fellow.

References

1. Cronan JE (2003) Bacterial membrane lipids: where do we stand? Annu Rev Microbiol 57(1):203–224
2. Nikaido H (2003) Molecular basis of bacterial outer membrane permeability revisited. Microbiol Mol Biol Rev 67(4):593–656
3. Okuda S, Tokuda H (2011) Lipoprotein sorting in bacteria. Annu Rev Microbiol 65(1): 239–259
4. Silhavy TJ, Kahne D, Walker S (2010) The bacterial cell envelope. Cold Spring Harb Perspect Biol 2:a000414
5. Dong C, Beis K, Nesper J et al (2006) Wza the translocon for *E. coli* capsular polysaccharides defines a new class of membrane protein. Nature 444(7116):226–229
6. Chandran V, Fronzes R, Duquerroy S et al (2009) Structure of the outer membrane complex of a type IV secretion system. Nature 462(7276):1011–1015
7. Wimley WC (2003) The versatile β-barrel membrane protein. Curr Opin Struct Biol 13(4):404–411
8. Nakamura K, Mizushima S (1976) Effects of heating in dodecyl sulfate solution on the conformation and electrophoretic mobility of isolated major outer membrane proteins from *Escherichia coli* K-12. J Biochem (Tokyo) 80(6):1411–1422
9. Fairman JW, Noinaj N, Buchanan SK (2011) The structural biology of β-barrel membrane proteins: a summary of recent reports. Curr Opin Struct Biol 21(4):523–531
10. Domene C, Bond PJ, Deol SS et al (2003) Lipid/protein interactions and the membrane/water interfacial region. J Am Chem Soc 125(49):14966–14967
11. Selkrig J, Leyton DL, Webb CT et al (2014) Assembly of β-barrel proteins into bacterial outer membranes. Biochim Biophys Acta 1843(8):1542–1550
12. Naveed H, Liang J (2013) Weakly stable regions and protein-protein interactions in beta-barrel membrane proteins. Curr Pharm Des 20:1268–1273
13. Celik N, Webb CT, Leyton DL et al (2012) A bioinformatic strategy for the detection, classification and analysis of bacterial autotransporters. PLoS One 7(8), e43245
14. Bitto E, McKay DB (2003) The periplasmic molecular chaperone protein SurA binds a peptide motif that is characteristic of integral outer membrane proteins. J Biol Chem 278(49): 49316–49322
15. Robert V, Volokhina EB, Senf F et al (2006) Assembly factor Omp85 recognizes its outer membrane protein substrates by a species-specific C-terminal motif. PLoS Biol 4(11), e377
16. Driessen AJM, Nouwen N (2008) Protein translocation across the bacterial cytoplasmic membrane. Annu Rev Biochem 77(1):643–667
17. Knowles TJ, Scott-Tucker A, Overduin M et al (2009) Membrane protein architects: the role of the BAM complex in outer membrane protein assembly. Nat Rev Micro 7(3):206–214
18. de Cock H, Struyvé M, Kleerebezem M et al (1997) Role of the carboxy-terminal phenylalanine in the biogenesis of outer membrane protein PhoE of *Escherichia coli* K-12. J Mol Biol 269(4):473–478
19. Struyvé M, Moons M, Tommassen J (1991) Carboxy-terminal phenylalanine is essential for the correct assembly of a bacterial outer membrane protein. J Mol Biol 218(1):141–148
20. Gawarzewski I, DiMaio F, Winterer E et al (2014) Crystal structure of the transport unit of the autotransporter adhesin involved in diffuse adherence from *Escherichia coli*. J Struct Biol 187(1):20–29
21. Gessmann D, Chung YH, Danoff EJ et al (2014) Outer membrane β-barrel protein

folding is physically controlled by periplasmic lipid head groups and BamA. Proc Natl Acad Sci U S A 111(16):5878–5883

22. Paramasivam N, Habeck M, Linke D (2012) Is the C-terminal insertional signal in Gram-negative bacterial outer membrane proteins species-specific or not? BMC Genomics 13(1):510
23. Volokhina EB, Grijpstra J, Beckers F et al (2013) Species-specificity of the BamA component of the bacterial outer membrane protein-assembly machinery. PLoS One 8(12), e85799
24. Rizzitello AE, Harper JR, Silhavy TJ (2001) Genetic evidence for parallel pathways of chaperone activity in the periplasm of *Escherichia coli*. J Bacteriol 183(23):6794–6800
25. Sklar JG, Wu T, Kahne D et al (2007) Defining the roles of the periplasmic chaperones SurA, Skp, and DegP in *Escherichia coli*. Genes Dev 21(19):2473–2484
26. Schäfer U, Beck K, Müller M (1999) Skp, a molecular chaperone of Gram-negative bacteria, is required for the formation of soluble periplasmic intermediates of outer membrane proteins. J Biol Chem 274(35):24567–24574
27. Vertommen D, Ruiz N, Leverrier P et al (2009) Characterization of the role of the *Escherichia coli* periplasmic chaperone SurA using differential proteomics. Proteomics 9(9):2432–2443
28. Harms N, Koningstein G, Dontje W et al (2001) The early interaction of the outer membrane protein PhoE with the periplasmic chaperone Skp occurs at the cytoplasmic membrane. J Biol Chem 276(22):18804–18811
29. Ureta AR, Endres RG, Wingreen NS et al (2007) Kinetic analysis of the assembly of the outer membrane protein LamB in *Escherichia coli* mutants each lacking a secretion or targeting factor in a different cellular compartment. J Bacteriol 189(2):446–454
30. Ruiz-Perez F, Henderson IR, Nataro JP (2010) Interaction of FkpA, a peptidyl-prolyl cis/trans isomerase with EspP autotransporter protein. Gut Microbes 1(5):339–344
31. Qu J, Mayer C, Behrens S et al (2007) The trimeric periplasmic chaperone Skp of *Escherichia coli* forms 1:1 complexes with outer membrane proteins via hydrophobic and electrostatic interactions. J Mol Biol 374(1):91–105
32. Hagan CL, Kim S, Kahne D (2010) Reconstitution of outer membrane protein assembly from purified components. Science 328(5980):890–892
33. Hagan CL, Kahne D (2011) The reconstituted *Escherichia coli* Bam complex catalyzes multiple rounds of β-barrel assembly. Biochemistry (Mosc) 50(35):7444–7446
34. Xu X, Wang S, Hu Y-X et al (2007) The periplasmic bacterial molecular chaperone SurA adapts its structure to bind peptides in different conformations to assert a sequence preference for aromatic residues. J Mol Biol 373(2):367–381
35. Walton TA, Sandoval CM, Fowler CA et al (2009) The cavity-chaperone Skp protects its substrate from aggregation but allows independent folding of substrate domains. Proc Natl Acad Sci U S A 106(6):1772–1777
36. Burmann BM, Hiller S (2012) Solution NMR studies of membrane-protein-chaperone complexes. Chima (Aarau) 66(10):759–763
37. Ge X, Wang R, Ma J et al (2014) DegP primarily functions as a protease for the biogenesis of β-barrel outer membrane proteins in the Gram-negative bacterium *Escherichia coli*. FEBS J 281(4):1226–1240
38. CastilloKeller M, Misra R (2003) Protease-deficient DegP suppresses lethal effects of a mutant OmpC protein by its capture. J Bacteriol 185(1):148–154
39. Webb CT, Heinz E, Lithgow T (2012) Evolution of the β-barrel assembly machinery. Trends Microbiol 20(12):612–620
40. Voulhoux R, Bos MP, Geurtsen J et al (2003) Role of a highly conserved bacterial protein in outer membrane protein assembly. Science 299(5604):262–265
41. Genevrois S, Steeghs L, Roholl P et al (2003) The Omp85 protein of *Neisseria meningitidis* is required for lipid export to the outer membrane. EMBO J 22(8):1780–1789
42. Heinz E, Lithgow T (2014) A comprehensive analysis of the Omp85/TpsB protein superfamily structural diversity, taxonomic occurrence and evolution. Front Microbiol 5:370
43. Clantin B, Delattre A-S, Rucktooa P et al (2007) Structure of the membrane protein FhaC: a member of the Omp85-TpsB transporter superfamily. Science 317(5840):957–961
44. Gruss F, Zähringer F, Jakob RP et al (2013) The structural basis of autotransporter translocation by TamA. Nat Struct Mol Biol 20(11):1318–1320
45. Noinaj N, Kuszak AJ, Gumbart JC et al (2013) Structural insight into the biogenesis of β-barrel membrane proteins. Nature 501(7467):385–390
46. Fan E, Fiedler S, Jacob-Dubuisson F et al (2012) Two-partner secretion of Gram-negative bacteria: a single β-barrel protein enables transport across the outer membrane. J Biol Chem 287(4):2591–2599
47. Guédin S, Willery E, Tommassen J et al (2000) Novel topological features of FhaC, the outer membrane transporter involved in the secretion of the *Bordetella pertussis* filamentous

hemagglutinin. J Biol Chem 275(39): 30202–30210
48. Rigel NW, Ricci DP, Silhavy TJ (2013) Conformation-specific labeling of BamA and suppressor analysis suggest a cyclic mechanism for β-barrel assembly in *Escherichia coli*. Proc Natl Acad Sci U S A 110(13):5151–5156
49. Delattre A-S, Clantin B, Saint N et al (2010) Functional importance of a conserved sequence motif in FhaC, a prototypic member of the TpsB/Omp85 superfamily. FEBS J 277(22): 4755–4765
50. Leonard-Rivera M, Misra R (2012) Conserved residues of the putative L6 loop of *Escherichia coli* BamA play a critical role in the assembly of β-barrel outer membrane proteins, including that of BamA itself. J Bacteriol 194(17): 4662–4668
51. Ni D, Wang Y, Yang X et al (2014) Structural and functional analysis of the β-barrel domain of BamA from *Escherichia coli*. FASEB J 28(6):2677–2685
52. Noinaj N, Kuszak AJ, Balusek C et al (2014) Lateral opening and exit pore formation are required for BamA function. Structure 22(7):1055–1062
53. Selkrig J, Mosbahi K, Webb CT et al (2012) Discovery of an archetypal protein transport system in bacterial outer membranes. Nat Struct Mol Biol 19(5):506–510
54. Kim S, Malinverni JC, Sliz P et al (2007) Structure and function of an essential component of the outer membrane protein assembly machine. Science 317(5840):961–964
55. Knowles TJ, Jeeves M, Bobat S et al (2008) Fold and function of polypeptide transport-associated domains responsible for delivering unfolded proteins to membranes. Mol Microbiol 68(5):1216–1227
56. Gatzeva-Topalova PZ, Warner LR, Pardi A et al (2010) Structure and flexibility of the complete periplasmic domain of BamA: the protein insertion machine of the outer membrane. Structure 18(11):1492–1501
57. Gatzeva-Topalova PZ, Warner LR, Pardi A et al (2008) Structure and flexibility of the complete periplasmic domain of BamA: the protein insertion machine of the outer membrane. Structure 18(11):1492–1501
58. Gatzeva-Topalova PZ, Walton TA, Sousa MC (2008) Crystal structure of YaeT: conformational flexibility and substrate recognition. Structure 16(12):1873–1881
59. Bennion D, Charlson ES, Coon E et al (2010) Dissection of β-barrel outer membrane protein assembly pathways through characterizing BamA POTRA 1 mutants of *Escherichia coli*. Mol Microbiol 77(5):1153–1171
60. Anwari K, Webb CT, Poggio S et al (2012) The evolution of new lipoprotein subunits of the bacterial outer membrane BAM complex. Mol Microbiol 84(5):832–844
61. Volokhina EB, Beckers F, Tommassen J et al (2009) The β-barrel outer membrane protein assembly complex of *Neisseria meningitidis*. J Bacteriol 191(22):7074–7085
62. Gatsos X, Perry AJ, Anwari K et al (2008) Protein secretion and outer membrane assembly in Alphaproteobacteria. FEMS Microbiol Rev 32(6):995–1009
63. Stenberg F, Chovanec P, Maslen SL et al (2005) Protein complexes of the *Escherichia coli* cell envelope. J Biol Chem 280(41):34409–34419
64. Wu T, Malinverni J, Ruiz N et al (2005) Identification of a multicomponent complex required for outer membrane biogenesis in *Escherichia coli*. Cell 121(2):235–245
65. Malinverni JC, Werner J, Kim S et al (2006) YfiO stabilizes the YaeT complex and is essential for outer membrane protein assembly in *Escherichia coli*. Mol Microbiol 61(1):151–164
66. Vuong P, Bennion D, Mantei J et al (2008) Analysis of YfgL and YaeT interactions through bioinformatics, mutagenesis, and biochemistry. J Bacteriol 190(5):1507–1517
67. Webb CT, Selkrig J, Perry AJ et al (2012) Dynamic association of BAM complex modules includes surface exposure of the lipoprotein BamC. J Mol Biol 422(4):545–555
68. Ricci DP, Hagan CL, Kahne D et al (2012) Activation of the *Escherichia coli* β-barrel assembly machine (Bam) is required for essential components to interact properly with substrate. Proc Natl Acad Sci U S A 109(9):3487–3491
69. Noinaj N, Fairman JW, Buchanan SK (2011) The crystal structure of BamB suggests interactions with BamA and its role within the BAM complex. J Mol Biol 407(2):248–260
70. Dong C, Yang X, Hou H-F et al (2012) Structure of *Escherichia coli* BamB and its interaction with POTRA domains of BamA. Acta Crystallogr D Biol Crystallogr 68(9): 1134–1139
71. Charlson ES, Werner JN, Misra R (2006) Differential effects of yfgL mutation on *Escherichia coli* outer membrane proteins and lipopolysaccharide. J Bacteriol 188(20): 7186–7194
72. Ruiz N, Falcone B, Kahne D et al (2005) Chemical conditionality: a genetic strategy to probe organelle assembly. Cell 121(2):307–317
73. Heuck A, Schleiffer A, Clausen T (2011) Augmenting β-augmentation: structural basis of how BamB binds BamA and may support

folding of outer membrane proteins. J Mol Biol 406(5):659–666

74. Jansen KB, Baker SL, Sousa MC (2012) Crystal structure of BamB from *Pseudomonas aeruginosa* and functional evaluation of its conserved structural features. PLoS One 7(11), e49749
75. Ieva R, Tian P, Peterson JH et al (2011) Sequential and spatially restricted interactions of assembly factors with an autotransporter β domain. Proc Natl Acad Sci U S A 108(31):E383–E391
76. Hagan CL, Westwood DB, Kahne D (2013) Bam lipoproteins assemble BamA in vitro. Biochemistry (Mosc) 52(35):6108–6113
77. Misra R, Stikeleather R, Gabriele R (2014) In vivo roles of BamA, BamB and BamD in the biogenesis of BamA, a core protein of the β-barrel assembly machine of *Escherichia coli*. J Mol Biol. doi:10.1016/j.jmb.2014.1004.1021
78. Sklar JG, Wu T, Gronenberg LS et al (2007) Lipoprotein SmpA is a component of the YaeT complex that assembles outer membrane proteins in *Escherichia coli*. Proc Natl Acad Sci U S A 104(15):6400–6405
79. Albrecht R, Zeth K (2011) Structural basis of outer membrane protein biogenesis in bacteria. J Biol Chem 286(31):27792–27803
80. Kim KH, Aulakh S, Paetzel M (2011) Crystal structure of β-barrel assembly machinery BamCD protein complex. J Biol Chem 286(45):39116–39121
81. Sandoval CM, Baker SL, Jansen K et al (2011) Crystal structure of BamD: an essential component of the β-barrel assembly machinery of Gram-negative bacteria. J Mol Biol 409(3): 348–357
82. Dong C, Hou H-F, Yang X et al (2012) Structure of *Escherichia coli* BamD and its functional implications in outer membrane protein assembly. Acta Crystallogr D Biol Crystallogr 68(2):95–101
83. D'Andrea LD, Regan L (2003) TPR proteins: the versatile helix. Trends Biochem Sci 28(12):655–662
84. Leyton DL, Sevastsyanovich YR, Browning DF et al (2011) Size and conformation limits to secretion of disulfide-bonded loops in autotransporter proteins. J Biol Chem 286(49): 42283–42291
85. Knowles TJ, Browning DF, Jeeves M et al (2011) Structure and function of BamE within the outer membrane and the β-barrel assembly machine. EMBO Rep 12(2): 123–128
86. Rigel NW, Schwalm J, Ricci DP et al (2012) BamE modulates the *Escherichia coli* beta-barrel assembly machine component BamA. J Bacteriol 194(5):1002–1008

Chapter 2

Yeast Mitochondria as a Model System to Study the Biogenesis of Bacterial β-Barrel Proteins

Thomas Ulrich, Philipp Oberhettinger, Ingo B. Autenrieth, and Doron Rapaport

Abstract

Beta-barrel proteins are found in the outer membrane of Gram-negative bacteria, mitochondria, and chloroplasts. The evolutionary conservation in the biogenesis of these proteins allows mitochondria to assemble bacterial β-barrel proteins in their functional form. In this chapter, we describe exemplarily how the capacity of yeast mitochondria to process the trimeric autotransporter YadA can be used to study the role of bacterial periplasmic chaperones in this process.

Key words BAM complex, β-Barrel proteins, Chaperones, Evolutionary conservation, Mitochondria, Skp, TOB complex

1 Introduction

Integral β-barrel proteins are exclusively found in the outer membrane of Gram-negative bacteria and in the outer membranes of eukaryotic organelles derived from prokaryotic ancestors namely, mitochondria and chloroplasts. Although most of the proteins in the bacterial outer membrane are members of this protein class, only five mitochondrial β-barrel proteins were identified in yeast so far [1]. Like the vast majority of mitochondrial genes they have undergone gene transfer to the host genome [2, 3]. Hence, precursors of β-barrel proteins are synthesized on cytosolic ribosomes and therefore have to contain all the information required to ensure an efficient and specific targeting to the mitochondrial outer membrane (MOM). Furthermore, translocases in the MOM had to evolve or adapt in order to facilitate the post-translational import and assembly of precursor β-barrel proteins.

1.1 Biogenesis of β-Barrel Proteins in Bacteria

Bacterial β-barrel proteins are synthesized on cytoplasmic ribosomes with N-terminal signal sequences. Upon their appearance at the exit channel of the ribosome they can be stabilized by the highly

Susan K. Buchanan and Nicholas Noinaj (eds.), *The BAM Complex: Methods and Protocols*, Methods in Molecular Biology, vol. 1329, DOI 10.1007/978-1-4939-2871-2_2, © Springer Science+Business Media New York 2015

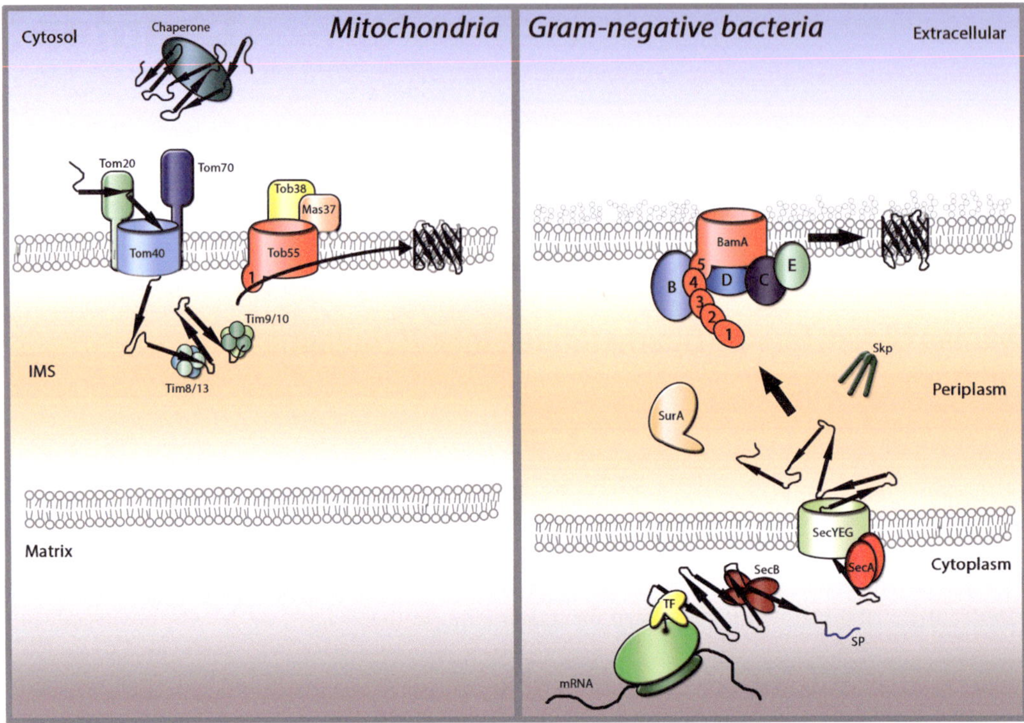

Fig. 1 Evolutionary conservation in the biogenesis pathways of β-barrel proteins between mitochondria and gram-negative bacteria

conserved trigger factor [4, 5]. Subsequently, the chaperone SecB is proposed to bind the nascent polypeptide chain and direct it to the Sec translocase (Fig. 1) [6]. Proteins destined for the outer membrane are translocated across the inner membrane through the Sec translocase in a process dependent on the hydrolysis of ATP by SecA [7]. Reaching the periplasmic side of the inner membrane, the signal peptide is cleaved off and the precursor proteins are escorted by periplasmic chaperones to the BAM complex [8, 9]. The precise roles of the chaperones SurA, Skp, and DegP are still under debate and seem to differ depending on the substrate and the organism [10–13]. The subsequent insertion into the outer bacterial membrane is facilitated by the BAM complex. In *Escherichia coli*, this complex is composed of the central β-barrel protein BamA (Omp85/YaeT) associated with four lipoproteins (BamB, BamC, BamD, and BamE) [14–16]. Despite remarkable progress in characterizing the various components in the biogenesis pathway of β-barrel proteins in bacteria, the exact mechanism by which the proteins are assembled into the lipid bilayer still remains unresolved.

1.2 Biogenesis of β-Barrel Proteins in Mitochondria

The requirement for N-terminal signal sequences in the sorting of β-barrel proteins was lost in the evolutionary transformation from bacteria to mitochondria. Upon their synthesis on cytosolic ribosomes, mitochondrial β-barrel proteins are recognized at the organelle's surface by import receptors of the translocase of the outer membrane (TOM) complex and transferred across the MOM through Tom40, the general entry gate of the TOM complex, a β-barrel protein itself (Fig. 1) [1, 17, 18]. Within the intermembrane space (IMS), the precursor proteins are then protected from misfolding and aggregation by the small heterohexameric Tim chaperone complexes Tim8/13 and Tim9/10. Finally, assembly of the precursor proteins into the MOM occurs with the help of a dedicated protein complex termed topogenesis of outer-membrane β-barrel proteins (TOB complex) or sorting and assembly machinery (SAM complex) [19–21]. This complex is composed of the central highly conserved β-barrel protein Tob55/Sam50 and the peripheral subunits Tob38/Sam35/Tom38 and Mas37/Sam37. The latter two are located on the cytosolic side of the membrane and do not show any sequence similarity with their bacterial counterparts of the BAM complex [20, 22–24]. The essential subunit Tob38 was previously shown to be involved in intramitochondrial substrate recognition through the so called β-signal, whereas Mas37 acts at a late stage of β-barrel protein assembly with a putative role in precursor release from the TOB complex [25–27].

1.3 Evolutionary Conservation in the Biogenesis of β-Barrel Proteins

The incorporation of an ancestral α-proteobacterium into the eukaryotic cell led to a major transfer of DNA to the host genome [2, 28]. Thereby, the developing organelle had to adapt in order to ensure post-translational import of proteins. However, functional expression of bacterial β-barrel proteins in eukaryotic cells suggests that during this adaptation process the ability of mitochondria to facilitate the assembly of prokaryotic β-barrel proteins was conserved [29–31]. In a reciprocal approach, the mitochondrial VDAC could also be assembled into the bacterial outer membrane upon its expression in *Escherichia coli* [32]. A closer look at the biogenesis pathways of β-barrel proteins reveals that many characteristics are shared among Gram-negative bacteria and mitochondria. In both cases the precursor proteins are initially translocated across a membrane and prevented from misfolding and aggregation in the intermembranal space by specialized soluble chaperones. Insertion into the outer membrane occurs in each instance from the inner side of the membrane. The most striking similarity, however, is the sequential and functional homology in the central components of the assembly machineries, Tob55/Sam50 and BamA, both being members of the Omp85 superfamily. Homologs of this family are present in all Gram-negative bacteria and in the outer membranes of the eukaryotic organelles, mitochondria, and chloroplasts [14, 21, 33–35]. A common feature of all proteins belonging to this family is the presence of N-terminal polypeptide-transport-associated (POTRA) domains followed by a 16-stranded C-terminal β-barrel domain. However, the number of POTRA domains can range from one in the mitochondrial Tob55/Sam50 and three in the chloroplast homolog Toc75-V up to seven in Omp85 from *Myxococcus xanthus* [36, 37]. It seems that the POTRA domains facilitate the transfer of the β-barrel precursors from the soluble chaperones to the membrane-embedded part of the translocase although, at least for mitochondria, a role in the release of the precursor from the TOB/SAM complex was also suggested [38–41]. Apart from the aforementioned similarities, the assembly processes differ in terms of accessory proteins and the requirement of N-terminal signal sequences. Whereas in mitochondria, the two accessory subunits are located at the cytosolic side of the MOM, all accessory lipoprotein subunits of the BAM complex reside on the internal side of the membrane similar to the N-terminal POTRA domains.

Due to this evolutionary conservation, yeast mitochondria provide a powerful system to study the biogenesis of prokaryotic β-barrel proteins. By using this system we investigated, for example, the involvement of periplasmic chaperones in the biogenesis pathway of the trimeric autotransporter protein Yersinia adhesin A (YadA) (Fig. 2). Mature YadA is composed of an internal passenger domain (also called the effector domain), and a relatively short C-terminal β-domain that anchors the protein to the OM. This anchor is made by a single 12-stranded β-barrel structure to which each monomer is contributing four β-strands. In this chapter we describe the expression of YadA in yeast cells, how to target periplasmic chaperones to the mitochondrial IMS and illustrate a method to examine the functionality of correctly assembled YadA by monitoring its ability to induce the secretion of IL8 from HeLa cells [42].

2 Materials

2.1 Yeast Transformation

1. Lithium acetate (LiOAc) (100 mM) (sterile). Store at room temperature (RT).
2. Lithium acetate (LiOAc) (1 M) (sterile). Store at RT.
3. Polyethylene glycol (PEG) (50 %), sterile: 50 g PEG-3350 in 100 mL water. Store at RT.
4. Salmon sperm DNA (10 mg/mL), store in small aliquots at −20 °C.
5. Vector to be transformed.

2.2 Targeting of Proteins to the Mitochondrial IMS

1. Yeast expression vector containing the protein of interest fused to the first 228 amino acids of *S. cerevisiae* Mgm1 lacking the first transmembrane domain (*see* **Note 1**).

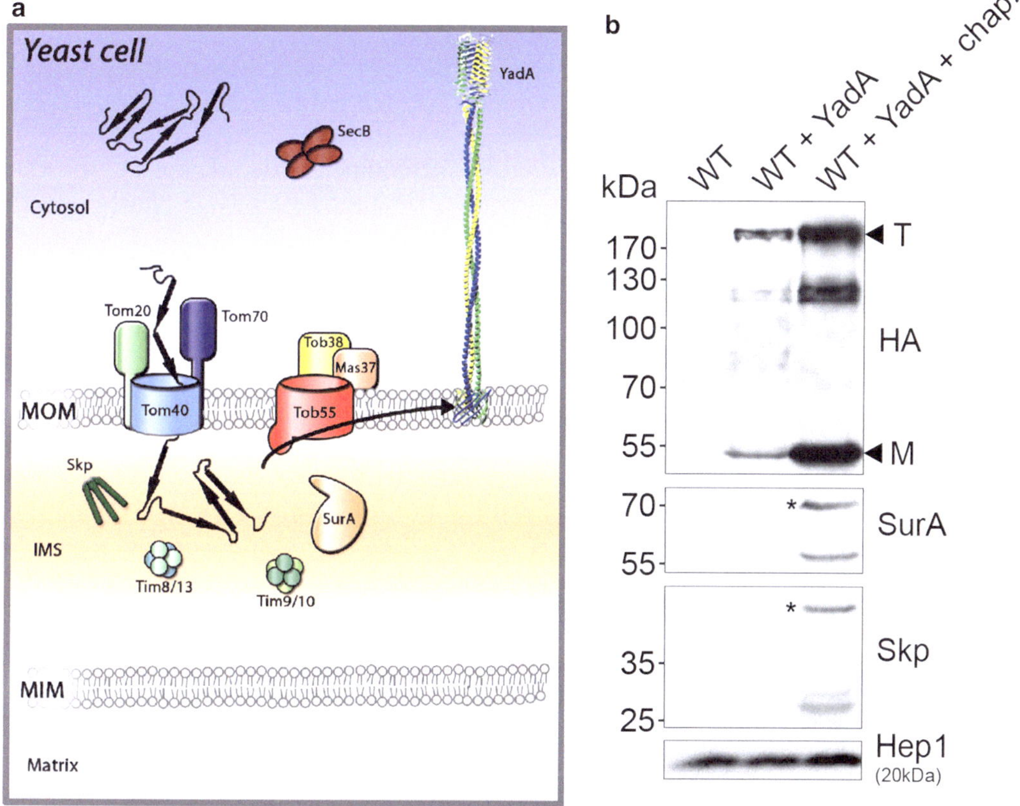

Fig. 2 The biogenesis of YadA in mitochondria. (**a**) Schematic representation of the co-expression of YadA and the bacterial chaperones SecB, SurA and Skp in yeast cells. (**b**) Comparison of the mitochondrial steady-state levels of YadA-HA in yeast cells expressing either empty plasmid, YadA-HA alone or co-expressing YadA-HA and the bacterial chaperones SurA, Skp, and SecB. Crude mitochondria were analyzed by SDS-PAGE and immunoblotting with antibodies against the HA-tag, Skp, SurA, and Hep1 (a mitochondrial matrix protein that serves as a loading control). Asterisks indicate unprocessed forms of SurA and Skp. The positions of monomeric and trimeric YadA-HA are indicated with M and T, respectively

2.3 Analysis of Steady-State Levels of Bacterial Proteins Expressed in Yeast Cells

2.3.1 Isolation of Crude Mitochondria by Lysis with Glass Beads

1. SEM buffer: 250 mM sucrose, 10 mM MOPS, 1 mM EDTA in water, pH 7.2 (adjusted with KOH) (*see* **Note 2**).
2. Phenylmethylsulfonylfluoride (PMSF) (200 mM) in isopropanol.
3. Reaction tubes (2 mL).
4. Glass beads, 0.25–0.5 mm.
5. Bradford solution.
6. Laemmli buffer (2×): 16 mL 1 M Tris–HCl, pH 6.8, 4 g SDS, 20 mL glycerol, 20 mg bromophenol blue, 5 % 2-mercaptoethanol. Add water to 100 mL. Store at RT.

7. S-medium: 0.17 % (w/v) yeast nitrogen base, 0.5 % (w/v) ammonium sulfate, 0.3 μM adenine and 0.5 μM uracil in water, pH 5.5 (adjusted with KOH). Add amino acids separately as 100× stock solution (200 mg arginine, 400 mg tryptophan, 1 g leucine, 400 mg lysine, 200 mg histidine, 600 mg phenylalanine, and 200 mg methionine in 100 mL water). Autoclave carbon source separately and add to 2 % (w/v) final D-glucose (SD), 2 % (w/v) final D-galactose (SGal), or 2 % (w/v) glycerol (SG). For selection media leave out amino acids corresponding to the auxotrophic marker of the employed plasmid.

2.3.2 SDS-PAGE

1. Bottom gel (15 %): 2.39 mL water, 3.75 mL 1 M Tris–HCl, pH 8.8, 3.75 mL acrylamide/*bis*-acrylamide (40 %/0.8 %), 0.1 mL 10 % ammonium persulfate (APS), 8 μL tetramethylethylenediamine (TEMED). Mix thoroughly before use.
2. Separating gel (10 %): 4.55 mL distilled water, 4.69 mL 1 M Tris–HCl, pH 8.8, 3.13 mL acrylamide/*bis*-acrylamide (40 %/0.8 %), 0.125 mL 10 % APS, 10 μL TEMED. Mix thoroughly before use.
3. Stacking gel (4 %): 3.76 mL distilled water, 0.625 mL 1 M Tris–HCl, pH 6.8, 0.563 mL acrylamide/*bis*-acrylamide (40 %/0.8 %), 0.05 mL 10 % APS, 4 μL TEMED. Mix thoroughly before use.
4. Running buffer (5×): 30 g Tris–HCl, 145 g glycine, 5 g sodium dodecyl sulfate (SDS), add distilled water to 1 L. Store at RT. Working concentration is 1×.

2.3.3 Western Blotting

1. Filter papers (thickness 0.35 mm).
2. Nitrocellulose membrane (pore size 0.2 μm).
3. Blotting buffer (10×): 60.57 g Tris–HCl, 281.51 g glycine, 5 g SDS, add distilled water to 2.5 L, store at 4 °C. Working concentration is 1×: 100 mL 10× blotting buffer, 200 mL methanol (20 % final), add distilled water to 1 L.
4. Ponceau S solution: 0.4 g Ponceau S, 8.5 mL 72 % trichloroacetic acid (TCA), add distilled water to 200 mL. Store at RT.
5. Blocking buffer (5 %): 5 g skim milk powder in 100 mL TBS-buffer.
6. TBS (20×): 121.16 g Tris–HCl, 900 g NaCl, add distilled water to 5 L, pH 7.5 (adjust with HCl). Working concentration is 1×. Store at room temperature.
7. TBS-T (1×): add 500 μL Tween-20 to 1 L 1× TBS solution. Store at RT.
8. Enhanced chemiluminescence (ECL) solution: 25 mL cold 1 M Tris–HCl, pH 8.5, 1.25 mL 4.4 % luminol, 0.55 mL 1.5 %

p-coumaric acid, add cold distilled water to 250 mL. Store light protected at 4 °C.

9. Secondary antibody: horseradish peroxidase conjugated immunoglobulin G specific for the corresponding primary antibody.

2.4 Cell Culture

1. HeLa cells, ATCC number CCL-2 (*see* **Note 3**).
2. RPMI-1640 medium (e.g. Biochrom).
3. Penicillin/Streptomycin (Pen/Strep, e.g. Gibco).
4. Fetal calf serum (FCS; e.g. Gibco) (*see* **Note 4**).
5. Phosphate buffered saline (PBS; e.g. Gibco) (*see* **Note 5**).
6. Trypsin EDTA (0.05 %; e.g. Gibco) (*see* **Note 6**).
7. Cell culture flask (e.g. Nunc).
8. 24-Well plate (e.g. Greiner).
9. Tabletop centrifuge (e.g. Heraeus Multifuge 3S-R).
10. Microcentrifuge tubes (1.5 mL).
11. Neubauer counting chamber.
12. Microscope (e.g. Zeiss Axiovert 25).
13. Cell culture incubator.

2.5 ELISA to Determine Amounts of IL8

1. RPMI-1640 medium (Biochrom), supplemented with 10 % FCS and 1 % Pen/Strep.
2. 96-Well plate (e.g. Nunc) (*see* **Note 7**).
3. Binding solution: 0.1 M Na_2HPO_4 (pH 9.0): Weigh 14.17 g of Na_2HPO_4 and dissolve in 900 mL ultra-pure H_2O. Adjust pH to 9.0 with NaOH and add ultra-pure H_2O to 1000 mL. For long-term storage, autoclave the solution and store at RT.
4. Blocking buffer: 1× PBS/10 % FCS. For one 96-well plate, freshly prepare 18 mL PBS supplemented with 2 mL FCS.
5. Blocking/Tween buffer: 1× PBS, 10 % FCS, 0.05 % Tween20. For one 96-well plate, mix 9 mL 1× PBS with 1 mL 10 % FCS and add 5 μL Tween20.
6. Substrate buffer: 0.05 M Na_2CO_3, 0.05 M $NaHCO_3$, 1 mM $MgCl_2$. Weight 5.3 g of Na_2CO_3, 4.2 g of $NaHCO_3$ and 0.094 g of $MgCl_2$ in 900 mL ultra-pure H_2O. Adjust pH to 9.8 with NaOH and fill up to 1000 mL with ultra-pure H_2O. Store the substrate solution at 4 °C.
7. Washing buffer: 1× PBS/0.05 % Tween20. Store the washing buffer at 4 °C.
8. Capture antibody: purified anti-human IL-8 (Becton Dickinson).
9. Detection antibody: biotin mouse anti-human IL-8 (Becton Dickinson).

10. Conjugate: Streptavidin alkaline phosphatase (Roche).
11. Substrate: 5 mg p-Nitrophenyl-phosphate (PNPP) tablets.
12. Recombinant human IL8 (Becton Dickinson) as standard.
13. Multichannel pipet.
14. Microplate washer (e.g. TECAN HydroFlex™).
15. Microplate reader (e.g. TECAN sunrise™).

3 Methods

3.1 Yeast Transformation

1. Scrape yeast cells from agar plate and wash in 1 mL sterile water (1.5 mL reaction tube).
2. Pellet the cells by centrifugation (5 s, top speed, tabletop centrifuge).
3. Discard the water and resuspend the cells in 1 mL 100 mM LiOAc. Incubate the cells at 30 °C for 5 min. In the meantime incubate salmon sperm carrier DNA for 5 min at 95 °C and place immediately on ice.
4. Pellet the cells by centrifugation (5 s, top speed, tabletop centrifuge).
5. Discard the supernatant and add in the following order: 240 μL PEG 3350 (50 %), 36 μL 1 M LiOAc, 10 μL salmon sperm carrier DNA (denatured), 60 μL sterile water, and 5 μL plasmid DNA (*see* **Note 8**).
6. Mix thoroughly and incubate for 20 min at 42 °C.
7. Collect the cells by centrifugation (10 s, top speed, tabletop centrifuge)
8. Discard supernatant and resuspend pellet in 100 μL sterile water.
9. Plate cells on selective SD-agar plates and incubate at 30 °C. Depending on the yeast strain, colonies appear after approximately 2 days.

3.2 Analysis of Steady-State Levels of Bacterial Proteins Expressed in Yeast Cells

3.2.1 Isolation of Crude Mitochondria by Lysis with Glass Beads

1. Inoculate one yeast colony into 30 mL of liquid SGal-medium (*see* **Note 9**) and grow overnight at 30 °C while shaking (120 rpm).
2. Dilute the overnight culture in fresh medium to 200 mL (OD_{600} = 0.2) and grow to OD_{600} = 0.8–1.5 (*see* **Note 10**).
3. Harvest cells by centrifugation (3000 × *g*, 5 min, RT).
4. Discard the supernatant and wash cells in 50 mL water.
5. Recollect cells by centrifugation (3000 × *g*, 5 min, RT) (*see* **Note 11**).

6. Resuspend the cells pellet in 2 mL SEM buffer + 2 mM PMSF (final conc.) and distribute into four 2 mL reaction tubes containing each 600 mg glass beads.
7. Vortex five times for 30 s each at max speed and cool cells for 30 s on ice in between.
8. To pellet down nuclei, unbroken cells and cell debris, spin the samples (1000 × *g*, 3 min, 4 °C).
9. Pool the supernatants of the four reaction tubes for each strain and measure protein concentration by Bradford method. In case of yeast cells grown on glucose, a rough estimation is that 7 % of total cellular proteins can be considered as mitochondria. If the cells are grown on galactose, ca. 15 % can be estimated to be mitochondrial proteins.
10. Collect crude mitochondria by centrifugation (13,200 × *g* for 10 min, 4 °C).
11. The pellet is the crude mitochondrial fraction. Supernatant contains proteins from the cytosolic fraction. Resuspend the mitochondrial fraction in 2× Laemmli buffer to a concentration of 2 μg/μL and boil samples for 5 min at 95 °C.

3.2.2 SDS-PAGE

1. Carefully remove the comb and wash the wells with 1× running buffer.
2. Prior to loading, centrifuge the samples shortly at 1000 × *g* in a tabletop centrifuge.
3. Load 15–30 μL (30–60 μg) of the samples per each well.
4. Run at 20 mA for each gel until the dye front reaches the bottom gel.
5. Remove the gel from the electrophoresis chamber and discard the stacking and bottom gel.

3.2.3 Western Blotting

For electrophoretic protein transfer from the SDS-gel to the nitrocellulose membrane, we employ the semi-dry western blotting.

1. Prior to the assembly of the blotting sandwich, incubate six filter papers (depending on the thickness of the filter papers), the nitrocellulose membrane and the SDS-gel in 1× blotting buffer.
2. Assemble the blotting sandwich in the following order:
   (a) Three wet filter papers.
   (b) Nitrocellulose membrane.
   (c) SDS-gel.
   (d) Three wet filter papers.
3. After assembly, carefully role a glass pipette over the sandwich to get rid of residual air bubbles.

4. Connect the blotting apparatus to the power supply and run for 1 h at 1 mA/cm^2 at RT.
5. Disassemble the blotting sandwich and transfer the nitrocellulose membrane to an incubation chamber (avoid touching the membrane without gloves). To control for successful blotting, incubate the membrane with Ponceau S solution for 2 min.
6. Wash the membrane with distilled water until protein bands appear. If decoration with different antibodies is required, cut the membrane into respective slices with a scalpel.
7. Block the membrane in 5 % blocking buffer for 1 h at RT under agitation.
8. Discard the blocking buffer and wash the membrane once with 1× TBS.
9. Incubate the membrane with the primary antibody for 1 h at RT while shaking (*see* **Note 12**).
10. Remove the primary antibody and wash three times 5 min with 1× TBS, once with TBS-T and again with 1× TBS.
11. Incubate the membrane for 1 h at RT with a secondary antibody that was raised against your first antibody and is conjugated to horseradish peroxidase.
12. Remove the secondary antibody and wash at least three times for 5 min with 1× TBS.
13. After brief incubation with ECL solution (add 1:1000 fresh H_2O_2 before use), chemiluminescence can be detected.

3.3 Cell Culture

1. Quickly thaw at 37 °C in a water bath human HeLa cervical epithelial cells out of liquid nitrogen. Seed cells subsequently in 25 mL RPMI-1640 medium containing 10 % FCS and 1 % Pen/Strep provided in a sterile cell culture flask with 175 cm^2 growth area.
2. Incubate cells in a cell culture incubator at 37 °C with 5 % CO_2 and 95 % air humidity. Check growth behavior of the cells under the microscope every day. If cell growth on the bottom of the flask is nearly confluent, cells can be split, counted and seeded for further assays.
3. Extract old medium from the flask and add 8 mL of prewarmed trypsin to remove HeLa cells from the bottom of the flask. Incubate for 5 min at 37 °C.
4. Afterwards add 20 mL of RPMI-1640 medium to dilute trypsin and transfer cells into a 50 mL Falcon tube.
5. Centrifuge cells for 5 min at 400 × g and discard supernatant. Resuspend cell pellet carefully in 5 mL pre-warmed RPMI-1640 medium and avoid air bubbles.

6. Prepare a 1:10 dilution (10 μL of HeLa cells in 90 μL trypan blue) and count cells under the microscope with the help of a Neubauer counting chamber.
7. Provide 1 mL of pre-warmed RPMI-1640 medium supplemented with 10 % FCS and 1 % Pen/Strep in each slot of a 24-well plate and add 1.5×10^5 HeLa cells per well (*see* **Note 13**).
8. Grow HeLa cells overnight at 37 °C in a cell culture incubator.
9. Next day, 1 h before adding either bacteria or isolated mitochondria, wash cells twice with 1 mL pre-warmed PBS and append 1 mL RPMI-1640 medium with 10 % FCS, but without antibiotics.
10. Afterwards add bacteria at a multiplicity of infection (MOI) of 100 (1.5×10^7) or mitochondria (100 μg) to desired wells and centrifuge at $400 \times g$ for 5 min (*see* **Note 14**).
11. After 1 h of incubation in the cell culture incubator, add 100 μg gentamicin (10 μL from a 10 mg/mL stock solution in cell culture PBS) to avoid further growth of bacteria in each well.
12. Incubate cells for further 5 h at 37 °C in the cell culture incubator and subsequently collect supernatant in 1.5 mL microcentrifuge tubes. Store tubes at −20 °C until further analysis.

3.4 IL8-ELISA Assay

1. Coat a 96-well plate with capture antibody diluted in binding solution. Then, mix 30 μL of capture antibody (ab) with 5 mL binding solution and pipet 50 μL per well with a multichannel pipette (*see* **Note 15**).
2. Store the 96-well plate at 4 °C overnight.
3. Wash the plate four times with a microplate washer. Afterwards add 200 μL blocking buffer to each well. Incubate at room temperature (RT) for 2 h.
4. Thaw on ice cell culture supernatants to be analyzed (from Subheading 3.3, **step 12**). Additionally, prepare IL8 standard by a serial twofold dilution of known protein concentrations from 800 to 12.5 pg/mL. Prepare one microcentrifuge tube containing medium without protein.
5. Wash the ELISA plate again four times. Pipet 100 μL of prepared standard in duplicates in the first two columns starting with 0 pg/mL (blank), followed by increasing IL8 concentrations from 12.5 to 800 pg/mL.
6. Distribute 100 μL of each cell culture supernatant sample in the 96-well plate and incubate for 2 h at RT or overnight at 4 °C. Perform technical duplicates.
7. Dilute 20 μL detection antibody in 10 mL blocking/tween buffer.

8. Wash ELISA plate four times and add 100 μL of the prepared detection antibody solution in each well. Incubate for 1 h at RT.
9. Prepare conjugate solution: add 10 μL conjugate in 10 mL blocking/tween buffer.
10. After washing the plate four times, add 100 μL conjugate solution per well and incubate for 1 h at RT.
11. Add a PNPP pill to 5 mL chilled substrate buffer and be aware that the pill is dissolved completely. After another washing step of the 96-well plate, add 50 μL of the substrate solution to each well and incubate again for 10–60 min at 37 °C in the dark. Check reaction intensity of the alkaline phosphatase every 10 min (*see* **Note 15**).
12. If a yellowish staining is visible, put the plate in a microplate reader and measure reaction intensity of the alkaline phosphatase at 405 nm.

4 Notes

1. Such a construct contains a bipartite targeting and sorting signal that targets the protein first to the mitochondrial inner membrane where the protein is then being processed by a specific peptidase and a soluble moiety is released to the IMS [43]. We used the yeast expression plasmid pYX113 but any other plasmid can be employed.
2. We always refer to purified water when "water" is mentioned.
3. HeLa cells can be stored as single use aliquots at −80 °C. Thaw an aliquot quickly at 37 °C and transfer the cells subsequently to a cell culture flask filled with pre-warmed 1640 medium supplemented with 10 % FCS and appropriate antibiotics.
4. If applicable, FCS can be heat-inactivated for 30 min at 56 °C in a water bath.
5. PBS contains $CaCl_2$ and $MgCl_2$.
6. All cell culture media and chemicals have to be stored at 4 °C after opening. Before usage and contact with cells, pre-warm all reagents to 37 °C. Sterile working is essential when dealing with cell culture.
7. Use a maxisorb ELISA plate with a flat top.
8. It is important to keep this exact order as direct contact of yeast cells with 1 M LiOAc can severely harm them.
9. The yield of mitochondria increases when yeast cells are grown in media with non-fermentable carbon sources such as glycerol, ethanol, or lactate. Growth on galactose combines a

decent yield with a moderate doubling time. The use of glucose as a carbon source should be avoided since in fungi glucose represses the expression of numerous mitochondrial genes. For selection, media should be prepared lacking the corresponding auxotrophic marker(s).

10. Try to harvest the cells in the mid-logarithmic growth phase. Cells from the stationary phase are harder to lyse with glass beads and differ from cells in the logarithmic phase in the composition and amount of mitochondrial proteins.
11. Yeast cell pellets can be kept at −20 °C for several days.
12. In some cases overnight incubation with the primary antibody at 4 °C is beneficial.
13. Carefully agitate the 24-well plate by tracing an hourglass shape to spread the cells over the whole well.
14. As a positive control for the assay, you can add 10 μg purified TNFα in one well which results in a strong IL8 secretion into the supernatant. Biological duplicates are recommended.
15. Check if every well is covered completely and remove all air bubbles.

Acknowledgments

Our work is supported by the Deutsche Forschungsgemeinschaft (SFB766/TP B11 and RA 1028/7-1 to D.R. and SFB766/TP B1 to I.A.) and by the UKT fortüne program (F1433253 to P.O.).

References

1. Paschen SA, Neupert W, Rapaport D (2005) Biogenesis of beta-barrel membrane proteins of mitochondria. Trends Biochem Sci 30: 575–582
2. Gray MW, Burger G, Lang BF (1999) Mitochondrial evolution. Science 283: 1476–1481
3. Esser C, Ahmadinejad N, Wiegand C et al (2004) A genome phylogeny for mitochondria among alpha-proteobacteria and a predominantly eubacterial ancestry of yeast nuclear genes. Mol Biol Evol 21:1643–1660
4. Ferbitz L, Maier T, Patzelt H et al (2004) Trigger factor in complex with the ribosome forms a molecular cradle for nascent proteins. Nature 431:590–596
5. Bos MP, Robert V, Tommassen J (2007) Biogenesis of the gram-negative bacterial outer membrane. Annu Rev Microbiol 61:191–214
6. Bechtluft P, Nouwen N, Tans SJ et al (2010) SecB-a chaperone dedicated to protein translocation. Mol BioSys 6:620–627
7. Zimmer J, Nam Y, Rapoport TA (2008) Structure of a complex of the ATPase SecA and the protein-translocation channel. Nature 455:936–943
8. De Keyzer J, Van Der Does C, Driessen AJ (2003) The bacterial translocase: a dynamic protein channel complex. Cell Mol Life Sci 60: 2034–2052
9. Paetzel M (2013) Structure and mechanism of *Escherichia coli* type I signal peptidase. Biochim Biophys Acta. Biochim Biophys Acta 1843:1497–1508.
10. Sklar JG, Wu T, Kahne D et al (2007) Defining the roles of the periplasmic chaperones SurA, Skp, and DegP in *Escherichia coli*. Genes Dev 21:2473–2484

11. Knowles TJ, Scott-Tucker A, Overduin M et al (2009) Membrane protein architects: the role of the BAM complex in outer membrane protein assembly. Nat Rev Microbiol 7:206–214
12. Patel GJ, Behrens-Kneip S, Holst O et al (2009) The periplasmic chaperone Skp facilitates targeting, insertion, and folding of OmpA into lipid membranes with a negative membrane surface potential. Biochemistry 48:10235–10245
13. Volokhina EB, Grijpstra J, Stork M et al (2011) Role of the periplasmic chaperones Skp, SurA, and DegQ in outer membrane protein biogenesis in *Neisseria meningitidis*. J Bact 193: 1612–1621
14. Voulhoux R, Bos MP, Geurtsen J et al (2003) Role of a highly conserved bacterial protein in outer membrane protein assembly. Science 299:262–265
15. Wu T, Malinverni J, Ruiz N et al (2005) Identification of a multicomponent complex required for outer membrane biogenesis in *Escherichia coli*. Cell 121:235–245
16. Hagan CL, Silhavy TJ, Kahne D (2011) β-Barrel membrane protein assembly by the BAM complex. Annu Rev Biochem 80:189–210
17. Pfanner N, Wiedemann N, Meisinger C et al (2004) Assembling the mitochondrial outer membrane. Nat Struct Mol Biol 11:1044–1048
18. Endo T, Yamano K (2009) Multiple pathways for mitochondrial protein traffic. Biol Chem 390:723–730
19. Paschen SA, Waizenegger T, Stan T et al (2003) Evolutionary conservation of biogenesis of β-barrel membrane proteins. Nature 426: 862–866
20. Wiedemann N, Kozjak V, Chacinska A et al (2003) Machinery for protein sorting and assembly in the mitochondrial outer membrane. Nature 424:565–571
21. Gentle I, Gabriel K, Beech P et al (2004) The Omp85 family of proteins is essential for outer membrane biogenesis in mitochondria and bacteria. J Cell Biol 164:19–24
22. Ishikawa D, Yamamoto H, Tamura Y et al (2004) Two novel proteins in the mitochondrial outer membrane mediate β-barrel protein assembly. J Cell Biol 166:621–627
23. Milenkovic D, Kozjak V, Wiedemann N et al (2004) Sam35 of the mitochondrial protein sorting and assembly machinery is a peripheral outer membrane protein essential for cell viability. J Biol Chem 279:22781–22785
24. Waizenegger T, Habib SJ, Lech M et al (2004) Tob38, a novel essential component in the biogenesis of β-barrel proteins of mitochondria. EMBO Rep 5:704–709
25. Chan NC, Lithgow T (2008) The peripheral membrane subunits of the SAM complex function codependently in mitochondrial outer membrane biogenesis. Mol Biol Cell 19:126–136
26. Kutik S, Stojanovski D, Becker L et al (2008) Dissecting membrane insertion of mitochondrial beta-barrel proteins. Cell 132:1011–1024
27. Dukanovic J, Dimmer KS, Bonnefoy N et al (2009) Genetic and functional interactions between the mitochondrial outer membrane proteins Tom6 and Sam37. Mol Cell Biol 29:5975–5988
28. Gray MW (2011) The incredible shrinking organelle. EMBO Rep 12:873
29. Walther DM, Papic D, Bos MP et al (2009) Signals in bacterial β-barrel proteins are functional in eukaryotic cells for targeting to and assembly in mitochondria. Proc Natl Acad Sci U S A 106:2531–2536
30. Kozjak-Pavlovic V, Ott C, Gotz M et al (2011) Neisserial Omp85 protein is selectively recognized and assembled into functional complexes in the outer membrane of human mitochondria. J Biol Chem 286:27019–27026
31. Müller JE, Papic D, Ulrich T et al (2011) Mitochondria can recognize and assemble fragments of a beta-barrel structure. Mol Biol Cell 22:1638–1647
32. Walther DM, Bos MP, Rapaport D et al (2010) The mitochondrial porin, VDAC, has retained the ability to be assembled in the bacterial outer membrane. Mol Biol Evol 27:887–895
33. Reumann S, Davila-Aponte J, Keegstra K (1999) The evolutionary origin of the protein-translocating channel of chloroplastic envelope membranes: identification of a cyanobacterial homolog. Proc Natl Acad Sci U S A 96: 784–789
34. Gentle IE, Burri L, Lithgow T (2005) Molecular architecture and function of the Omp85 family of proteins. Mol Microbiol 58: 1216–1225
35. Moslavac S, Mirus O, Bredemeier R et al (2005) Conserved pore-forming regions in polypeptide-transporting proteins. FEBS J 272:1367–1378
36. Sanchez-Pulido L, Devos D, Genevrois S et al (2003) POTRA: a conserved domain in the FtsQ family and a class of beta-barrel outer membrane proteins. Trends Biochem Sci 28:523–526
37. Arnold T, Zeth K, Linke D (2010) Omp85 from the thermophilic cyanobacterium *Thermosynechococcus elongatus* differs from proteobacterial Omp85 in structure and domain composition. J Biol Chem 285:18003–18015

38. Habib SJ, Waizenegger T, Niewienda A et al (2007) The N-terminal domain of Tob55 has a receptor-like function in the biogenesis of mitochondrial beta-barrel proteins. J Cell Biol 176:77–88
39. Kim S, Malinverni JC, Sliz P et al (2007) Structure and function of an essential component of the outer membrane protein assembly machine. Science 317:961–964
40. Koenig P, Mirus O, Haarmann R et al (2010) Conserved properties of polypeptide transport-associated (POTRA) domains derived from cyanobacterial Omp85. J Biol Chem 285: 18016–18024
41. Stroud DA, Becker T, Qiu J et al (2011) Biogenesis of mitochondrial beta-barrel proteins: the POTRA domain is involved in precursor release from the SAM complex. Mol Biol Cell 22:2823–2833
42. Schmid Y, Grassl GA, Buhler OT et al (2004) *Yersinia enterocolitica* adhesin A induces production of interleukin-8 in epithelial cells. Infect Immun 72:6780–6789
43. Herlan M, Bornhovd C, Hell K et al (2004) Alternative topogenesis of Mgm1 and mitochondrial morphology depend on ATP and a functional import motor. J Cell Biol 165: 167–173

Chapter 3

Experimental Methods for Studying the BAM Complex in *Neisseria meningitidis*

Martine P. Bos, Ria Tommassen-van Boxtel, and Jan Tommassen

Abstract

Neisseria meningitidis is a human pathogen. It is intensively studied for host–pathogen interactions and vaccine development. However, its favorable growth properties, genetic accessibility, and small genome size also make it an excellent model organism for studying fundamental biological processes, such as outer membrane biogenesis. Indeed, the first component of the assembly machinery for outer-membrane proteins, the BAM complex, was identified in *N. meningitidis*. Here, we describe protocols to inactivate chromosomal genes and to express genes from a well-controlled promoter on a plasmid in *N. meningitidis*. Together, these protocols can be used, for example, to deplete cells from essential components of the BAM complex. We also describe a simple, gel-based assay to assess the proper functioning of the BAM complex in vivo.

Key words BAM complex, Expression vector, Gene cloning, Mutagenesis, *Neisseria meningitidis*, Outer membrane protein, Semi-native SDS-PAGE

1 Introduction

Neisseria meningitidis is a strictly human pathogenic gram-negative diplococcus. It normally resides as a commensal in the nasopharynx in up to 20 % of the population. However, occasionally, it crosses the epithelial cell layers to enter the blood stream and it can cross the blood–brain barrier. The resulting sepsis and meningitis are diseases with high morbidity and mortality. One of the main virulence factors is the capsule, which protects the bacteria against phagocytosis and the complement system. *N. meningitidis* isolates can be either encapsulated or unencapsulated. Only the encapsulated forms cause disease. Based on the structure of the capsular polysaccharide, *N. meningitidis* isolates have been classified into several serogroups. The polysaccharides of serogroups A, C, W, and Y have successfully been employed for vaccine development [1]. However, the capsular polysaccharide of serogroup B strains, the most prevalent serogroup in the industrialized countries, is not

Susan K. Buchanan and Nicholas Noinaj (eds.), *The BAM Complex: Methods and Protocols*, Methods in Molecular Biology, vol. 1329, DOI 10.1007/978-1-4939-2871-2_3, © Springer Science+Business Media New York 2015

immunogenic. Therefore, much research has been focused in the past decades on the identification of suitable subcapsular antigens, including outer membrane proteins (OMPs).

N. meningitidis was the first organism in which a component of the machinery for the assembly of β-barrel OMPs, the BAM complex, was identified [2]. Reports that homologs of a protein designated Omp85, which was under investigation as a vaccine candidate, are essential proteins in *Haemophilus ducreyi* and *Synechocystis* sp. [3, 4] and the observation that the corresponding gene was ubiquitously present in all available gram-negative bacterial genome sequences suggested an important role for this protein and triggered research on its function. It was found that Omp85 is an essential protein also in *N. meningitidis* and that it is required for the proper assembly of all OMPs investigated into the outer membrane [2]. This protein is now called BamA in *Escherichia coli*, where it was found to be in a complex with four lipoproteins, now called BamB-E [5, 6] of which only BamD, along with BamA, is essential [7]. Subsequent studies demonstrated that the BAM complex is different in *N. meningitidis* in that the BamB component is lacking [8]. Furthermore, an additional protein was found to be a constituent of the complex; this protein, RmpM, has no direct role in OMP assembly but stabilizes the BAM complex [8]. Interestingly, like in *E. coli*, the BamD component was found to be essential in *N. meningitidis* [8], although its homolog in the closely related species *Neisseria gonorrhoeae*, ComL, was earlier reported to be a dispensable protein [9].

N. meningitidis has many advantages as a model organism for studying fundamental biological processes and in particular those related to the cell envelope (*see* **Note 1**). The organism is naturally competent to take up DNA and is recombination proficient, which facilitates the construction of mutants. Also, plasmid systems for regulated gene expression have been developed [2]. With 2000–2200 genes, the genome is relatively small, which facilitates the identification of the important ones for a specific process, and the genome sequences of many strains are publicly available [10]. In contrast to *E. coli*, *N. meningitidis* is viable without lipopolysaccharides (LPS), a major lipid component of the outer membrane [11]. This property facilitates research into the mechanism of LPS transport, since also the components of the LPS transport machinery are dispensable and, thus, their genes can be knocked out. Indeed, *N. meningitidis* was also the organism in which the first component of the Lpt machinery, required for the transport of LPS from the inner to the outer membrane, was identified [12]. With respect to studies on OMP assembly, it is important to notice that *N. meningitidis* lacks the σ^E-dependent stress response [13], which is induced in *E. coli* upon OMP assembly defects. In *E. coli*, this response results in the degradation of unassembled OMPs by the periplasmic protease DegP and in the inhibition of OMP synthesis by small regulatory RNAs [14]. Consequently, in contrast to *E. coli*, even slight OMP assembly defects result in *N. meningitidis* in the accumulation of unassembled OMPs, which can readily be detected by semi-native sodium dodecyl sulfate–polyacrylamide gel electrophoresis (SDS-PAGE). This method makes use of the extraordinary stability of native OMPs in SDS at room temperature and of the different electrophoretic mobility of native and unfolded OMPs in SDS-PAGE [2, 15]. Here, we describe the genetic tools that are available to study OMP assembly in *N. meningitidis*. Protocols that can be used to knock out chromosomal *bam* genes and to express *bam* genes from a well-controlled promoter, thus allowing to deplete cells of essential Bam components, are provided. In addition, we describe the semi-native SDS-PAGE method that can be used to detect misfolded OMPs that accumulate when the BAM complex is not properly functioning.

2 Materials

2.1 Bacterial Culture

1. Humidified CO_2 incubator (*see* **Note 2**).
2. Inoculating loops.
3. Spectrophotometer.
4. Laminar flow cabinet.
5. Incubator shaker.
6. GC-agar plates for growth of *N. meningitidis*: Stir 18 g of GC-agar base (Oxoid) in 500 mL of H_2O (*see* **Note 3**). The agar base will not dissolve. Autoclave the suspension for 15 min at 120 °C and 2.7 kg/cm^2. After cooling to 60 °C, add one bottle of Vitox supplement SR0090A (Oxoid) and isopropyl β-D-1-thiogalactopyranoside (IPTG) and antibiotics as required.
7. Lysogeny broth (LB) for growth of *E. coli*: dissolve 10 g of tryptone, 5 g of yeast extract, and 5 g of NaCl in 1 L of H_2O. Adjust pH to 7.0 with NaOH and autoclave as above. To prepare solid medium, add 1.5 % (w/v) agar before autoclaving.
8. Tryptic soy broth (TSB) for liquid culture of *N. meningitidis*: dissolve 15 g of TSB in 500 mL of H_2O and autoclave as above.
9. Antibiotic stock solutions: 100 mg/mL ampicillin (Amp) in H_2O, filter-sterilized through a 0.2-μm filter; 80 mg/mL kanamycin (Kan) in H_2O, filter-sterilized; 10 mg/mL chloramphenicol (Cam) in 96 % (v/v) ethanol. All stocks can be stored in aliquots at −20 °C.
10. IPTG (1 M): dissolve 238 mg IPTG in 1 mL H_2O. Sterilize by filtration using a 0.2-μm filter. For long-term storage, place at −20 °C. Stocks can be kept at 4 °C for several weeks.

2.2 Agarose Gel Electrophoresis

1. Microwave.
2. Horizontal gel electrophoresis system.
3. Power supply.
4. Gel imaging system with a UV lamp (λ254 nm).
5. Diaminoethane tetraacetic acid (EDTA) (0.5 M): add 18.6 g EDTA to 100 mL of H_2O and adjust the pH to 8.0 with NaOH to dissolve the EDTA.
6. Tris–Borate–EDTA (TBE) solution, 5× stock: add 54 g Tris base, 27.5 g boric acid, and 20 mL of 0.5 M EDTA, pH 8, to 800 mL of H_2O. After all compounds are dissolved, adjust the volume to 1 L with H_2O.
7. Agarose D1 LEEO (Hispanagar).
8. Loading buffer (6×) (Fermentas).
9. DNA ladder.
10. Ethidium bromide solution: for a stock solution, dissolve 10 mg of ethidium bromide in 1 mL of H_2O. Before application, dilute the stock solution 1:10,000 in 0.5× TBE buffer (*see* **Note 4**).

2.3 Isolation of DNA

1. Tabletop centrifuge.
2. Thermocycler.
3. Sterile toothpicks: sterilize household toothpicks by autoclaving (*see* Subheading 2.1).
4. Tris-EDTA (TE) solution: mix 10 mL of 1 M Tris–HCl buffer, pH 7.5, and 2 mL of 0.5 M EDTA with 988 mL of H_2O.
5. Commercial plasmid isolation kit.
6. Razor blade.
7. Gel and PCR cleanup kit, e.g., Wizard SV gel and PCR cleanup kit (Promega).
8. SpeedVac concentrator.

2.4 Polymerase Chain Reaction (PCR)

1. Thermocycler.
2. PCR tubes.
3. Commercial PCR mixes, such as Taq DNA polymerase with buffer (Fermentas) or the Expand High Fidelity PCR system (Roche).
4. Stock solution containing 2 mM of each deoxynucleotide (dNTPs).
5. Primers: 10-μM stock solutions.
6. DNA template (*see* Subheadings 3.3.1 and 3.3.2).

2.5 Plasmid Constructions

1. SpeedVac concentrator.
2. Cloning vector pCRII-TOPO (Invitrogen).
3. Chemically competent *E. coli* TOP10F' cells (Invitrogen).
4. Neisserial expression plasmid pEN11 [12].
5. Plasmid pMB25 [12].
6. Restriction enzymes SalI, EcoRI, HindIII, NdeI, and AatII and T4 DNA ligase with corresponding buffers.
7. Designed primers plus primers M13For(-20) (5′-GTAAAACGACGGCCAGT-3′) and M13Rev(-20) (5′-GATAACAATTTCACACAGG-3′) that anneal on either side of the insert in pCRII-TOPO.

2.6 Transformation of N. meningitidis

1. GC-agar plates with and without the appropriate antibiotics.
2. Leica S6D Stereozoom binocular Microscope (Leica Microsystems GmbH).
3. Sterile glass tubes, 0.85 × 15.5 cm (Beldico).
4. Humidified CO_2 incubator (*see* **Note 2**).
5. $MgCl_2$ solution (30 mM): dissolve 28 mg $MgCl_2$ in 10 mL of H_2O and filter-sterilize using a 0.2-μm filter.

2.7 Cell Envelope Preparation

1. Refrigerated benchtop centrifuge.
2. Spectrophotometer.
3. Freezer, −80 °C.
4. Branson sonifier 450 ultrasonic disintegrator (Branson Ultrasonics Corporation).
5. Ultracentrifuge.
6. Lysate buffer: add 10 mL of a solution of 0.5 M EDTA to 50 mL of 1 M Tris-base, adjust to pH 8 with HCl, and add H_2O to a final volume of 1 L.
7. Tris–HCl, pH 7.6 (2 mM): add 2 mL of 1 M Tris-base to 800 mL of H_2O, adjust to pH 7.6 with HCl, and add H_2O to a final volume of 1 L.

2.8 Semi-native SDS-PAGE

1. Mini-Protean III gel system with glass plates and combs (BioRad).
2. Power supply.
3. Running gel (RG) buffer: dissolve 90.9 g Tris-base in 1 L of H_2O and adjust the pH to 8.8 with HCl.
4. Stacking gel (SG) buffer: dissolve 30.3 g Tris-base in 1 L of H_2O and adjust the pH to 6.8 with HCl.
5. Acrylamide–bis solution 37.5:1 (40 %).
6. Acrylamide–bis solution 37.5:1 (30 %).

7. Ammonium persulfate (APS) (1 % (v/v)): dissolve 1 g APS in 100 mL of H_2O.
8. SDS (10 % (v/v)): dissolve 10 g SDS in 100 mL of H_2O (*see* **Note 5**).
9. Tetramethylethylenediamine (TEMED).
10. Isopropanol ≥99.5 % pure.
11. Glycine (1 M): dissolve 75.07 g glycine in 1 L of H_2O.
12. Running buffer: mix 25 mL of 1 M Tris-base with 192 mL of 1 M glycine and 10 mL of 10 % SDS (*see* **Note 6**). Adjust the volume to 1 L with H_2O; this should yield a pH of 8.3.
13. Tris–HCl, pH 6.8 (1 M): dissolve 12.114 g of Tris-base in 80 mL of H_2O and adjust the pH to 6.8 with HCl and the volume to 100 mL with H_2O.
14. Sample buffer (2×): dissolve 2 mg bromophenol blue in 1.25 mL of 1 M Tris–HCl (pH 6.8), 2 mL of glycerol, and 4 mL of 10 % SDS (*see* **Note 6**). Adjust the volume to 10 mL with H_2O.
15. Molecular weight marker, e.g., a prestained marker.

2.9 Coomassie Brilliant Blue Staining

1. Orbital shaker.
2. Microwave.
3. Bradford reagent: dissolve 100 mg of Coomassie Brilliant Blue (CBB) G250 in 50 mL of ethanol, adjust the volume to 900 mL with H_2O, and add 100 mL of an aqueous solution of 85 % phosphoric acid.
4. Acetic acid solution (10 % (v/v)): add 100 mL acetic acid to 900 mL H_2O.

3 Methods

3.1 Bacterial Culture

1. Scrape some material from the surface of a frozen stock of *N. meningitidis* onto GC-agar plates using an inoculating loop. Put the plates at 37 °C in a humidified incubator with 5 % (v/v) CO_2 (*see* **Note 2**). For the selection of transformants, add appropriate antibiotics at the following concentrations to the plates: 100 μg/mL Kan or 10 μg/mL Cam. For growth of meningococci in liquid culture, inoculate TSB medium with bacteria at a starting optical density at λ550 nm (OD_{550}) of 0.1 (*see* **Note 7**) and grow in an incubator shaker at 37 °C. Shaking speed is not crucial, as long as the suspension does not start to foam. For induction of expression of plasmid-encoded genes under *lac* promoter control, add 1 mM IPTG.

2. Grow *E. coli* on LB-agar plates at 37 °C in a dry incubator. For selection or maintenance of plasmids, the medium should be supplemented with Kan, Cam, or Amp at 50 μg/mL, 25 μg/mL or 100 μg/mL, respectively.

3.2 Agarose Gel Electrophoresis

1. Add 1.5 g agarose D1 LEEO to 100 mL of 0.5× TBE solution and boil the suspension in a microwave oven until the solution is completely clear.
2. Add 50 mL of 0.5× TBE solution and pour the gel into a horizontal gel electrophoresis system tray. Solidification at room temperature usually takes ~30 min.
3. Immerse the agarose gel fully in 0.5× TBE solution in the horizontal gel electrophoresis system.
4. Mix the DNA samples and 6× loading buffer in a 5:1 ratio.
5. Load the samples on the gel and carry out electrophoresis for 1 h at 100 V.
6. Impregnate the gel in a solution of 1 μg/mL ethidium bromide in 0.5× TBE buffer (*see* **Note 4**).
7. Stain for 15–30 min and visualize the DNA with UV light using a gel imaging system.

3.3 Isolation of DNA

3.3.1 Isolation of Chromosomal Template DNA

1. Use a toothpick to transfer a small amount of bacteria, grown overnight on plates, into 50 μL of H_2O in a PCR tube.
2. Incubate the tube for 5 min in a thermocycler at 95 °C.
3. Spin for 5 min at 20,000 × *g* in a tabletop centrifuge.
4. The supernatant can be used as a source of genomic template DNA in PCR reactions.

3.3.2 Plasmid Isolation

1. Use a sterile cotton swab to collect overnight grown bacteria from half an agar plate and transfer them into 1.5 mL of TSB for meningococci or 1.5 mL of LB for *E. coli* in an Eppendorf tube
2. Centrifuge the suspension for 5 min at 6000 × *g*.
3. Isolate plasmid DNA from the pellet using a commercially available mini-prep plasmid isolation kit in a final volume of 50–100 μL following the manufacturer's instructions.
4. If necessary, plasmids can be fivefold concentrated in approximately 30 min in a SpeedVac device.

3.3.3 Purification of DNA Fragments

1. Separate DNA mixtures on agarose gels (*see* Subheading 3.2) and cut PCR products or restriction fragments of the expected size from the gels using a razor blade while illuminating the gel with UV light (*see* **Note 8**).

2. Use a commercially available PCR clean-up kit or gel purification kit to obtain DNA fragments in a volume of 30–50 μL following the manufacturer's instructions.

3.4 Polymerase Chain Reaction (PCR)

Download the sequenced genome of the relevant *N. meningitidis* strain into a bioinformatics software program, such as CloneManager. Design primers using the primer design feature of the program.

1. Add in this order into the PCR tubes: 16.8 μL of H_2O; 2.5 μL of 10× PCR buffer supplied with the polymerase; 2.5 μL of a 2-mM stock solution of dNTPs; 1 μL of 10-μM stock solutions of each primer; 1 μL of template DNA and, finally, 0.2 μL of DNA polymerase (*see* **Note 9**).
2. Perform the following PCR steps in the thermocycler:
   (a) DNA melting for 5 min at 95 °C.
   (b) DNA melting for 30 s at 95 °C.
   (c) Primer annealing for 30 s at a temperature that is 5 °C below the melting temperature (T_m) of the primer with the lowest T_m.
   (d) Extension by the polymerase for a time period that depends on the size of the expected product. Use 1 min per 1000 bp. Use the extension temperature recommended by the manufacturer of the polymerase, which is usually 72 °C.
   (e) Repeat **steps b–d** for 29 cycles.
   (f) Finalization of products for 5 min at 72 °C.

3.5 Plasmid Constructions

3.5.1 Construction of Neisserial Expression Plasmids

Most cloning and expression vectors, designed for use in *E. coli*, do not propagate in *N. meningitidis*. Pagotto et al. used the origin of replication of a plasmid naturally occurring *N. gonorrhoeae* to construct a shuttle vector, pFP10, which replicates extrachromosomally in both *Neisseria* spp. and *E. coli* [16]. We further adapted this plasmid to create a vector capable of regulated gene expression. To that end, we introduced a *lac*-derived promoter into this plasmid, which contains two promoter and operator sequences to provide tight promoter control, and a *lacI*^{Q} gene encoding the *lac* repressor [17]. Next, we introduced the ribosome-binding site of the well-expressed *bamA* gene [2], and NdeI and AatII sites allowing for easy subcloning of any ORF into this construct [18]. This plasmid is called pEN11 [12] (Fig. 1). Genes can be cloned into this plasmid as follows:

1. Amplify the gene of interest by PCR (*see* Subheading 3.4) with a forward primer containing an NdeI site, such that the ATG present in the NdeI recognition site (CATATG) represents the start codon of the gene, and a reverse primer containing an AatII site after the stop codon.

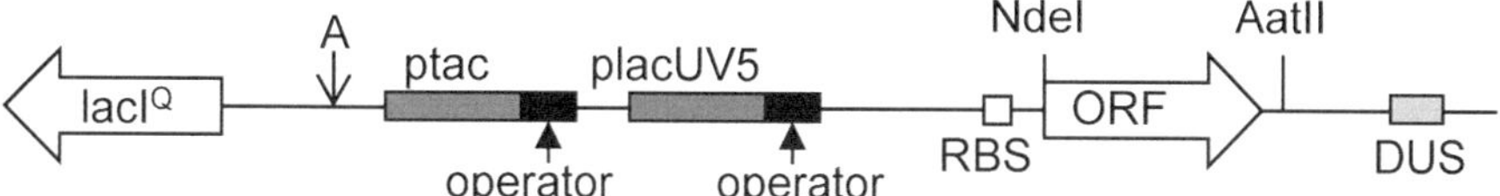

Fig. 1 Schematic representation of the cloning region of pEN11. RBS, ribosome-binding site; ORF, open reading frame; DUS, DNA uptake sequence. The *arrow* points to the annealing site of primer A (*see* **Note 19**). The elements shown are not drawn to scale. ptac and placUV5 are two engineered versions of the original *lac* promoter

2. Ligate the PCR product directly from the PCR mix into pCRII-TOPO. Transform *E. coli* TOP10F' with the ligation mixture and select on LB plates with Amp (*see* **Note 10**).
3. Purify plasmid from Amp-resistant transformants and verify the sequence of the inserted fragment using primers M13For(-20) and M13Rev(-20) (*see* **Note 11**).
4. Cut the plasmid with NdeI and AatII and ligate the released DNA fragment into pEN11 digested with the same enzymes. Transform *E. coli* TOP10F' with the ligation mixture and select on LB plates containing Cam.
5. Isolate the resulting plasmid (*see* Subheading 3.3.2) and bring the volume of the preparation down to 10 μL using a SpeedVac.

3.5.2 Generation of Constructs to Create Chromosomal Knockout Mutants

Gene inactivation is accomplished by replacing all or most of the target gene by an antibiotic resistance marker via homologous recombination. To that end, two DNA segments flanking the target gene (Up- and Down-region) are cloned with an antibiotic resistance cassette in between them. The size of the flanking segments should be at least 100 bp but preferably longer (up to 500 bp or longer) to enable more efficient recombination.

1. Design primers and include a SalI site (CTGGAC) at the 5′ end of both the reverse primer of the Up-region and the forward primer of the Down-region.
2. Amplify the flanking DNA segments by PCR (*see* Subheading 3.4) using 1 μL of a 1:100 dilution of chromosomal DNA as template (*see* Subheading 3.3.1) and clone the PCR products separately into pCRII-TOPO.
3. Determine the orientation of the flanks in pCRII-TOPO. To this end, perform a PCR on the plasmid DNA with one of the primers used to amplify the flank from the chromosome and the other being the M13for or M13rev primer. Determine the size of the PCR products obtained by agarose gel electrophoresis (*see* Subheading 3.2) to deduce the orientation of the insert and continue with the plasmids that have both flanks in the same orientation.

4. Combine the two flanks into one plasmid by restricting each flank-containing plasmid with HindIII and SalI. Ligate the fragment released from one plasmid into the other restricted plasmid.
5. Verify the sequence of the insert in the resulting plasmid (*see* **Note 11**)
6. Release the Kan-resistance cassette from pMB25 (*see* **Note 12**) with SalI and insert this cassette into the SalI-restricted plasmid containing both flanks.
7. Determine the orientation of the cassette by PCR using an internal primer in the cassette and a primer in one of the flanks (*see* **Note 13**).
8. Restrict the resulting plasmid with EcoRI and purify the fragment containing the flanks and the antibiotic-resistance cassette (*see* Subheading 3.3.3) to obtain linear knockout DNA for transformation into *N. meningitidis* (*see* Subheading 3.6.2).

3.6 Transformation of N. meningitidis

3.6.1 Selection and Maintenance of Piliated N. meningitidis

N. meningitidis uses type IV pili to take up DNA. However, the capacity to express these pili is occasionally lost, e.g., upon repeated or prolonged (i.e., >48 h) subculture on agar plates [19]. The piliation status can be assessed by growing the bacteria on GC-agar plates at 30 °C and judging their colony morphology by using a binocular microscope: piliated colonies form small domed colonies with well-defined edges due to their auto-agglutinability, while non-piliated organisms form flat, spreading colonies [20]. If the capacity to express pili is lost, piliated derivatives can be selected by growing the meningococci in TSB at 30 °C without shaking for 16 h in sterile glass tubes. Piliated cells can then be collected from the air–water interface (*see* **Note 14**). For highest transformation efficiencies, take bacteria from a −80 °C stock that is known to be piliated and grow these only once on GC-agar for no longer than 18 h (*see* **Note 15**).

3.6.2 Transformation of N. meningitidis

1. Grow meningococci overnight (~16 h) on a GC-agar plate.
2. Prepare transformation DNA: add 2 μL of 30 mM $MgCl_2$ to 10 μL of the concentrated pEN11 preparation (*see* Subheading 3.5.1) or to 10 μL of the linear knockout DNA (*see* Subheading 3.5.2)
3. Draw a square of 1 × 1 cm on the back of a fresh GC-agar plate using a marker and streak a small amount of the overnight-grown bacteria onto the plate within this square: the streak should be barely visible.
4. Drop 12 μL of transformation DNA onto the bacteria within the square. Incubate the plate at 37 °C with 5 % (v/v) CO_2 for 6–8 h.
5. Bacterial growth should be seen within the square after this period. Scrape all the bacteria with a sterile cotton stick from the plate

and transfer them onto a fresh GC-agar plate containing the appropriate antibiotics and, if necessary, IPTG (*see* **Note 16**).

6. Pick single colonies over the next 24–48 h and streak them onto fresh selection GC-agar plates (*see* **Note 17**).
7. Verify chromosomal knockout mutants by PCR (*see* Subheading 3.4) using chromosomal DNA as template and the forward primer of the upstream flanking region and the reverse primer of the downstream flanking region (*see* Subheading 3.5.2). Correct transformants should produce a PCR product of a different size than wild type cells, which should be taken along as control.
8. Verify the presence of pEN11-derived plasmid in transformants by testing for IPTG-inducible expression of the protein (*see* **Notes 18–20**).

3.7 Cell Envelope Isolation

1. Grow meningococci for 5 h in 25 mL of TSB and measure the OD_{550}.
2. Harvest the bacteria from the culture by centrifugation in a benchtop centrifuge (5000 × *g*, 10 min, 4 °C).
3. Resuspend the cells in 15 mL of ice-cold lysate buffer.
4. Incubate the resulting suspension for at least 30 min in the −80 °C freezer (*see* **Note 21**).
5. After thawing, ultrasonically disintegrate the bacteria in a Branson sonifier 450 for 5 min (duty cycle: 40 %, output control: 7) using a macrotip. For optimal cooling, leave the tube with the bacterial suspension in a holder containing melting ice during the sonication.
6. Remove unbroken cells from the lysate by centrifugation in a benchtop centrifuge (12,000 × *g*, 15 min, 4 °C).
7. Harvest the cell envelopes from the supernatant by ultracentrifugation (100,000 × *g*, 8 min, 4 °C).
8. Dissolve the pellet in 2 mM Tris–HCl, pH 7.6. Store the cell envelopes until further use at −20 °C (*see* **Note 22**).

3.8 Semi-native SDS-PAGE

1. Prepare running gel solution for two gels by mixing 5 mL of RG buffer with 2.5 mL of 40 % acrylamide–bis solution, 2.2 mL of H_2O, 0.25 mL of 1 % APS, and, finally, 0.02 mL of TEMED (*see* **Note 23**). Immediately after the addition of TEMED, pour the running gel solution in between the glass plates until approximately 3 cm from the top. Fill the remaining space with isopropanol to prevent evaporation and contact of the gel solution with oxygen, which retards polymerization. In addition, this step creates a straight gel border.
2. Polymerization of the gel requires 10–20 min at room temperature.

3. Meanwhile, prepare the stacking gel solution for two gels by mixing 2.5 mL of SG buffer with 0.5 mL of 30 % acrylamide–bis solution, 1.88 mL of H_2O, 0.2 mL of 1 % APS, and, finally, 0.02 mL of TEMED.
4. Discard the layer of isopropanol from the polymerized gel and pour the stacking gel solution on top of the gel.
5. Immediately add a comb into the stacking gel solution (in between the two glass plates).
6. The stacking gels will polymerize in 20–30 min.
7. Assemble the gels in a Mini-Protean III gel system and fill the buffer chambers with the running buffer.
8. Prepare native samples by mixing cell envelope preparations with an equal volume of 2× sample buffer and leave the samples at room temperature.
9. Prepare denatured samples by mixing preparations with an equal volume of 2× sample buffer supplemented with 5 % β-mercaptoethanol and boil the samples for 10 min.
10. Load the samples and the molecular weight marker to the slots of the gels. Electrophoresis should be performed at 12 mA at 4 °C in a cold room or by putting the gel unit in a bucket filled with ice (*see* **Note 24**) until the blue dye front reaches the bottom of the running gel.

3.9 CBB Staining

Most CBB-staining methods require staining of the gel, followed by extensive destaining. The method presented below does not require destaining and the results can, therefore, be evaluated faster (after ~50 min). In addition, because no destaining is required, the method is cheaper and produces less organic waste.

1. After finishing semi-native SDS-PAGE, dismantle the gel unit and rinse the gel three times for 5 min in H_2O at room temperature while shaking.
2. Rinse the gel for a fourth time in H_2O while heating in a microwave until boiling.
3. Remove H_2O and add 10 mL Bradford reagent while the gel is still hot. Incubate for 30 min at room temperature while shaking. All major bands are visible now with hardly any background. The next steps can increase the intensity of staining and reveal additional minor bands (*see* **Note 25**).
4. Add 40 mL H_2O and continue incubation at room temperature overnight.
5. Wash two to five times with H_2O for 5 min until you are satisfied with the intensity of the color and the clearness of the background.
6. For storage, incubate the gel for 10 min in 10 % acetic acid solution to fix the results (*see* **Note 26**).

4 Notes

1. *N. meningitidis* is a biosafety-level II organism. Since the capsule is regarded as a major virulence factor, it is advisable to use unencapsulated strains for fundamental research projects. For example, in our laboratory, we use strain HB-1, an unencapsulated derivative of the commonly used serogroup B strain H44/76 [21], for fundamental research. However note that in the absence of a suitable animal model for *N. meningitidis* infection, the safety of such strains has never been rigorously proven. Therefore, to eliminate the risk of infection for laboratory workers, *N. meningitidis* should always be grown in closed containers, which should only be opened in a laminar flow cabinet.
2. *N. meningitidis* grows better with elevated levels of CO_2. When a standard CO_2 incubator is not available, candle jars can be used.
3. In all materials described in this chapter, we use distilled or demineralized water.
4. Caution: as ethidium bromide is a known mutagen and toxic, wear gloves and protective clothing.
5. Caution: wear a dust mask for protection against breathing SDS powder.
6. The porins, PorA and PorB, are resistant to denaturation in the SDS concentrations routinely used in SDS-PAGE if the samples are left unheated before electrophoresis. These proteins are by far the most abundant OMPs of *N. meningitidis* and are, therefore, ideal markers to evaluate correct OMP assembly. If correctly assembled, these porins form homotrimeric complexes that are associated with another major OMP, RmpM [22]. These complexes migrate with an apparent molecular weight of ~150 kDa during semi-native SDS-PAGE, whilst the unassembled monomeric forms migrate at ~35–45 kDa (depending on the specific porin species and the strain from which they are derived) (*see* Fig. 2). Other OMPs may be more sensitive to SDS. Therefore, when interested in the proper assembly of specific OMPs, the concentration of SDS in the sample and running buffers may need to be adapted, and the optimal concentrations have to be empirically determined for each OMP.
7. *N. meningitidis* is very sensitive to detergents. Therefore, it is recommend not to grow these bacteria in washed glassware that may still contain traces of soap. Growing them in plastic disposables is also advisable for biosafety reasons.

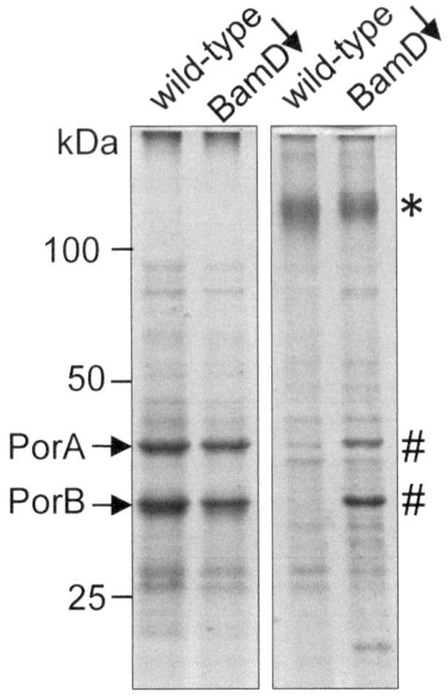

Fig. 2 Assessment of porin assembly by semi-native SDS-PAGE. Cell envelopes of wild type and BamD-depleted cells were subjected to denaturing SDS-PAGE (*left panel*) or semi-native SDS-PAGE (*right panel*) and stained with CBB. The positions of porin oligomers and of unassembled porin monomers in the semi-native gel are indicated by an asterisk and by #, respectively

8. Wear glasses and gloves and work quickly to avoid exposure of eye lenses and skin to UV light. Working as quickly as possible also avoids damage to the DNA.
9. Use a proofreading DNA polymerase such as Expand High Fidelity PCR system from Roche for cloning, and any cheaper DNA polymerase for testing plasmids and transformants.
10. PCR products can also be cut with NdeI and AatII and ligated directly into pEN11. However, in our experience, initial cloning in pCRII-TOPO, followed by digestion of the resulting plasmid with NdeI and AatII and subcloning of the released fragment in pEN11 is generally more efficient.
11. Sequencing can be done by Sanger sequencing at any commercial company.
12. Any antibiotic-resistance cassette could work. However, *N. meningitidis* selectively takes up its own DNA, which is recognized by the presence of neisserial DNA uptake sequences (5′-GCCGTCTGAA-3′), during natural transformation. The Kan-resistance cassette of pMB25 includes such an uptake sequence, which warrants the efficient uptake of the generated knockout construct for subsequent gene exchange. If an

antibiotic-resistance marker from another source is chosen, it should be verified that an uptake sequence is present in the cloned flanks of the target gene.

13. A knockout construct should be selected with the gene conferring kanamycin resistance in the same transcriptional orientation as the target gene to avoid potential polar effects on expression of downstream genes.
14. Piliated strains of *N. meningitidis* and other gram-negative species form a pellicle at the air–water interface when cultured in static broth medium, and such conditions can be used to selectively enrich for piliated organisms.
15. If all attempts to isolate piliated derivatives of a non-piliated strain fail, an alternative transformation protocol for non-piliated strains is available [23]. However, this procedure is far less efficient than natural transformation. For fundamental research projects, such as on the mechanism of OMP assembly, we recommend the use of an efficiently transformable strain, e.g., HB-1, as the model organism.
16. If the gene to be inactivated is essential, such as *bamA*, use a strain containing another copy of the gene on a pEN11-derived plasmid, and use IPTG-containing plates for pre-growth, transformation, and selection of the bacteria. The resulting strains can be depleted of BamA by growth in the absence of IPTG.
17. It is important to streak the colonies from the first selection plate onto a new one before testing them by PCR. Colonies on the first selection plate may still contain non-incorporated DNA that was used for the transformation, which may yield false-positive PCR results. Also, regrowth on a fresh selection plate indicates a stable resistance phenotype.
18. Grow transformants in TSB in the presence and absence of 1 mM IPTG for 4–6 h. Collect the cells and test for presence of the protein of interest by SDS-PAGE and CBB staining or Western blotting if appropriate antiserum is available.
19. Occasionally, we obtained transformants that did not show IPTG-inducible expression of the gene cloned into pEN11. This was often due to a recombination event that had taken place between the two *lac*-operator sequences present on the plasmid (Fig. 1) resulting in an internal deletion of the double *lac* promoter. Such an event can be demonstrated by PCR using primer A (5′-TCTGGATAATGTTTTTTGCGCCGAC-3′) annealing just upstream of the promoter region (Fig. 1) and a reverse primer annealing in the 5′ end of the cloned ORF. In case of a promoter deletion, the size of the PCR product will be 70 base pairs smaller than expected.

20. In our experience, the recombinant plasmids obtained according to this procedure are stably maintained once introduced in *N. meningitidis*. However during the transformation, recombination can occur if the cloned gene is a mutant allele of a gene that is also present on the chromosome. In such a case, transformants should be carefully checked, e.g., by sequence analysis, to verify that the mutant allele is still on the plasmid and the wild type allele on the chromosome. In most studies, we introduce mutant alleles of a gene into a strain with a complete deletion of the chromosomal copy of the gene. This strategy, however, is not possible in the case of essential genes, such as *bamA*, where the chromosomal copy can only be removed after introduction of a (partially) functional allele on the plasmid.
21. This step will both kill *N. meningitidis* and facilitate disruption of the cells during subsequent ultrasonication.
22. Resuspend the pellet of cell envelopes at 4 °C in 0.5–1 mL of 2 mM Tris–HCl depending on the input amount of cells (e.g., cell envelopes of a wild type strain grown to an OD_{550} of 3.5 can be resuspended in 1 mL and the cell envelopes from other strains in a volume in proportion to the OD_{550} measured at **step 1** of the procedure). To facilitate resuspension, use a small magnetic stirrer or leave the pellet in the buffer for several hours without stirring. Pipetting up and down to speed up resuspension often results in foaming and poor solubilisation.
23. The resulting 10 % acrylamide gels are suitable for the analysis of porins. For the analysis of smaller or larger OMPs, the concentration of acrylamide may need to be adapted. Always choose a concentration of acrylamide that allows the OMP of interest to migrate a fair distance into the gel. It is advisable to use the lowest possible concentration as this was shown to increase the difference in migration between folded and non-folded OMPs [24].
24. Gels heat up during standard electrophoresis conditions. Running the gels at low amperage with cooling prevents heat denaturation of the proteins during electrophoresis.
25. The porins, which are ideal markers for evaluating the proper functioning of the BAM complex (*see* **Note 6**), are present in large quantities in cell envelope preparations and can be readily detected on gels stained for 30 min with Bradford reagent. Minor OMPs, including BamA, may require further incubation as described or Western blotting can be used provided that antisera are available.
26. The acetic acid solution can be reused.

References

1. Zahlanie YC, Hammadi MM, Ghanem ST et al (2014) Review of meningococcal vaccines with updates on immunization in adults. Hum Vaccin Immunother 10:995–1007
2. Voulhoux R, Bos MP, Geurtsen J et al (2003) Role of a highly conserved bacterial protein in outer membrane protein assembly. Science 199:262–265
3. Thomas KL, Leduc I, Olsen B et al (2001) Cloning, overexpression, purification, and immunobiology of an 85-kDa outer membrane protein from *Haemophilus ducreyi*. Infect Immun 69:4438–4446
4. Reumann S, Davilla-Aponte J, Keegstra K (1999) The evolutionary origin of the protein-translocating channel of chloroplastic envelope membranes: identification of a cyanobacterial homolog. Proc Natl Acad Sci U S A 96:784–789
5. Wu T, Malinverni J, Ruiz N et al (2005) Identification of a multicomponent complex required for outer membrane biogenesis in *Escherichia coli*. Cell 121:235–245
6. Sklar J, Wu T, Gronenberg LS et al (2007) Lipoprotein SmpA is a component of the YaeT complex that assembles outer membrane proteins in *Escherichia coli*. Proc Natl Acad Sci U S A 104:6400–6405
7. Malinverni JC, Werner J, Kim S et al (2006) YfiO stabilizes the YaeT complex and is essential for outer membrane protein assembly in *Escherichia coli*. Mol Microbiol 61:151–164
8. Volokhina EB, Beckers F, Tommassen J et al (2009) The β-barrel outer membrane protein assembly complex of *Neisseria meningitidis*. J Bacteriol 191:7074–7085
9. Fussenegger M, Facius D, Meier J et al (1996) A novel peptidoglycan-linked lipoprotein (ComL) that functions in natural transformation competence of *Neisseria gonorrhoeae*. Mol Microbiol 19:1095–1105
10. Meningitis Research Foundation Meningococcus Genome Library at http://www.meningitis.org/research/genome
11. Steeghs L, den Hartog R, den Boer A et al (1998) Meningitis bacterium is viable without endotoxin. Nature 392:449–450
12. Bos MP, Tefsen B, Geurtsen J et al (2004) Identification of an outer membrane protein required for lipopolysaccharide transport to the bacterial cell surface. Proc Natl Acad Sci U S A 101:9417–9422
13. Bos MP, Robert V, Tommassen J (2007) Biogenesis of the Gram-negative bacterial outer membrane. Annu Rev Microbiol 61:191–214
14. Ruiz N, Silhavy TJ (2005) Sensing external stress: watchdogs of the *Escherichia coli* cell envelope. Curr Opin Microbiol 8:122–126
15. Nakamura K, Mizushima S (1976) Effects of heating in dodecyl sulfate solution on the conformation and electrophoretic mobility of isolated major outer membrane proteins from *Escherichia coli* K-12. J Biochem 80:1411–1422
16. Pagotto FJ, Salimnia H, Totten PA et al (2000) Stable shuttle vectors for *Neisseria gonorrhoeae*, *Haemophilus* spp. and other bacteria based on a single origin of replication. Gene 244:13–19
17. Seifert HS (1997) Insertionally inactivated and inducible *recA* alleles for use in *Neisseria*. Gene 188:215–220
18. van Ulsen P, van Alphen L, ten Hove J et al (2003) A neisserial autotransporter NalP modulating the processing of other autotransporters. Mol Microbiol 50:1017–1030
19. McGee ZA, Street CH, Chappell CL et al (1979) Pili of *Neisseria meningitidis*: effect of media on maintenance of piliation, characteristics of pili, and colonial morphology. Infect Immun 24:194–201
20. Blake MS, MacDonald CM, Klugman KP (1989) Colony morphology of piliated *Neisseria meningitidis*. J Exp Med 170:1727–1736
21. Bos MP, Tommassen J (2005) Viability of a capsule- and lipopolysaccharide-deficient mutant of *Neisseria meningitidis*. Infect Immun 73:6194–6197
22. Jansen C, Wiese A, Reubsaet L et al (2000) Biochemical and biophysical characterization of *in vitro* folded outer membrane porin PorA of *Neisseria meningitidis*. Biochim Biophys Acta 1464:284–298
23. Bogdan JA, Minetti CA, Blake LS (2002) A one-step method for genetic transformation of non-piliated *Neisseria meningitidis*. J Microbiol Methods 49:97–101
24. Heller KB (1978) Apparent molecular weights of a heat-modifiable protein from the outer membrane of *Escherichia coli* in gels with different acrylamide concentrations. J Bacteriol 134:1181–1183

Chapter 4

Heat Modifiability of Outer Membrane Proteins from Gram-Negative Bacteria

Nicholas Noinaj, Adam J. Kuszak, and Susan K. Buchanan

Abstract

β-barrel membrane proteins are somewhat unique in that their folding states can be monitored using semi-native SDS-PAGE methods to determine if they are folded properly or not. This property, which is commonly referred to as heat modifiability, has been used for many years on both purified protein and on whole cells to monitor folded states of proteins of interest. Additionally, heat modifiability assays have proven indispensable in studying the BAM complex and its role in folding and inserting β-barrel membrane proteins into the outer membrane. Here, we describe the protocol our lab uses for performing the heat modifiability assay in our studies on outer membrane proteins.

Key words BamA, Heat modifiability, OMP, Outer membrane protein, β-barrel membrane protein, BAM complex, Protein folding

1 Introduction

β-barrel outer membrane proteins (OMPs) from gram-negative bacteria often exhibit a useful characteristic when analyzed by semi-native SDS-PAGE methods [1–5]. Unboiled protein samples will run differently than samples that have been boiled. This behavior is due to the extensive hydrogen bonding network holding the beta strands together into the barrel shape, a structural feature that is often resistant to denaturation by SDS alone. As a result, an OMP that has been solubilized with a mild detergent-like DDM retains a relatively compact globular shape on semi-native SDS-PAGE which allows it to migrate further along the gel than the same sample that has been denatured (unfolded) by being boiled (heat). This migration difference between the folded and unfolded states of an OMP is the basis of the heat modifiability assay. Here we present a protocol our lab uses when determining whether OMPs are folded properly or not, either (1) after purification from recombinant expression or (2) after refolding from inclusion

Susan K. Buchanan and Nicholas Noinaj (eds.), *The BAM Complex: Methods and Protocols*, Methods in Molecular Biology, vol. 1329, DOI 10.1007/978-1-4939-2871-2_4,

bodies, for structure determination. We also show some real examples of OMPs that we have recently worked on in our lab and determined their crystal structures.

2 Materials

Reagents and Equipment

1. Protein sample(s): *Ec*BamA, *Hd*BamAΔ3, ProtX, *Ec*TamA, and *Nm*TbpA are samples from our lab used here.
2. UV–Vis spectrometer for determining protein concentration.
3. SDS sample loading buffer (2×): 20 % glycerol, 120 mM Tris–HCl, pH 6.8, 2 % SDS, 0.02 % bromophenol blue (*see* **Note 1**). The concentration of SDS is varied from 0 to 1 % in our assays.
4. Microcentrifuge tubes (1.5 mL).
5. Microliter pipettes and tips.
6. Benchtop microcentrifuge for 1.5 mL microcentrifuge tubes.
7. Native gels (*see* **Note 2**).
8. MES-SDS Running Buffer (20×): 1 M MES, 1 M Tris base, 2 % (w/v) SDS, 20 mM EDTA, pH 7.3.
9. Polyacrylamide gel electrophoresis (PAGE) system.
10. Heat block set to 95 °C.
11. Instant Blue protein stain (Expedeon).
12. Gel staining box.
13. Ice bucket and ice.

3 Methods

Heat Modifiability of OMPs to Assess Folding

1. Determine the concentration of all protein samples using the absorbance at $\lambda = 280$ nm, the calculated protein extinction coefficient, and the path length (Beer's law).
2. Adjust the concentration of the protein samples to ~2 mg/mL by either concentrating or diluting the samples (*see* **Note 3**).
3. Prior to preparing the samples for analysis, assemble the gel running apparatus so that the gel may be loaded immediately upon sample preparation. Place the gel apparatus tank into a bed of ice within an ice bucket. Insert the gel and then fill the tank completely with cold 1× MES running buffer. It is important to keep the buffer within the tank cold during the entire

experiment (*see* **Note 4**). Once assembled, you can move to the next step.

4. Place 1 μL of the protein sample into two 1.5 mL microcentrifuge tubes, labeling one as "boiled" and the other as "unboiled." Repeat for each sample.
5. Add 9 μL of sample buffer (i.e. the buffer the protein sample is in) to each tube.
6. Add 10 μL of 2× SDS loading buffer to each tube and mix by gently pipetting up and down a few times (*see* **Note 5**).
7. Place the "boiled" samples in a heating block set to 95 °C for 5 min while leaving the "unboiled" samples at room temperature (*see* **Note 6**).
8. Centrifuge the boiled samples using a microcentrifuge at full speed for 30 s and then load the samples onto the native gel assembled in **step 3** (*see* **Note 7**).
9. Run the gel for 60 min at constant 150 V.
10. Remove the gel and place it into a gel staining box. Add enough Instant Blue gel stain to cover the gel and place on a rotating or rocking platform for 5 min (*see* **Note 8**).
11. Visualize the gel to determine if the samples show heat modifiability (Fig. 1) (*see* **Notes 9** and **10**).

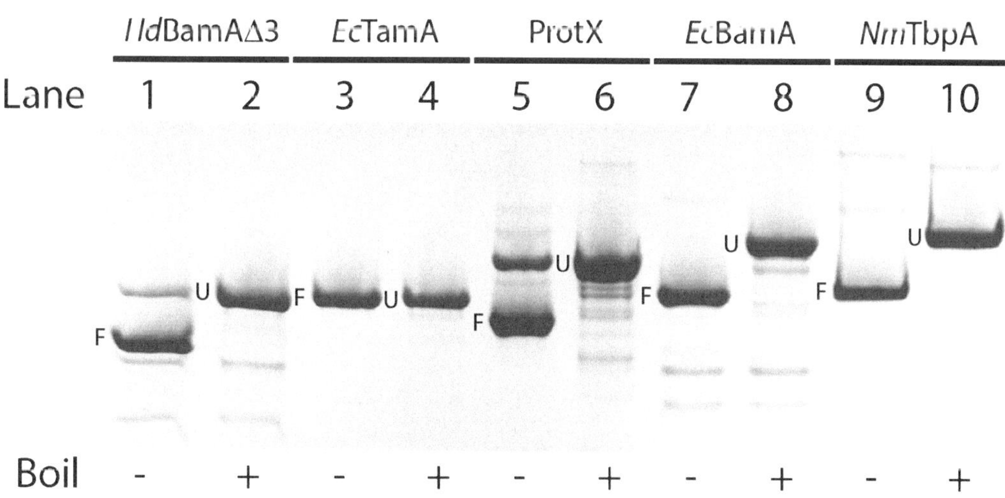

Fig. 1 Heat modifiability assay of outer membrane proteins from gram-negative bacteria. Shown is a stained semi-native SDS-PAGE gel for various OMPs that were either boiled (+) or left at room temperature for 5 min (–) prior to loading onto the gel for analysis. *Lanes 1* and *2* are of *Hd*BamAΔ3, *lanes 3* and *4* are of *Ec*TamA, *lanes 5* and *6* are of a sample we are referring to as ProtX, *lanes 7* and *8* are of *Ec*BamA, and *lanes 9* and *10* are of *Nm*TbpA. All samples were "heat modifiable" (i.e. they showed a gel-shift in this assay) except for *Ec*TamA in these assays. However, knowing that *Ec*TamA was indeed folded properly, further investigations revealed that EcTamA is indeed heat modifiable as well after a few optimization of the assay conditions (*see* Fig. 2)

4 Notes

1. For the dye, we advise against using coomassie G250 here since we have experienced odd results when substituted for bromophenol blue. We hypothesize that the coomassie G250, which is in fact used for blue native PAGE methods, interacts with the samples due to its properties while that is not the case for the bromophenol blue.
2. Most native gels, either fixed or gradient, should work here. For the examples shown here, we used precast NativePAGE Novex 4–12 % Bis–Tris protein gels (Life Technologies).
3. For convenience, it is easier to maintain all samples at the same sample concentration. However, it is acceptable to also use your protein samples at varying concentrations as long as you ensure ~2 μg of protein is being loaded and you must ensure you adjust the volume of buffer required to maintain constant volume of 10 μL. Also, 2 μg is a starting point but you may find that you need to add more or less protein for best visualization.
4. Since we keep our electrophoresis equipment at room temperature, it was easier for us to do everything at room temperature and put the gel apparatus in ice to keep it cool. However, a gel apparatus set up in a cold room would be just as good and would not require the ice bucket or ice.
5. Here, you can also use other concentrations of SDS. As you see in our examples, *Ec*TamA appeared to be quite sensitive to this variable (Fig. 2).
6. When boiling samples, sometimes the tops on the microcentrifuge tubes can pop up due to the pressure. To prevent this, poke a single small hole into the tops of microcentrifuge tubes to relieve the pressure buildup.
7. If desired, protein standards can also be loaded to help indicate the first lane. The protein standards can also be used to align results from several gels run separately.
8. If no protein standards are used to indicate the first lane, be sure to mark the gel to indicate the first lane. One method is to just cut off the corner on the side of the gel next to the first lane.
9. While 5 min will allow you to view the bands in the gel, one should leave the gel to stain overnight and then wash in water for 1 h at least three times and then leave in water overnight. If desired, the gel can then be imaged and densitometry performed to determine the percentage of folded versus unfolded states.

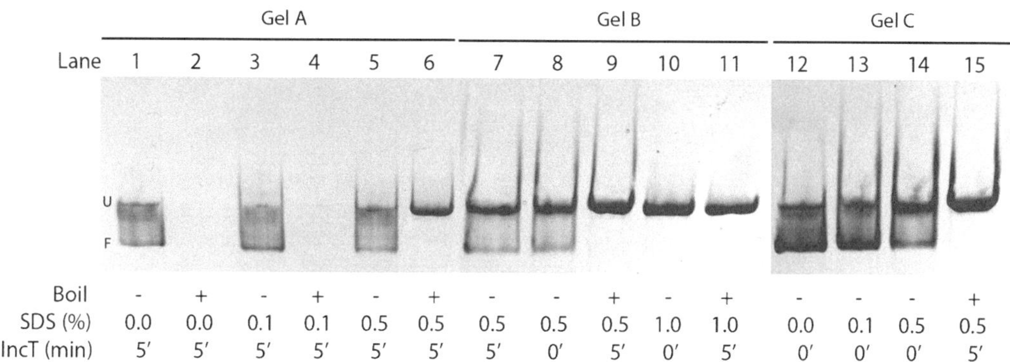

Fig. 2 Optimizing the heat modifiability assay for SDS-sensitive *Ec*TamA. *Ec*TamA did not show heat modifiability in our standard assays, however, knowing it was folded properly prompted further investigation to quell our curiosity. Here, we found that *Ec*TamA was more sensitive to SDS than the other OMPs and that reducing the SDS in our sample loading buffer and incubation time (IncT) was enough to detect predominantly the folded form in our assays. Gel A shows a preliminary stained semi-native SDS-PAGE gel varying the percent of SDS in our sample loading buffer. While we could observe a smear from the folded (F) to unfolded (U) states for all unboiled (–) samples (*lanes 1, 3*, and *5*), at least 0.5 % SDS was required to allow any migration of the unfolded/boiled sample (*lanes 2, 4*, and *6*). Gel B shows another gel where the incubation time (IncT) was varied as well, yet no difference was observed (*lanes 7, 8*, and *9*) indicating the incubation time was contributing minimally. However, the percentage of SDS was increased and we found that 1 % was sufficient to fully unfold the protein even with no heat or incubation time (*lanes 10* and *11*). Following up on this in Gel C, the concentration of SDS in the sample loading buffer was varied. It was found that no SDS in the sample loading buffer yielded the largest band for the folded state and that an increase in the percent of SDS in the sample loading buffer was accompanied by an increase in the percentage of unfolded protein (*lanes 12–15*). As shown in Gel A, boiled samples do not migrate into the gel in the absence of at least 0.5 % SDS in the sample loading buffer. However, folded proteins do not need the SDS to migrate in the gel (compare *lanes 2* and *4* of Gel A to *lanes 12* and *13* of Gel C)

10. While the folded state of most OMPs will migrate faster than the unfolded state, it is also common for the folded state to migrate slower than the unfolded state. This is often observed for OMPs that are found as oligomers in their fully folded states.

Acknowledgements

We would like to thank Matthew Belousoff and Christine Jao for providing the *Ec*TamA and ProtX samples, respectively. This research presented here was supported by the Intramural Research Program of the NIH, The National Institute of Diabetes and Digestive and Kidney Diseases (NIDDK).

References

1. Noinaj N, Kuszak AJ, Balusek C et al (2014) Lateral opening and exit pore formation are required for BamA function. Structure 22(7): 1055–1062
2. Noinaj N, Kuszak AJ, Gumbart JC et al (2013) Structural insight into the biogenesis of beta-barrel membrane proteins. Nature 501(7467): 385–390
3. Burgess NK, Dao TP, Stanley AM et al (2008) Beta-barrel proteins that reside in the *Escherichia coli* outer membrane in vivo demonstrate varied folding behavior in vitro. J Biol Chem 283(39): 26748–26758
4. Heller KB (1978) Apparent molecular weights of a heat-modifiable protein from the outer membrane of *Escherichia coli* in gels with different acrylamide concentrations. J Bacteriol 134(3):1181–1183
5. Stegmeier JF, Andersen C (2006) Characterization of pores formed by YaeT (Omp85) from *Escherichia coli*. J Biochem 140(2):275–283

Chapter 5

The Role of a Destabilized Membrane for OMP Insertion

Ashlee M. Plummer,* Dennis Gessmann,* and Karen G. Fleming

Abstract

Here we describe the procedures used in our laboratory for the in vitro investigation of the apparent folding kinetics as well as the folding efficiencies of outer membrane proteins (OMPs). Because microbial OMPs display a change in their gel migration upon folding, the usage of traditional gel electrophoresis is a standard method of folding analysis. Additional aspects of the method we detail herein include the preparation and storage of OMP stocks, the setup procedures for a folding reaction, and the analysis of fraction folded from scanned gel images.

Key words Extrusion, Kinetics, Large unilamellar vesicles, Membrane protein folding, BamA, OMP

1 Introduction

In vitro folding studies of outer membrane proteins have been useful for interrogating the lipid requirements for folding, for finding folding conditions under which folded proteins can be generated with high yields, and to test ideas about how BamA catalyzes folding of client OMPs. The main methods utilized in these studies are kinetic folding assays followed by gel electrophoresis and image analysis to determine populations of folded and unfolded proteins. There are protocols for two types of procedures we utilize.

The first protocol concerns intrinsic folding kinetics. In this set of experiments we examine the apparent folding kinetics for an outer membrane protein given a particular phospholipid environment. This protocol was used in 2008 to demonstrate that a thinner phospholipid bilayer accelerates folding [1]. These findings lead to the conclusion that a destabilized membrane accelerated OMP insertion. The conditions for folding reactions using *di*C_{10}PC, *di*C_{11}PC, or *di*C_{12}PC are included to address this issue.

The original version of this chapter was revised. The erratum to this chapter is available at: DOI 10.1007/978-1-4939-2871-2_23

*Authors contributed equally to this manuscript.

Susan K. Buchanan and Nicholas Noinaj (eds.), *The BAM Complex: Methods and Protocols*, Methods in Molecular Biology, vol. 1329, DOI 10.1007/978-1-4939-2871-2_5, © Springer Science+Business Media New York 2015

The second protocol describes a method to observe the folding acceleration of a client OMP by BamA [2]. In this protocol, BamA, which is itself a transmembrane β-barrel, is prefolded into LUVs. The apparent folding kinetics of a client OMP are then measured in the presence of prefolded BamA using a protocol similar to that of the intrinsic folding reaction described first. The conditions for folding into *di*C_{10} lipid mixtures containing 20:80 PE:PC or 20:80 PG:PC or 20:20:60 PE:PG:PC are provided. Aside from adjusting these lipid mole ratios, the basic procedures are identical.

2 Materials

Prepare all solutions using ultrapure water. Prepare and store all reagents at room temperature unless otherwise indicated. Follow all waste disposal regulations when disposing waste materials. We do not add sodium azide to the reagents. We recommend and use Amresco high purity grade urea, CAS # 57-13-6.

1. Lysis buffer: 50 mM Tris–HCl, 40 mM EDTA, pH = 8.0.
2. Brij L23: 30 % w/v, Sigma, CAS 9002-92-0.
3. Wash buffer: 10 mM Tris–HCl, 1 mM EDTA, pH = 8.0.
4. Urea buffer: 8 M urea, 20 mM Tris–HCl, pH = 8.0.
5. Borate buffer: 20 mM borate, pH = 10.0.
6. Elution Buffer (20 %): 100 mM NaCl, 8 M urea, 20 mM Tris–HCl, pH = 8.0.
7. Elution Buffer (30 %): 150 mM NaCl, 8 M urea, 20 mM Tris–HCl, pH = 8.0.
8. Elution Buffer (100 %): 500 mM NaCl 8 M urea, 20 mM Tris–HCl, pH = 8.0.
9. SDS gel-loading buffer (4×): 200 mM Tris–HCl pH = 6.8, 8 % SDS w/v, 40 % glycerol, 0.4 % bromophenol blue.
10. Syringe (10 mL) with needle removed.
11. Syringe filter (0.45 μm): Millex HV Durapore PVDF membrane.
12. Q Sepharose Fast Flow matrix: GE Healthcare, 17-0510-01.
13. Amicon Ultra-15: 10,000 MW cutoff, UFC901024.
14. Chloroform-dissolved lipids with desired head group and chain lengths: Avanti Polar Lipids (Table 1).
15. Glass vials: Fisher 15 × 45 mm 1 Dram.
16. Mini-extruder with 0.1 μm filters: Avanti.
17. EDTA stock: 100 mM EDTA.
18. Glass cuvettes (1 cm).
19. Micro Stirbar: 7 mm × 2 mm, Sci. Gear, SBM-0702-MIC.

Table 1
Lipid amounts and volumes for some desired lipid compositions

Desired lipid composition	mg per vial	μL* per vial
PC-*di*C_{10}	5.66	226.3
PC-*di*C_{11}	5.94	237.6
PC-*di*C_{12}	6.22	248.7
20 % PE-*di*C_{10} 80 % PC-*di*C_{10}	1.05 4.53	41.9 181.0
20 % PG-*di*C_{10} 80 % PC-*di*C_{10}	1.15 4.53	46.1 181.0
20 % PE-*di*C_{10} 20 % PG-*di*C_{10} 60 % PC-*di*C_{10}	1.05 1.15 3.39	41.9 46.1 135.8

*The volumes are based on the standard Avanti lipid concentrations of 25 mg/mL

20. Stirring Incubator: Aviv Biomedical, Inc. 10 sample-thermoelectric temperature incubator, Model T-10 or stir plate.
21. Eppendorf tubes (1.5 mL).
22. Timer.
23. Heat block: VWR scientific.
24. Protein ladder: We use PageRuler Prestained Protein Ladder, but any commercially available protein ladder could be used.
25. Precast gel (10 %): We use Mini-PROTEAN TGX available from Bio-Rad.
26. Coomassie Blue stain.
27. SDS Running Buffer (1×): 3.02 g Tris–HCl, 18.8 g glycine, 1.7 g SDS in 1 L.
28. Destain: 10 % acetic acid, 30 % methanol, 60 % water.
29. Epson 4490 Photo Scanner.

3 Methods

Carry out all procedures at room temperature unless otherwise specified.

3.1 OMP Expression and Preparation

1. The open reading frames for mature OMP proteins (lacking the N-terminal single peptides) were cloned and expressed into inclusion bodies as described in Burgess et al. [1]. Harvest cells by centrifugation at 4400×g for 15 min at 4 °C. Either store pellets at −20 °C or resuspend pellets in 25 mL Lysis buffer for each 500 mL of growth.

2. Isolate the OMP inclusion bodies by lysing cells. Add 83.3 μL Brij L23 (30 % w/v, Sigma, CAS 9002-92-0) to lysed cell resuspension.
3. Collect inclusion bodies by centrifugation at 5300 × *g* for 30 min at 4 °C.
4. Wash pellets two times by suspending in 25 mL Wash buffer followed by centrifuging for 5300 × *g* for 30 min at 4 °C. Discard supernatant.
5. Resuspend pellet in 25 mL Wash buffer. Aliquot equal volumes into four 15 mL Falcon tubes.
6. Collect inclusion bodies in pellets by centrifugation 6300 × *g* for 30 min at 4 °C. Discard supernatant and store at −20 °C.

3.2 Inclusion Body Purification and Aliquot Storage

1. Thaw OMP inclusion bodies on the bench top. Resuspend by adding 6 mL urea buffer and gently pipetting until no clumps are visible. Let sit for 15 min on the bench top.
2. Aliquot ~1 mL of volume into each 1.5 mL Eppendorf tube.
3. Centrifuge at 9600 × *g* for 5 min in benchtop centrifuge to remove insoluble debris.
4. Combine supernatants in a new 15 mL Falcon tube.
5. Remove plunger and needle from a 10 mL Luer-Lok syringe. Dispose of needle, and attach a 0.45 μm syringe filter. Pour combined supernatant into the syringe. Insert plunger and filter into a new 15 mL Falcon tube.
6. Pour a 6 mL Q Sepharose Fast Flow column (GE Healthcare, 17-0510-01) into a 1 inch diameter clean, gravity flow column.
7. Gently apply 30 mL of sterile water to remove ethanol from the Q Sepharose Fast Flow column (GE Healthcare, 17-0510-01).
8. Pre-equilibrate Q Sepharose column with 40 mL urea buffer.
9. Gently apply filtered supernatant to pre-equilibrated column and collect flow through in 10 mL fractions.
10. Wash three times with 10 mL urea buffer and collect flow through in 10 mL fractions.
11. For all OMPs except BamA, elute protein with three applications of 10 mL of 20 % elution buffer. Collect 10 mL fractions. The protein will normally elute in the first two fractions. 30 % elution buffer is required to elute BamA.
12. Regenerate column by the gentle application of 100 % elution Buffer. Wash column with 50 mL sterile water. Store in 20 % ethanol.
13. Mix 15 μL of each collected fraction with 5 μL of 4× SDS gel-loading buffer. Load 5 μL protein ladder (PageRuler Prestained Protein Ladder) into lane 1, and load 10 μL of sample into each subsequent lane of a 10 % SDS-PAGE gel.

Run for 30 min at 200 V. Stain with Coomassie Brilliant Blue to visualize protein bands.

14. Combine protein-containing fractions. Concentrate the samples by centrifugation using an Amicon Ultra-15 (10,000 MW cutoff, UFC901024). Centrifuge at 4400×*g* for 20 min at 4 °C. Refill the Amicon filter with urea buffer and repeat until the salt concentration is calculated to be below 1 mM and the final protein concentration is 50 μM as determined by an absorbance scan.
15. Calculate A_{260}:A_{280} ratio to evaluate purity of the protein sample with regard to DNA contamination. Pure protein should have a A_{260}:A_{280} ratio of ≤0.6.
16. Aliquot protein into 100 μL OMP stocks in 1.5 mL Eppendorf tubes and store at −80 °C (*see* **Note 1**).

3.3 Large Unilamellar Vesicle (LUV) Stock Preparation

1. Work in a hood and use Hamilton syringes for all lipid manipulations involving chloroform.
2. Aliquot lipids dissolved in chloroform into glass vials according to desired lipid compositions given in Table 1. To demonstrate that thinner bilayers accelerate folding, three separate intrinsic folding experiments using the *di*C_{10}PC, *di*C_{11}PC, and *di*C_{12}PC lipids should be performed. To conduct intrinsic folding controls for BamA-catalyzed folding, the *di*C_{10}PE, or *di*C_{10}PG mixtures should be used. Each lipid condition will constitute one set of kinetics reactions.
3. Evaporate chloroform under a gentle stream of Ar or N_2.
4. Lyophilize under vacuum overnight to remove any residual organic solvent. Dried lipids should be stored under Ar or N_2 gas at −20 °C or used immediately.
5. Before each experiment, reconstitute lipids by the addition of 1 mL of borate buffer followed by gentle vortexing. This results in a final lipid concentration of 10 mM.
6. Extrude lipids 25 times through a 0.1 μm filter using a mini-extruder (Avanti) following the manufacturer's instructions (*see* **Note 2**) (Fig. 1).

3.4 Intrinsic Folding Kinetics

1. Pre-label 1.5 mL Eppendorf tubes for the desired time points. Aliquot 5 μL of 4× SDS gel-loading buffer into each tube. Close lid to avoid evaporation.
2. Mix 200.6 μL of borate buffer, 120 μL LUV stock, 16.9 μL urea buffer, and 7.5 μL of EDTA stock in 1 cm square glass cuvette with a Micro Stirbar. The total volume should be 345 μL.
3. Pre-incubate cuvette in Aviv Model T-10 Stirring Incubator with rigorous stirring at 37 °C for 5 min. Alternatively, tape cuvette to stir plate to hold in place and conduct experiments at room temperature.

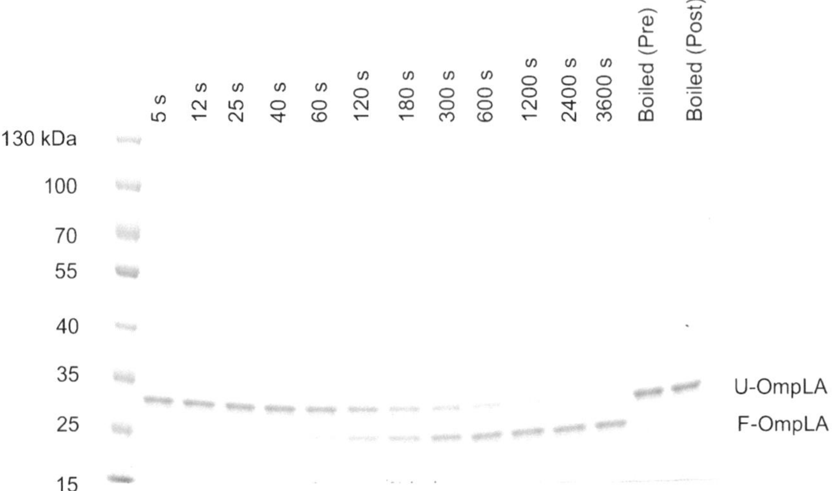

Fig. 1 Typical Intrinsic kinetics OMP folding gel. The time points in seconds for this kinetic assay and the folded (F) and unfolded (U) bands for OmpLA are labeled. The lipid conditions for this experiment were *di*C_{10}PC. The gel for a BamA-catalyzed folding experiment would look similar except for the addition of the BamA bands

4. Thaw 100 μL OMP stock. Mix gently then centrifuge to collect all volume at bottom of tube.
5. Add 30 μL of 50 μM OMP to start folding reaction. Start timer immediately upon OMP addition to reaction. This folding reaction is completed at 37 °C with vigorous stirring if using the Aviv Model T-10 stirring incubator. The final concentrations for the OMP folding reaction are: 4 μM OMP, 3200 μM lipid, 1 M urea, 2 mM EDTA. The total volume is 375 μL.
6. At each desired time point, remove 15 μL of the reaction from the cuvette and add to 4× SDS gel-loading buffer in pre-labeled 1.5 mL Eppendorf tube. If desired take an additional sample in the first 5 min termed "Boiled (Pre)" sample. Store at 4 °C until all time points to be loaded on the same gel have been collected. Analyze immediately by electrophoresis (see instructions below). Place the lid on the cuvette between time points to prevent sample evaporation.
7. At the end of the experiment, take an additional time point, termed "Boiled (Post)" sample, to be boiled in parallel with the "Boiled (Pre)" sample tube. Incubate these samples at 95 °C for 5 min in a heat block prior to electrophoresis (VWR scientific, High Setting) (*see* **Note 3**).

3.5 BamA-Catalyzed Folding Kinetics

1. Pre-label 1.5 mL Eppendorf tubes for 14 desired time points. Aliquot 5 μL of 4× SDS gel-loading buffer into each tube. Close lid to avoid evaporation.

2. Mix 200.6 μL of borate buffer, 120 μL LUV stock, 16.9 μL urea buffer, and 7.5 μL of EDTA stock in 1 cm square glass cuvette with a Micro Stirbar. The total volume should be 345 μL.
3. Pre-incubate cuvette in Aviv Model T-10 Stirring Incubator with rigorous stirring at 37 °C for 5 min. Alternatively, tape cuvette to stir plate to hold in place and conduct experiments at room temperature.
4. Thaw 100 μL BamA stock. Mix gently then centrifuge to collect all volume at bottom of tube.
5. Add 30 μL of 50 μM BamA stock to folding reaction. Cover cuvette with lid and Parafilm to seal. Allow BamA folding reaction to proceed for 2–3 h with stirring. At this point, the final volume is 375 μL, and the reaction concentrations are as follows: 4 μM BamA, 3200 μM LUVs, 1 M urea, and 2 mM EDTA.
6. In a 1.5 mL Eppendorf tube, premix 320.6 μL borate buffer, 16.9 μL urea buffer, and 7.5 μL EDTA stock.
7. After 2–3 h, remove the Parafilm and lid from the cuvette containing the BamA folding reaction. Add the 345 μL of premixed solution from **step 6** directly to the BamA reaction.
8. Remove 360 μL from the BamA reaction and place into a new 1 cm square glass cuvette with a Micro Stirbar (*see* **Note 4**).
9. Thaw 100 μL client OMP stock. Mix gently then centrifuge to collect all volume at bottom of tube.
10. Add 15 μL of 50 μM client OMP to start OMP folding reaction. Start timer immediately upon OMP addition to reaction. This folding reaction is completed at 37 °C with vigorous stirring, if using the Aviv Model T-10 stirring incubator. The final concentrations for the OMP folding reaction are: 2 μM BamA, 1600 μM lipid, 1 M urea, 2 mM EDTA, and 2 μM client OMP.
11. At each desired time point, remove 15 μL of the reaction from the cuvette and add to 4× SDS gel-loading buffer in pre-labeled 1.5 mL Eppendorf tube. Take an additional sample in the first 5 min termed "Boiled (Pre)" sample. Store at 4 °C until all time points to be loaded on the same gel have been collected. Analyze immediately by electrophoresis (below). Place the lid on the cuvette between time points to prevent sample evaporation.
12. At the end of the experiment, take an additional time point, termed "Boiled (Post)" sample, to be boiled in parallel with the "Boiled (Pre)" sample tube. Incubate these samples at 95 °C for 5 min in a heat block (VWR scientific, High Setting) (*see* **Note 3**).

3.6 Gel Separation and Quantification of Folding Reactions

1. Load 5 μL protein ladder (PageRuler Prestained Protein Ladder) and 10 μL of each sample into the lanes of a 10 % precast gel (Mini-PROTEAN TGX, Bio-Rad).
2. Run gel at constant voltage of 150 mV for 55 min at room temperature.
3. Remove gel from casing and stain with Coomassie Brilliant Blue. Destain gel to remove excess Coomassie Brilliant Blue.
4. Scan gel in transmission mode with the following settings:
   (a) Settings: Home.
   (b) Document Type: Positive Film.
   (c) Image Type: Color.
   (d) Resolution: 800 dpi.
5. Save images as jpeg format.

3.7 ImageJ Analysis of Scanned Gels

1. Analyze bands with ImageJ (NIH) to determine band intensities.
2. Select File→Open.
3. Select image to open it.
4. Use rectangular tool to draw a rectangle around one lane.
5. Use Ctrl 1 (PC) to save this lane selection.
6. Use the mouse to translate the rectangle to the second lane.
7. Use Ctrl 2 (PC) to save this lane selection.
8. Repeat **steps 5** and **6** for all lanes.
9. From ImageJ tool bar select Analyze→Gels→Plot Lanes. This will open a window with x-y plots of the lane intensity on the *y*-axis and the position within the rectangle along the *x*-axis.
10. Use "Straight" tool to draw a baseline for the bands. Repeat for each lane.
11. Select "Wand" tool. Click inside the area of each band. A popup will appear with a numerical intensity. Copy these numbers into a spreadsheet file or list for analysis.
12. Calculate fraction folded by dividing the intensity of each band by the intensity of the boiled control band. That is:

$$\text{Fraction Folded} = \frac{\text{Density of Folded Band}}{\text{Density of Boiled Band}}$$

4 Notes

1. The quality of the OMP inclusion body preparation can degrade over time if stored at room temperature or even at −20 °C. OMP stocks in urea should therefore be stored in

small volume (<100 μL) aliquots at −80 °C. Once thawed, proteins should be used immediately for experiments or discarded.

2. These experiments predominantly use lipids with fully saturated acyl chains, which are more resistant to oxidation than lipids with unsaturated acyl chains. Still, care should be taken to avoid oxidative conditions and not to unnecessarily heat lipid samples.
3. The intensity of the boiled samples "Boiled (Pre)" and "Boiled (Post)" should be identical.
4. Each BamA folding reaction yields two OMP folding reaction setups.

Acknowledgements

This work was supported by grants from the National Science Foundation (MCB0919868), the National Institutes of Health (R01 GM079440, T32 GM008403) to KGF and NSF Fellowship DGE-1232825 to AMP.

References

1. Burgess NK, Dao TP, Stanley AM et al (2008) Beta-barrel proteins that reside in the *Escherichia coli* outer membrane *in vivo* demonstrate varied folding behavior *in vitro*. J Biol Chem 283: 26748–26758
2. Gessmann D, Chung YH, Danoff EJ et al (2014) Outer membrane β-barrel protein folding is physically controlled by periplasmic lipid head groups and BamA. Proc Natl Acad Sci U S A 111:5878–5883

Chapter 6

Treponema pallidum in Gel Microdroplets: A Method for Topological Analysis of BamA (TP0326) and Localization of Rare Outer Membrane Proteins

Amit Luthra, Arvind Anand, and Justin D. Radolf

Abstract

The noncultivable spirochete *Treponema pallidum* subspecies *pallidum* (*T. pallidum*) is the etiological agent of venereal syphilis. In contrast to the outer membranes (OMs) of gram-negative bacteria, the OM of *T. pallidum* lacks lipopolysaccharide, contains a paucity of integral membrane proteins, and is extremely labile. The lability of the *T. pallidum* OM greatly hinders efforts to localize the bacterium's rare outer membrane proteins (OMPs). To circumvent this problem, we developed the gel microdroplet method in which treponemes are encapsulated in porous agarose beads and then probed with specific antibodies in the absence or presence of low concentrations of the non-ionic detergent Triton X-100. To demonstrate the general utility of this method for surface localization of any *T. pallidum* antigen, herein we describe a protocol for immunolabeling of encapsulated treponemes using antibodies directed against the β-barrel and POTRA domains of TP0326, the spirochete's BamA ortholog.

Key words *Treponema pallidum*, Syphilis, Rare outer membrane protein, BamA (TP0326), Gel microdroplets, Surface immunolabeling

1 Introduction

Syphilis is a multistage, sexually transmitted illness caused by the spirochetal pathogen *Treponema pallidum*, an extracellular, highly motile bacterium [1]. While syphilis is one of the oldest recognized sexually transmitted diseases, the mechanisms underlying its pathogenesis are poorly understood in large part because *T. pallidum* cannot be cultivated in vitro and, therefore, is not amenable to genetic manipulation [1]. Like gram-negative bacteria, *T. pallidum* has both outer and cytoplasmic membranes; however, the composition and physical properties of its cell envelope differ markedly from those of prototypical diderms [2, 3]. In addition to lacking lipopolysaccharide, the OM of *T. pallidum* has a much lower density of membrane-spanning proteins, often referred to collectively as rare OMPs, and it is easily disrupted by

Susan K. Buchanan and Nicholas Noinaj (eds.), *The BAM Complex: Methods and Protocols*, Methods in Molecular Biology, vol. 1329, DOI 10.1007/978-1-4939-2871-2_6,

manipulations such as centrifugation and resuspension [4, 5]. The paucity of OMPs is the ultrastructural basis for the syphilis spirochete's extraordinary capacity for immune evasion and, consequently, its well-earned reputation as a "stealth pathogen" [6]. Recently, we reported that TP0326, the only protein encoded by the *T. pallidum* genome with sequence homology to known gram-negative β-barrel OMPs, is a rare OMP as well as an ortholog of BamA [7]. TP0326 consists of a soluble N-terminal portion containing five polypeptide transport-associated (POTRA) domains and a C-terminal β-barrel domain and is expressed at very low copy number (approximately 100 molecules per cell) [7]. Interestingly, although TP0326 exists as part of a high molecular mass complex, *T. pallidum* lacks orthologs for any of the other known Bam subunits [7]. Thus, evolution appears to have tailored a unique variant of the Bam molecular machine in order to mediate biogenesis of *T. pallidum's* highly unusual OM.

Over the years, the lability of the *T. pallidum* OM has been a major obstacle to efforts by investigators to identify the syphilis spirochete's surface-exposed antigens [5]. To circumvent this problem, we developed the gel microdroplet method in which treponemes are encapsulated in porous agarose beads (Fig. 1) and then probed with specific antibodies in the absence or presence of low concentrations of the non-ionic detergent Triton X-100 [5]. Although we have demonstrated that the gel microdroplet method can detect surface labeling of treponemes with sera from patients with secondary syphilis [8], we had not proven that it possesses the requisite sensitivity to detect surface labeling by antibodies directed against individual OMPs. Taking advantage of the bipartite topology of TP0326 (Fig. 2), herein we demonstrate that the technique represents a general protocol for definitive surface localization of *T. pallidum* rare OMPs.

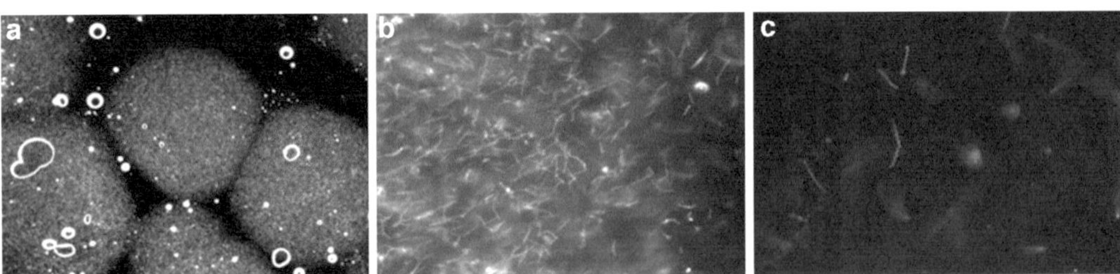

Fig. 1 *T. pallidum* encapsulated in gel microdroplets visualized by dark-field microscopy. Total magnification: ×200 (**a**), ×400 (**b**), ×1000 (**c**)

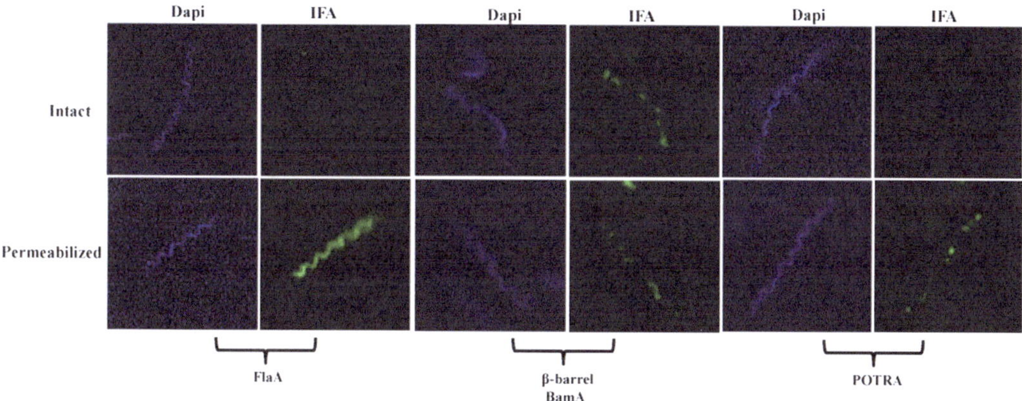

Fig. 2 *T. pallidum* encapsulated in gel microdroplets. Gel microdroplets were probed with rat anti-TP0326 β-barrel, anti-TP0326 Potra, or anti-FlaA (TP0249) antisera in the absence or presence of 0.10 % Triton X-100. Antibody binding was detected with goat anti-rat Alexa Fluor 488 (*Green*)

2 Materials

2.1 Gel Microdroplets (GMDs)

1. Phosphate-buffer saline (PBS) buffer pH 7.4 (1×): Add 800 mL water to a 1 L glass beaker. Weigh 8 g sodium chloride (NaCl), 0.2 g potassium chloride (KCl), 1.44 g of disodium hydrogen phosphate (Na_2HPO_4), 0.24 g of potassium hydrogen phosphate and transfer to the beaker. Dissolve the reagents and adjust the pH to 7.4 with HCl, and then add H_2O to 1 L. Dispense the solution into aliquots and sterilize by autoclaving for 20 min at 15 psi (1.05 kg/cm^2) on liquid cycle or by filter sterilization. Store at room temperature.
2. Water bath at 42 °C.
3. Stock solution of 2-hydroxyethyl agarose Type VII low-melting-temperature (LMT) agarose (2 %): Weigh 2 g LMT agarose and transfer to a 200 mL beaker. Add PBS (1X) to a volume of 100 mL.
4. Light mineral oil.
5. Triton X-100 (TX-100) (10 %): Add 1 mL of 100 % TX-100 solution to 9 mL of PBS (1X).
6. Round-bottomed glass screw-cap tube.
7. Polystyrene conical centrifuge tubes (15 mL).
8. CMRL-1066 medium (Invitrogen).
9. Fetal bovine serum (FBS).
10. Bovine serum albumin (BSA): BSA is used to prevent nonspecific binding in immunoassays and immunolabeling protocols (*see* **Note 1**).

11. Blocking buffer: Dissolve 1 g of BSA in 1 mL of PBS (100 % stock).
12. *T. pallidum* culture medium (TpCM) [9]: Add 20 mL of heat-inactivated FBS (10 mL) to 80 mL of CMRL medium, mix thoroughly, and filter using a 0.22 μm filter unit (*see* **Note 2**).
13. Motile *T. pallidum*: Propagation and harvesting of *T. pallidum* as described by Lukehart and Marra [10]. The animal protocol for this work is approved by the University of Connecticut Health (UCH) Animal Care Committee and the Animal Care Committee of the Centers for Disease Control and Prevention (CDC) under the auspices of Animal Welfare Assurance numbers A3471-01 and A4365-01, respectively (*see* **Note 3**). *Briefly*, maintain male New Zealand White rabbits (approximately 3.5 kg) on antibiotic-free food and water at 16 °C (*see* **Note 4**). Propagate *T. pallidum* by intratesticular inoculation of each rabbit with 1×10^8 treponemes per testis (1 mL each). Ten days later, euthanize the animals and remove the testes aseptically. Make several lengthwise cuts in each testis (*see* **Note 5**) and extract the treponemes on a rotary shaker in TpCM [9]. Remove gross testicular contaminants by centrifugation at $500 \times g$ for 15 min (*see* **Note 6**).

2.2 Immunologic Reagents

1. Primary Antibodies: Rat polyclonal antisera directed against the β-barrel and POTRA regions of BamA (TP0326) [7] and rat polyclonal antisera directed against FlaA (TP0249) protein [11] (*see* **Note 7**).
2. Secondary antibody–fluorescent dye conjugates: The most commonly used fluorochromes are Alexa Fluor® 488 (green) or Alexa Fluor® 594 (red) (Molecular Probes) (*see* **Note 8**).

2.3 Microscopy

1. VECTASHIELD mounting medium with 4′6-diamidino-2-phenylindole (DAPI) (1.5 μg/mL) (Vector Laboratories).
2. Fluorescence microscope equipped with a dark-field condenser and DAPI, fluorescein, and rhodamine filters.
3. Glass slides and coverslips (*see* **Note 9**).

3 Methods

3.1 Preparation of Gel Microdroplets (GMDs)

1. Boil 2 % stock solution of low-melting-temperature (LMT) agarose in PBS using a microwave oven and then transfer to a 42 °C water bath (*see* **Note 10**).
2. Place the mineral oil in the 42 °C water bath (*see* **Note 11**).
3. Add approximately 0.5 mL of a *T. pallidum* suspension in TpCM to each 50 mL round-bottomed glass screw-cap tube.
4. Place glass screw-cap tubes in a 42 °C water bath.

5. After 45–60 s, remove the tubes from the water bath and add 0.5 mL of warm LMT agarose.
6. Vortex the suspension at low speed four or five times for a total vortex time of 15–20 s. Immediately return the tubes to the water bath for 30 s to prevent the agar from solidifying (*see* **Note 12**).
7. Add 5 mL of warm (42 °C) light mineral oil to the suspension of *T. pallidum*-LMT agarose and vortex vigorously for approximately 10–15 s (*see* **Note 13**).
8. Plunge immediately into an ice-filled container and let stand for 5 min.
9. Warm the tubes to 35 °C (3–5 min) to collect the beads. Add 5 mL of sterile TpCM to each tube.
10. Mix the samples vigorously three to four times on a vortex mixer. Phase-separate the mineral oil (top) and aqueous (bottom) layers by centrifugation at 500 × *g* for 5 min.
11. Carefully remove the bottom aqueous layer containing the GMDs and transfer to a 15 mL polystyrene conical centrifuge tube.
12. Extract the remaining GMDs from the mineral oil by adding another 5 mL of sterile TpCM followed by centrifugation at 500 × *g* for 5 min to phase-separate the mineral oil (top) and aqueous (bottom) layers.
13. Remove the aqueous (bottom) layer from the tubes and transfer to a 15 mL polystyrene conical centrifuge tube.
14. Collect the GMDs encapsulated with treponemes by centrifugation at 100 × *g* for 5 min.
15. Pool the GMDs in a clean 15 mL polystyrene conical centrifuge tube.
16. Wash them two to three times by centrifugation at 500 × *g* for 3 min with sterile TpCM.

3.2 Probing Beads with Antibodies

1. For each labeling experiment with specific antibody, label two 1.5 mL sterile Eppendorf tubes, "intact" (no TX-100) and "permeabilized" (TX-100) (*see* **Note 14**).
2. For each sample, Transfer 250 μL of the bead suspension (50–60 μL of packed beads) to an Eppendorf tube.
3. Add appropriate dilution of each primary Ab (e.g., rat anti-TP0326 β-barrel and rat anti-TP0326 POTRA antiserum) to each group (*see* **Note 15**).
4. Add 750 μL of TpCM with 10 μL of blocking buffer (leaving a small bubble in the tube) and place the cap back on the tube. The bubble aids in mixing.
5. Add 10 μL of 10 % TX-100 to permeabilized samples immediately after the addition of the primary antibody.

6. Set the tubes on a rocker at 34 °C for 2 h with slow gentle rocking. (2–5 cycles per min) (*see* **Note 16**).
7. Centrifuge at 500 × *g* for 3 min.
8. Decant the medium being careful not to disturb the bead pellet (*see* **Note 17**).
9. Add 1 mL of TpCM in the tubes; place the tubes on the rocker for 5 min.
10. Repeat **steps 7–9** three times.
11. Refill the tubes with 1 mL of TpCM with 10 μL of blocking buffer. Add 1 μL (1 mg/mL) of secondary Ab (Goat anti-rat Alexa Fluor ®488) (*see* **Notes 18, 19** and **20**).
12. Set on a rocker at 34 °C for 2 h with slow gentile rocking.
13. Centrifuge the tubes at 500 × *g* for 3 min.
14. Decant the medium, being careful not to disturb the pellet (*see* **Note 17**).
15. Add 1 mL of TpCM to the tubes; place the tubes on a rocker for 5 min.
16. Repeat **steps 13–15** three times.
17. After the last washing step, resuspend the beads in 1 mL of TpCM medium for viewing and storage (*see* **Notes 21** and **22**).

3.3 Observation

1. Mix the labeled beads carefully with a 200 μL pipette tip in the bottom of tubes.
2. Pipet 20 μL of gravity-sedimented, labeled beads from each sample on a glass slide and add 20 μL of mounting medium with DAPI onto the beads. Mix gently and transfer 10–15 μL beads onto another clean glass slide.
3. Drop the appropriate size cover slip onto beads and press down gently to push any trapped air bubbles to the outer edges of the cover slip.
4. The slide is now ready for viewing.
5. Observe and count treponemes using a epifluorescent Olympus BX-41 microscope under dark-field, and using DAPI (λ^{ext} 327 nm) fluorescein, (λ^{ext} 495, green) and rhodamine (λ^{ext} 594, red) filters (*see* **Notes 23** and **24**).

4 Notes

1. BSA should be IgG- and protease-free. Many commercial preparations of BSA, even some of the highest purity grades, contain IgG that may interfere with binding of secondary antibodies.

2. 20 % FBS in the CMRL helps to maintain the viability and integrity of organisms. Complement in the FBS must be heat-inactivated for 30 min at 55 °C before adding to CMRL.
3. The experimental manipulations for noncultivable treponemes should be done under BSL-2 laboratory conditions. The protocol for the propagation and harvesting of *T. pallidum* must be approved by the investigator's Institutional Animal Care Committee.
4. 16 °C is the optimum ambient temperature for propagation of *T. pallidum* in rabbit testes.
5. While rabbit testes are in the Petri dish, make multiple cuts in the testes with a small sterile scissor to help release treponemes from the tissue.
6. The centrifugation step (500 × *g* for 10 min) should be repeated twice to remove gross testicular debris and blood contaminants.
7. Because *T. pallidum* are extracted from rabbit testes, primary antibodies generated in rabbits should be avoided to minimize background fluorescence.
8. The fluorescent conjugated secondary antibodies are light sensitive. Always keep them wrapped in aluminum foil.
9. Glass slides should be cleaned with acetone and dried completely.
10. *T. pallidum* is extremely heat-sensitive. The LMT agarose used for the preparation of GMDs has a gelling temperature of approximately 30 °C and does not solidify at 42 °C.
11. All solutions should be equilibrated at the specific temperature at which they will be used.
12. The timescale for this step is very important. We recommend that the investigators practice this step with beads alone. Use 1 mL of CMRL and 5 mL of warm light mineral oil and mix them. The mix should still be fluid and grayish in color. If the mix turns white and is very thick or semisolid, it contains too much air.
13. All of the manipulations at 42 °C should be completed in less than 4 min to minimize heat stress to the treponemes.
14. The OM of *T. pallidum* is extremely fragile. To assess OM integrity, each experiment should include samples in which the encapsulated treponemes are incubated with rat anti-FlaA (flagellar sheath) antiserum as primary antibodies [5, 8] in the absence or presence of 0.1 % TX-100.
15. The dilution factor of the primary antibody depends upon antibody titer, the avidity of the antibodies, and the copy number of the antigenic targets. It is best determined empirically.

16. The period of incubation with primary antibody is also variable and can be extended at 4 °C.
17. After centrifugation, remove as much medium as possible from 1.5 mL polypropylene snap-cap tube.
18. Secondary antibodies are purchased commercially and should be used according to the recommendations of the manufacturer.
19. Every step after incubation with secondary antibody must be performed in the dark or under reduced ambient light. Always put samples in a dark box or wrap with aluminum foils when transporting to a lit area.
20. Each experiment should include a secondary antibody control (i.e., samples incubated with secondary antibody alone) in order to assess nonspecific binding. If nonspecific labeling occurs, it can frequently be eliminated by increasing the dilution factor of the secondary antibody.
21. The entire experiment typically takes 1–3 days from the extraction of *T. pallidum* from rabbit testis to the enumeration of labeling. Microscopy and analysis can be performed in the latter part of day 2 and the beginning of day 3.
22. Once the experiment is complete, the labeled beads are stable at 4 °C for up to 2 weeks with no loss of fluorescence as long as they are refrigerated and stored away from light.
23. Enumeration of immunofluorescent organisms must be done in complete darkness.
24. Before enumeration, each slide is scanned to identify representative fields. All of the treponemes in a particular field must be counted. An organism is considered to be labeled if it displays any discernible fluorescence. At least 100 organisms must be counted for each sample. The percentage of labeled treponemes is determined by dividing the number of fluorescent organisms by the number visualized by dark-field microscopy or DAPI staining and then multiplying by 100.

Acknowledgements

This work was partially supported by NIH grant AI-26756 (J.D.R.) and the Department of Research, Connecticut Children's Medical Center.

References

1. Lafond RE, Lukehart SA (2006) Biological basis for syphilis. Clin Microbiol Rev 19(1):29–49
2. Cameron CE (2006) The *T. pallidum* outer membrane and outer membrane proteins. In: Radolf JD, Lukehart SA (eds) Pathogenic Treponema: Molecular and Cellular Biology. Caister Academic Press, Norwich, UK, pp 237–266
3. Radolf JD, Hazlett KRO, Lukehart SA (2006) Pathogenesis of Syphilis. In: Radolf JD, Lukehart SA (eds) Pathogenic Treponemes: Cellular and Molecular Biology. Caister Academic Press, Norfolk, UK, pp 197–236
4. Cox DL, Chang P, McDowall AW et al (1992) The outer membrane, not a coat of host proteins, limits antigenicity of virulent *Treponema pallidum*. Infect Immun 60(3):1076–1083
5. Radolf JD (1995) *Treponema pallidum* and the quest for outer membrane proteins. Mol Microbiol 16(6):1067–1073
6. Salazar JC, Hazlett KR, Radolf JD (2002) The immune response to infection with *Treponema pallidum*, the stealth pathogen. Microbes Infect 4(11):1133–1140
7. Desrosiers DC, Anand A, Luthra A et al (2011) TP0326, a *Treponema pallidum* B-barrel assembly machinery A (BamA) orthologue and rare outer membrane protein. Mol Microbiol 80(6):1496–1515
8. Cox DL, Luthra A, Dunham-Ems S et al (2010) Surface immunolabeling and consensus computational framework to identify candidate rare outer membrane proteins of *Treponema pallidum*. Infect Immun 78(12):5178–5194
9. Cox DL (1994) Culture of *Treponema pallidum*. Methods Enzymol 236:390–405
10. Lukehart SA, Marra CM (2007) Isolation and laboratory maintenance of *Treponema pallidum*. Curr Protoc Microbiol Chapter 12:Unit 12A 11
11. Isaacs RD, Radolf JD (1990) Expression in *Escherichia coli* of the 37-kilodalton endoflagellar sheath protein of *Treponema pallidum* by use of the polymerase chain reaction and a T7 expression system. Infect Immun 58(7):2025–2034

Chapter 7

Analyzing the Role of Periplasmic Folding Factors in the Biogenesis of OMPs and Members of the Type V Secretion System

Gustavo Bodelón, Elvira Marín, and Luis Ángel Fernández

Abstract

The outer membrane (OM) of gram-negative bacteria is highly packed with OM proteins (OMPs) and the trafficking and assembly of OMPs in gram-negative bacteria is a subject of intense research. Structurally, OMPs vary in the number of β-strands and in the size and complexity of extra-membrane domains, with extreme examples being the members of the type V protein secretion system (T5SS), such as the autotransporter (AT) and intimin/invasin families of secreted proteins, in which a large extracellular "passenger" domain is linked to a β-barrel that inserts in the OM. Despite their structural and functional diversity, OMPs interact in the periplasm with a relatively small set of protein chaperones that facilitate their transport from the inner membrane (IM) to the β-barrel assembly machinery (BAM complex), preventing aggregation and assisting their folding in various aspects including disulfide bond formation. This chapter is focused on the periplasmic folding factors involved in the biogenesis of integral OMPs and members of T5SS in *E. coli*, which are used as a model system in this field. Background information on these periplasmic folding factors is provided along with genetic methods to generate conditional mutants that deplete these factors from *E. coli* and biochemical methods to analyze the folding, surface display, disulfide formation and oligomerization state of OMPs/T5SS in these mutants.

Key words Periplasmic chaperones, Disulfide bonds, Autotransporters, Intimin, OMP, Type V secretion systems, *E. coli*

1 Introduction

The biogenesis of OMPs begins in the bacterial cytoplasm and encompasses their translocation across the IM to the periplasm via the Sec pathway before final folding and assembly in the OM by the β-barrel assembly machinery (BAM complex) [1–4]. OMPs contain an N-terminal cleavable signal sequence that is recognized by the SecYEG translocon. In the cytoplasm, OMPs bind to SecB, which maintains them in a fully unfolded state until they reach the translocase that drives translocation of the polypeptide across the IM in an unfolded conformation [5]. Subsequently, in the periplasm

Susan K. Buchanan and Nicholas Noinaj (eds.), *The BAM Complex: Methods and Protocols*, Methods in Molecular Biology, vol. 1329, DOI 10.1007/978-1-4939-2871-2_7, © Springer Science+Business Media New York 2015

the polypeptides are assisted by different types of folding factors and escorted to the BAM complex for OM assembly [6, 7].

The identification of extra-cytoplasmic folding factors started in the early nineties with the discovery of disulfide bond formation by DsbA and the peptidylprolyl-cis-trans-isomerase PpiA [8, 9]. In 1996, as a result of the search for proteins that decreased the σE-dependent response constitutively induced by the accumulation of misfolded proteins in the periplasm, SurA, FkpA, and Skp proteins were identified as periplasmic folding factors [10]. In that same year, SurA and Skp were shown to bind and assist the folding of OMPs, confirming their function as molecular chaperones [11, 12]. Since these initial discoveries, the work of the past decades has led to the identification of numerous periplasmic folding factors, the elucidation of their crystallized structures and an understanding of the molecular mechanisms and events that participate in the biogenesis of OMPs [1–4, 13–17].

The periplasm contains numerous folding catalysts, that in some cases present overlapping functions: chaperones for preventing undesirable off-pathway interactions or aggregation of improperly folded polypeptides (e.g., Skp, SurA, DegP, HdeA, and Spy), peptidyl–prolyl cis/trans isomerases (PPIases) catalyzing cis/trans isomerization of proline peptide bonds in proteins (e.g., SurA, FkpA, PpiA, PpiD), oxidoreductases that mediate the formation and exchange of disulfide bonds (e.g., DsbA, DsbC), and proteases (e.g., DegP). These proteins assist the folding of OMPs but also other types of periplasmic and membrane-associated proteins. The structures and functions of the abovementioned folding factors have been reviewed recently [3, 13, 17, 18]. In addition to these proteins, the periplasm of gram-negative bacteria contains other types of folding factors that are specifically dedicated to assist the subunits of surface organelles (e.g., fimbria, pili). The function of dedicated periplasmic chaperones of fimbrial subunits (e.g., PapD, FimC) has been reviewed recently [19].

The periplasmic chaperones are also involved in the protection of OMPs under stress conditions. Envelope stress induces the accumulation of misfolded or aggregated OMPs in the periplasm, which is lethal to the cell. To avoid cellular damage, gram-negative bacteria have evolved signal transduction pathways (e.g., σE and Cpx systems) that sense and respond to stress, in part by upregulating periplasmic folding factors (e.g., SurA, Skp, FkpA) and proteases (DegP) [13, 20].

The SurA polypeptide is composed of four domains: an N-terminal domain, two central PPIase domains (PPIase 1 and 2) of the parvulin family, and a short C-terminal domain [21]. The N- and the C-terminal regions have chaperone activity, whereas the PPIase activity exclusively resides in the PPIase domain 2 [22], the PPIase domain 1 could be responsible for substrate selection because it was shown to selectively bind peptides that are rich in aromatic residues and characteristic for OMPs [23]. SurA has a strong preference for binding Ar-X-Ar tripeptide motifs, commonly found at the C-terminus of OMPs, with Ar being any aromatic and X being any residue [24]. SurA binds OMPs immediately after they leave the Sec translocon before signal sequence cleavage [25], and it has been shown to interact with the AT serine protease EspP from enterohemorrhagic *E. coli* (EHEC) O157:H7 [26]. It is also the main chaperone assisting folding of EHEC Intimin [27]. Interestingly, SurA was also detected in association with BamA, the central component of the BAM complex [26, 28, 29]. These and other experimental observations [30], have led to the proposal that SurA binds unfolded OMPs in the periplasm and transports them to BamA, where they are taken by the periplasmic polypeptide transport-associated (POTRA) domains through a β-augmentation mechanism. SurA and BamB have been also suggested to function in the same folding pathway of numerous OMPs [25, 31–33]. Recently, Ricci and collaborators have shown that defects in OMP assembly caused by mutations of BamA or BamB can be corrected by gain-of-function mutations in the PPIase 1 domain of SurA, suggesting that the activity of SurA could be regulated by interactions between its PPIase 1 domain and BamA/BamB [34].

Skp is a small protein of about 17 kDa that forms a homotrimer, which binds its substrate in a 1:1 stoichiometry. The crystallized trimer shows a structure with three α-helical tentacles protruding from a basal trimerization interface forming a central cavity that holds substrates, protecting them from aggregation [35, 36]. Recent structural and biophysical analyses showed that Skp forms transient interactions with OMP polypeptides that are weak and not sequence specific, thus allowing Skp to interact with a broad range of different substrates. The authors speculate that the flexibility of the interaction between Skp and the OMP substrate could also facilitate its release to a downstream receptor in the BAM complex bearing higher affinity for a particular sequence in the OMP [37]. In vivo, Skp binds unfolded OMPs while they are still engaged in translocation through the Sec translocon [38, 39]. Site-specific photo-cross-linking experiments have demonstrated that Skp interacts with the autotransporter EspP, which is at the same time engaged with components of the BAM complex [26, 40]. Intriguingly, although Skp was shown to interact with more than 30 envelope proteins of *E. coli* [40–42], to date no OMP appears to depend preferentially on Skp for folding in *E. coli* [43]. However, recent experiments in *Neisseria meningitidis* demonstrated that two major OMPs, PorA and PorB, use Skp preferentially [44], and Skp was demonstrated to be required for the efficient assembly of the autotransporter IcsA in *Shigella flexneri* [45].

DegP belongs to the serine proteases high temperature requirement (HtrA) family and is considered both a chaperone and a protease [46–48]. The DegP monomer is composed of a N-terminal trypsin-like protease domain and two C-terminal PDZ domains allowing DegP to recognize its substrates [47]. DegP exists in different oligomeric forms, ranging from a 6-mer to a 24-mer with a 3-mer as the fundamental building block. In the presence of the substrate, the homotrimers further oligomerize via PDZ1–PDZ2 interactions into oligomeric cage-like structures that exhibit both protease and chaperone activity [48–53]. The likely reason for the different oligomeric states of DegP is to provide functional control of its efficient but rather indiscriminate proteolytic activity [52]. DegP is upregulated by the σE and the Cpx systems in response to heat shock or to other envelope stresses performing both the antagonistic functions of protein repair and degradation [13]. DegP was proposed to act as a molecular chaperone mainly based on the observations that it captures folded OmpA and OmpC [48]. Interestingly, complementation experiments with a DegP protein mutated in its protease active site (DegP-S236A) rescued full expression of EspP in a *degP* mutant strain with severe reduction of EspP secretion [42]. In the case of EHEC Intimin, DegP was shown to act mainly as a protease to degrade unfolded polypeptide in the periplasm [27]. Recently, Ge and collaborators used genetically incorporated non-natural amino acids for the identification of DegP substrates by photo-cross-linking and co-purification in *E. coli* cells, and identified several OMPs including OmpC, F, A, X, W, and NmpC [54]. In this work, the authors stated that DegP mainly functions as a protease with hardly any appreciable chaperone function [54].

Two pathways have been shown to assist the biogenesis of OMPs: SurA and Skp/DegP pathways. Genetic analysis performed in *E. coli* indicated that bacterial cells that lack either SurA or Skp are viable, but cells that lack SurA and Skp or SurA and DegP are not viable (synthetic lethality). This discovery suggested that DegP/Skp and SurA operate in parallel pathways that are functionally redundant [55]. In *E. coli*, OMP assembly preferentially depends on SurA, suggesting that the DegP/Skp pathway functions as a backup to rescue OMPs that fall off of the SurA pathway, particularly under stressful conditions [27, 28, 43]. The major arguments in favor of this hypothesis are that whereas strains lacking *surA* have a decreased OM density [28] and are hypersensitive to detergents and hydrophobic antibiotics, an indicative of OM perturbations [43], deletion of the *skp* gene only leads to a moderate reduction in OMPs [12] and shows a much less severe phenotype than *surA* mutants [28]. In an alternative model, it has been proposed that Skp and SurA cooperate sequentially in the same pathway [56]. Interestingly, it has been reported that both SurA and Skp perform distinct roles in the biogenesis of the essential OMP LptD, indicating that both pathways might act in concert to efficiently assemble certain OMPs [57]. Despite the numerous findings, the exact role of the various players involved is far from clear [57]. It is still difficult to provide a detailed description of the role of the periplasmic folding factors that is universally applicable to all OMPs. The preference for periplasmic folding factors seems to be protein and potentially species-specific [14].

In addition to SurA, the periplasm of *E. coli* contains several other PPIases: PpiD and PpiA, whose function in OMP biogenesis remains to be clarified [17], and FkpA. FkpA belongs to the FKBP family of PPIases, presents general chaperone activity [58] and is upregulated by the σE system in response to envelope stress [59]. Structurally, FkpA is a V-shaped homodimer comprising two monomers, each containing a C-terminal domain with PPIase activity and an N-terminal dimerization domain constituted by three α-helices that accommodate the chaperone activity [60]. The activity of FkpA as a periplasmic chaperone was initially demonstrated using heterologous proteins expressed in *E. coli*. Although there is limited data regarding the natural substrates of this folding factor, FkpA was shown to bind unfolded EspP passenger domain with high affinity [61], and recently it has been proposed that this chaperone plays a role in the folding of LptD and FhuA [57, 61].

The periplasmic disulfide bond catalysts DsbA and DsbC that carry out disulfide bond formation and isomerization have been shown to improve both the speed of otherwise slow steps in protein folding and the stability of proteins. DsbA is a monomeric protein with a thioredoxin fold and has a CXXC catalytic motif with a redox-active cysteine pair (Cys30–Cys33) [62]. It has been suggested that DsbA is the main oxidant in the periplasm having more than 300 in vivo substrates in *E. coli* [63], including Imp [64], OmpA [65], or the passenger domains of intimin [27] and IcsA [66]. DsbA also forms disulfide bonds in LptD, the OMP translocon of LPS [67]. It has been shown that DsbA introduces disulfide bonds nonspecifically between consecutive cysteines into proteins entering the periplasm [68], without regard for the correct disulfide pairing that is found in the native conformation. The isomerization of incorrect disulfide bonds is catalyzed by the disulphide isomerase DsbC [69]. DsbC is a V-shaped homodimeric protein, and each monomer consists of two domains: an N-terminal dimerization domain and a C-terminal thioredoxin-like domain with a CXXC motif bearing with a redox-active cysteine pair (Cys98–Cys101) [70], which should be maintained in its reduced state for proper isomerase activity in a process mediated by the disulfide reductase DsbD [71]. Currently, it is becoming apparent the importance of Dsb systems for bacterial pathogenesis, and numerous gram-negative pathogens either encode several copies of their Dsb genes or possess functional Dsb paralogs [72]. The study of *dsb* mutants of various bacterial pathogens has shown an impaired assembly of different adhesive organelles [18] and attenuation [73, 74], revealing the implication of Dsbs in bacterial virulence and, importantly, that disulfide formation systems are interesting targets for the design of new antimicrobial drugs.

Overall, a possible picture for OMP biogenesis is that major OMPs and certain ATs and LptD may recruit Skp and/or FkpA at an early stage for preventing aggregation. SurA plays a role for targeting the OMP β-domain and the AT passenger domain of EspP to the BAM complex [3, 40, 56, 75]. Although the function of FkpA is still unclear, it has been reported that this protein assists Skp in the folding of certain OMPs [57]. FkpA may be a novel quality control factor under heat-shock conditions [76]. DsbA and DsbC catalyze the formation of disulfide bonds in cysteine-containing polypeptides. The role of DegP as a chaperone still remains to be elucidated but it clearly plays an important role as a protease in the quality control of OMP biogenesis. In general, translocation incompetent OMPs are degraded by DegP, relieving the cell of the toxic effects associated with the presence of misfolded polypeptides.

Despite the recent advances and the intense research in OMP biogenesis, we still have a relatively poor understanding of how OMPs and ATs, as structurally challenging substrates, are maintained in a translocation competent state and transported across the periplasm by the different folding factors. Also, defining the exact roles of the different players and how they interact and are regulated are important questions that remain unanswered. It is becoming apparent that even homologous chaperones in different species function quite differently. Hence, further studies are required to allow an in-depth understanding of the molecular mechanisms that govern the periplasmic transit and assembly of OMPs.

2 Materials

2.1 Bacterial Culture

1. Incubators and shakers set at 30 °C, 37 °C or 42 °C.
2. Spectrophotometer and cuvettes for measuring optical densities of bacterial cultures.
3. Culture media and plates. *LB medium*: 10 g tryptone, 5 g yeast extract, and 10 g NaCl. Low salt LB medium: 10 g tryptone, 5 g yeast extract, and 5 g NaCl. Combine the dry reagents above and add distilled water to 950 mL. Adjust pH to 7.5 with 1 N NaOH and bring the volume up to 1 L. *SOC medium*: Add the following to 900 mL of distilled H_2O, 20 g tryptone, 5 g yeast extract, 2 mL of 5 M NaCl, 2.5 mL of 1 M KCl, 10 mL of 1 M $MgCl_2$, 10 mL of 1 M $MgSO_4$. Adjust to 1 L with distilled H_2O. *BHI broth medium*: 37 g of powder with 1 L of distilled water (*see* **Note 1**). Autoclave all media on liquid cycle at 15 psi and 121 °C for 20 min and store at room temperature.

For *LB agar plates*, add 15 g agar to 1 L of LB, autoclave as above, and allow it to cool at room temperature for at least 55 °C. Add antibiotics as needed, and pour into 100 mm × 15 mm petri plates using 25–30 mL per plate.

4. Ampicillin (sodium salt): dissolve 1 g of sodium ampicillin in distilled deionized water to a final volume of 10 mL. Filter-sterilize the ampicillin solution through a 0.22 μm filter. Store the ampicillin in aliquots at −20 °C (or at 4 °C for 3 months).
5. Kanamycin (monosulfate): make kanamycin to a final concentration of 50 mg/mL in distilled deionized water. Filter-sterilize as above and store in aliquots at −20 °C.
6. Zeocin: the stock solution is dissolved at 100 mg/mL in distilled deionized water. Filter-sterilize as above and store in aliquots at −20 °C. Do not store at 4 °C for more than a week (*see* **Note 2**).
7. L-Arabinose stock (20 %): dissolve 10 g in 50 mL of distilled deionized water and filter-sterilize.
8. D-Glucose stock (20 %): dissolve 10 g in 50 mL of distilled deionized water and filter-sterilize.
9. TG1zeoRExBAD *E. coli* bacterial strain is a gift of Dr. Jean-Marc Ghigo. The zeoRExBAD (zeoR araC PBAD) promoter cassette is a modification of the catRExBAD cassette [77].
10. Bacterial strains used in all methods: UT5600 *E. coli* as wild type strain [78], UT5600*dsbA* [79] and EPEC E2348/69 strain [80].
11. pKD4 plasmid: template for gene disruption. OriR6Kgamma, (AmpR, KmR), rgnB(Ter). The Km resistance gene is flanked by FRT sites. Gene Bank: AY048743 [81].
12. pKD46 plasmid: lambda Red recombinase expression plasmid. repA101(ts), araBp-gam-bet-exo, oriR101, (AmpR). Gene Bank: AY048746 [81].
13. pCP20 plasmid: has the yeast Flp recombinase gene, ts-rep, [cI857](lambda)(ts), (AmpR, CmR) [82].
14. pCVD438 plasmid expressing intimin$_{\mathrm{EPEC}}$ (*eae*) [83]: this plasmid carries the *eaeA* gene from EPEC under the control of its natural promoter and was previously used for complementation of Δeae mutants of EPEC and *C. rodentium* [84].
15. Chloramphenicol (34 mg/mL): dissolve 340 mg chloramphenicol stock in 10 mL (100 %) ethanol. If necessary, vortex solution to ensure the antibiotic is fully dissolved. Filter-sterilize this solution using a 0.2 μm syringe filter. Store in aliquots at −20 °C until use.
16. β-mercaptoethanol (2-ME).

17. pInt550 and pNeae plasmids expressing intimin$_{EPEC}$ (eae) polypeptides [27]. ATs plasmids with C-terminal domain fused to 6xHistidine-tag and E-tag [85]. All the plasmids are CmR and derived from pAK-Not [78].
18. IPTG (1 M): dissolve 1 g in 5 mL dH_2O. Filter-sterilize the IPTG solution through a 0.22 μm filter. Store the IPTG in aliquots at −20 °C and one aliquot ready to use at 4 °C.

2.2 Recombineering Reagents

1. Thermocycler (e.g., T100 Thermal Cycler, Bio-Rad).
2. Bio-Rad MicroPulser Electroporator (#165-2100).
3. Electroporation cuvettes: sterile, 0.2 cm gap, package of 50 (Bio-Rad, 165-2086).
4. Agarose: use at 0.75–1.5 % for analysis of PCR products.
5. Vent DNA polymerase (NEB, M0254S): enzyme used for generating PCR recombineering substrates.
6. Taq DNA polymerase: enzyme used for colony PCR.
7. DpnI enzyme (NEB, R0176S): restriction enzyme used to cleave methylated DNA.
8. QIAprep Spin Miniprep kit (Qiagen): used for the isolation of plasmids.
9. QIAquick PCR purification kit (Qiagen) and Qiaex II gel extraction kit (Qiagen): used for the purification of PCR products to be used as substrates for recombineering.
10. Elution buffer: 10 mM Tris–HCl, pH 8.5.
11. dNTPs: 2.5 mM each of dATP, dCTP, dGTP, dTTP.
12. Sterile distilled water and sterile distilled deionized water.
13. Glycerol.
14. Electroporation washing buffer: dilute 50 mL of glycerol in 450 mL distilled deionized water and sterilize using 0.22 μm filters. Store at 4 °C.

2.3 Denaturing SDS-PAGE

1. Electrophoresis system: Bio-Rad Mini-PROTEAN 3 or equivalent. Electrophoresis power supply (PowerPac Basic Power Supply. Bio-Rad, 164-5050EDU) or equivalent.
2. Running gel buffer (1 M Tris–HCl, pH 8.8): dissolve 90.9 g Tris-base in 1 L of dH_2O, adjust to pH 8.8 with HCl.
3. Stacking gel buffer (1 M Tris–HCl, pH 6.8): dissolve 30.3 g Tris-base in 1 L of dH_2O, adjust to pH 6.8 with HCl.
4. Acrylamide solution (30 %): 29.2 % acrylamide–0.8 % bis-acrylamide (*see* **Note 3**).
5. Ammonium persulfate (APS) 25 %: dissolve 2.5 g APS in a total volume of 10 mL of distilled water. Filter (0.45 μm) and

make 1 mL aliquots. Store a working aliquot at 4 °C and the rest of the aliquots at −80 °C.

6. SDS (10 %): dissolve 10 g of SDS in 80 mL of distilled water, and then add distilled water to 100 mL. This stock solution is stable for 6 months at room temperature. Caution: Wear a dust mask for protection against breathing SDS powder.
7. Tetramethylethylenediamine (TEMED).
8. Running/ Electrophoresis buffer (5×): 15.1 g Tris base, 72 g glycine, and 5 g SDS in 1 L of distilled water.
9. Sample buffer (2×): dissolve 2 mg bromophenol blue in 1.25 mL of 1 M Tris–HCl (pH 6.8), 2 mL of glycerol, 4 mL of 10 % SDS, 0.5 mL of β-mercaptoethanol (2-ME), and 2.25 mL of dH_2O. Store in aliquots at −20 °C.
10. Molecular weight marker (e.g., a prestained protein markers, Bio-Rad, or equivalent).
11. Urea–SDS sample buffer (1×): 120 mM Tris–HCl, pH 6.8, 4 % (w/v) SDS, 8 M urea, 10 mM EDTA, 10 % (v/v) glycerol, 0.01 % (w/v) bromophenol blue, and 2 % (v/v) 2-ME (*see* **Note 4**). The urea–SDS sample buffer is used for achieving complete denaturation of intimin polypeptides.

2.4 Western Blotting

1. Semi-dry transfer apparatus (e.g., Trans-Blot SD, Bio-Rad, or equivalent).
2. Transfer buffer (5×): 15.1 g Tris and 72 g glycine in 1 L of distilled water.
3. Transfer buffer (1×): mix 300 mL of H_2O, 100 mL of 5× transfer buffer, 100 mL of methanol, and 1.9 mL of 10 % SDS.
4. PVDF membrane (e.g., Immobilon-P, Millipore).
5. Methanol.
6. Extra-thick blot paper (e.g., Bio-Rad); 2–3 mm thickness.
7. Antibodies used on Subheading 3.2. Primary antibody: anti-GroEL monoclonal antibody-peroxidase (POD) conjugate (1:5000; Sigma) and anti-OmpA (1:20,000; a gift of Hiroshi Nikaido), anti-Skp (1:1000; a gift of Matthias Mueller), anti-MBP-DegP (1:5000; a gift of Michael Ehrmann), anti-SurA (1:10,000; a gift of Roberto Kolter), anti-Fim D [33].
8. Secondary antibody: bound rabbit antibodies were detected with protein A-POD conjugate (1:8000; Zymed) (*see* **Note 5**).
9. PBS: dissolve 8 g NaCl, 0.2 g KCl, 1.44 g Na_2HPO_4, and 0.24 g KH_2PO_4 in 800 mL of distilled water. Adjust pH to 7.4 with HCl; add distilled water to 1 L and autoclave.
10. PBS-0.1 % Tween 20 (PBST): mix on magnetic stirrer 1 mL Tween 20 (Sigma, P-1379) in 1 L of PBS.

11. Blocking and antibody dilution solution: dissolve 5 g of nonfat dry milk in 100 mL of PBST.
12. X-ray film.
13. Antibodies used on Subheadings 3.2, 3.3 and 3.4. Primary antibody: rabbit anti-Int280$_{EPEC}$ polyclonal serum (1:200; a gift from Gad Frankel). Bound rabbit antibodies were detected with protein A-POD conjugate (1:8000; Zymed).
14. Antibodies used on Subheading 3.4. Primary antibody: E-tag mAb-POD conjugate (1:2000; GE Bioscience) and rabbit anti-Int280EPEC.

Secondary antibody: bound rabbit antibodies were detected with protein A-POD conjugate.

2.5 Protease Digestion

1. Sonicator.
2. Proteinase K (PK) (10 mg/mL) (Sigma) (*see* **Note 6**).
3. Phenyl-methylsulfonyl fluoride (PMSF) (100 mM): dissolve 17.4 mg PMSF in 1 mL of ethanol. Caution: Wear gloves as PMSF is highly toxic and carcinogenic.
4. PBS: dissolve 8 g NaCl, 0.2 g KCl, 1.44 g Na_2HPO_4, and 0.24 g KH_2PO_4 in 800 mL of distilled water. Adjust pH to 7.4 with HCl; add distilled water to 1 L and autoclave.

2.6 Alkylation Assay

1. Dithiothreitol (DTT) (1 M): dissolve 0.15 g in 1 mL H_20.
2. Alkylation buffer (AK): 80 mM Tris–HCl, pH 6.8, 150 mM NaCl.
3. mPEG-MAL (10 mM): dissolve 0.05 g in 0.25 mL AK buffer. Prepare immediately prior to use due to quick loss of alkylation activity (*see* **Note 7**).
4. Trichloroacetic acid (TCA) (100 % (w/v)): dissolve 500 g TCA (as shipped) into 350 mL dH2O, store at RT.
5. Acetone.
6. Glycerol.
7. PBS: dissolve 8 g NaCl, 0.2 g KCl, 1.44 g Na_2HPO_4, and 0.24 g KH_2PO_4 in 800 mL of distilled water. Adjust pH to 7.4 with HCl; add distilled water to 1 L and autoclave.

2.7 Non-reducing SDS-PAGE

Prepare materials as described in Subheading 2.3.

1. Non-reducing urea–SDS sample buffer (2×) (without 2-ME): 120 mM Tris–HCl, pH 6.8, 4 % (w/v) SDS, 8 M urea, 10 mM EDTA, 10 % (v/v) glycerol, and 0.01 % (w/v) bromophenol blue.

2.8 Blue Native (BN)-PAGE

1. Electrophoresis system: Bio-Rad Mini-PROTEAN 3 or equivalent.
2. Electrophoresis power supply (PowerPac Basic Power Supply. Bio-Rad, 164-5050EDU) or power supply capable of providing constant voltage of 150 V or higher.
3. Electrophoresis buffer: 50 mM Bis–Tris HCl, 500 mM 6-aminocaproic acid, 10 % (v/v) glycerol pH 7.0.
4. Cathode buffer: 50 mM Tricine, 15 mM Bis–Tris HCl, pH 7.0, and 0.002 % (w/v) Coomassie blue G250.
5. Anode buffer: 50 mM Bis–Tris HCl, pH 7.0.
6. Protein standards of high molecular mass (66–669 kDa) for native electrophoresis (GE Bioscience) were resuspended at a 2.5 mg/mL final concentration in 50 mM Bis–Tris HCl, pH 7.0, containing 750 mM 6-aminocaproic acid.
7. Sample buffer: 20 mM Tris–HCl, pH 8.0, 10 mM NaCl, 1 % (w/v) Zwittergent 3-14, 8.7 % (v/v) glycerol, and 0.5 % (w/v) Coomassie blue G-250.
8. Stocks solution: 10 % (w/v) Zwittergent 3-14, 87 % (v/v) glycerol, and 5 % (w/v) Coomassie blue G-250.
9. To prepare the separating (polyacrylamide gradient gel) and stacking gel use the following solutions:
   (a) AB solution: 49.5 % T and 3 % C
10. %T, total concentration of both monomers acrylamide and bis acrylamide.
11. %C, percentage of cross-linker relative to the total concentration.
    (a) Buffer 3×: 150 mM Bis–Tris HCl pH 7.0, 1.5 M 6-aminocaproic acid
    (b) Glycerol 87 % (v/v)
    (c) 10 % (w/v) ammonium persulfate (APS) solution. Store at −20 °C for longer time periods and at 4 °C for up to 2 weeks.
    (d) *N,N,N,N′*-tetramethyl-ethylenediamine (TEMED).

2.9 Cross-Linking with Dithiobis-Succinimidyl Propionate (DSP)

1. PBS: dissolve 8 g NaCl, 0.2 g KCl, 1.44 g Na_2HPO_4, and 0.24 g KH_2PO_4 in 800 mL of distilled water. Adjust pH to 7.4 with HCl; add distilled water to 1 L and autoclave.
2. Buffer: 1 M Tris–HCl, pH 7.5, to quench cross-linking reaction.
3. DSP cross-linker: must be dissolved in an organic solvent, such as DMSO, and then added to an aqueous cross-linking reaction. Stock solution is prepared at 250 mM on DMSO by rigorous vortex-mixing plus 1 min incubation at 37 °C. A few particles may remain still insoluble.

3 Methods

3.1 Genetic Strategy for Analyzing the Role of Periplasmic Folding Factors

Obtaining Conditional Depletion Mutants of Periplasmic Folding Factors in E. coli

The evidence for the functional role of periplasmic chaperones have been defined in vivo using genetically modified strains bearing either null mutations, or in conditional depletion strains, in which the expression of the chaperone gene can be regulated allowing control of protein levels [28, 55].

The procedure presented here describes the generation of a conditional mutant strain in which an inducible araC-P_{BAD} promoter cassette replaces the native promoter of *surA* on the chromosome (Fig. 1). The transcriptional activity of the P_{BAD}::*surA* allele can be controlled by the presence in the growth medium of the inducer L-arabinose or the repressor D-glucose. This strategy was used to obtain conditional mutants in *surA* [27, 33] and *bamA* [27]. This approach can be easily adapted to the periplasmic folding factor of interest in each case.

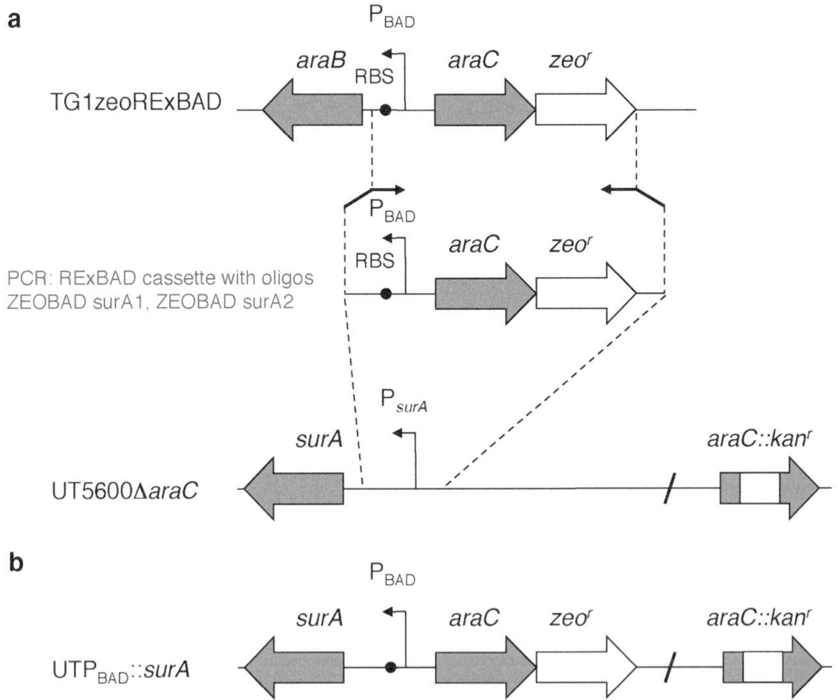

Fig. 1 Construction of SurA depletion *E. coli* mutant strain. (**a**) Scheme of the RExBAD casette amplified by PCR from the TG1zeoRExBAD strain using primers ZEOBAD SurA1 and ZEOBAD SurA2 and insertion of the RExBAD cassette by λred-driven homologous recombination after transformation with the PCR product containing the zeoRExBAD cassette and flanking DNA of the *surA* promoter region. (**b**) Schematic of P_{BAD} controlled *surA* allele in the UTP_{BAD}::surA strain

In this approach we adapted a method developed by the laboratory of Dr Jean-Marc Ghigo that combines the lambda-red linear DNA recombination method with site-directed insertion of a repression and expression (RExBAD) cassette that replaces a native promoter with a functional pBAD promoter [77]. The mutant strains described here were obtained by one-step inactivation of chromosomal genes with PCR products in *E. coli* cells bearing the bacteriophage lambda-red recombination system [81, 86].

To avoid recombination in the *araC* gene we generated an *E. coli* UT5600 knockout strain devoid of *araC* termed UT5601. Subsequently, this strain was used for generating the RExBAD conditional mutant with regulated *surA* expression termed UTP_{BAD}::s*urA*.

3.1.1 Creation of UT5600 ΔaraC (UT5601) E. coli Strain

1. The targeting substrate for recombineering is a PCR product consisting in a Kanamycin (K_m) drug marker flanked by upstream and downstream regions of *araC* target site amplified from pKD4 plasmid template. The primers for the PCR are shown below. The 20 bases on the 3′ end of each oligonucleotide (lower case) anneal to and amplify the Km cassette, the rest of the bases on the 5′ end of each primer (upper case) contain the upstream sequence and the reverse complement of the downstream sequence.

   AraC KO1 (5′- CCC TAT GCT ACT CCG TCA AGC CGT CAA TTG TCT GAT TCG TTA Cgt gta ggc tgg agc tgc ttc) and AraC KO2 (5′- CCG CCA AAG CTC GCA CAG AAT CAC TGC CAA AAT CGA GGC Cat atg aat atc ctc ctt agt) were used as primers for PCR on pKD4 template.
2. Prepare a PCR reaction (75 μL) as follows: 60.75 μL sterile deionized water, 7.5 μL 10× Vent PCR buffer (NEB), 1.5 μL dNTPs (10 mM), 1.5 μL primer AraC KO1 (20 μM), 1.5 μL primer AraC KO2 (20 μM), 1.5 μL pKD4 (10 ng), and 0.75 μL Vent polymerase (NEB) used as a high fidelity polymerase.
3. Perform standard PCR as follows: (step 1) 95 °C, 4 min; (step 2) 94 °C, 30 s; (step 3) 59 °C, 30 s; (step 4) 74 °C, 2 min. Repeat last three steps 29 times; (step 5) 74 °C, 5 min; (step 6) hold at 4 °C. The extension times should be increased for products expected to be longer than 1 kb. When completed, load 10 μL of the PCR on a 1 % agarose gel to check for correct size and purity of the PCR product. Clean the PCR product with QIAquick PCR purification kit (Qiagen) and elute DNA in 50 μL of elution buffer.
4. Degrade chromosomal DNA that may be present in the sample with DpnI (NEB), a methylation-dependent restriction enzyme that cleaves Gm6A^TC sites. Mix 43 μL of the PCR product with 1 μL of DpnI (NEB) and 5 μL of 10× CutSmart Buffer (NEB) and incubate at 37 °C for 2 h.

5. Gel-purify the PCR product using a gel extraction kit (Qiaex II, Qiagen) following manufacturer´s instructions and elute in 50 μL of sterile deionized water.
6. Transform UT5600 *E. coli* bacterial strain [78] with Red-recombineering plasmid pKD46 (AmpR) following a protocol suited for this purpose. Plate transformation at 30 °C on LB plates containing 100 μg/mL Amp overnight. Inoculate a fresh colony into 25 mL LB containing 100 μg/mL Amp and grow overnight at 30 °C.
7. In a 125 mL flask, inoculate 20 mL of LB containing 100 μg/mL ampicillin and 0.2 % (v/v) l-arabinose with 200 μL of the 5 mL overnight (ON) culture containing pKD46. Grow cells shaking (210 rpm) at 30 °C to an OD_{600} of 0.5 (~10^8 cells/mL).
8. Chill the culture on ice for 5 min and pour it into prechilled sterile 50 mL Falcon tubes. Collect cells by centrifugation in swinging bucket bench top centrifuge at 3800 × *g* for 20 min and 4 °C. Handle tubes gently so as not to disturb the cell pellet. Pour off supernatant slowly, resuspend the cells in 25 mL of ice-cold 10 % glycerol and collect the cells by centrifugation as above. Resuspend the cell pellet in 5 mL of ice-cold 10 % glycerol and centrifuge. Resuspend the cells in 100 μL of ice-cold 10 % glycerol by gently pipetting back and forth. Make sure no clumps are present. Place cells on ice and use within 30 min. This amount of cells is good for two to three trials using 50 μL of electrocompetent cells per electroporation.
9. Prechill the electroporation cuvettes (0.2 cm, Bio-Rad) in ice for 10 min. In a prechilled sterile Eppendorf tube, mix 50 μL of electrocompetent cells with 3 μL of DNA containing 0.1–0.5 ng of PCR product. Do not exceed this volume of DNA-containing sample since it will cause arcing during electroporation.
10. In a Micropulser electroporation apparatus (Bio-Rad) select EC2, a preset protocol for transformation of *E. coli* cells using 0.2 cm cuvette.
11. Transfer the DNA-bacteria mixture to a prechilled cuvette and incubate on ice for 1 min. Quickly dry the cuvette with Kimwipes, place the cuvette into the electroporation chamber, and release charge. Immediately add 1 mL of SOC medium to the cuvette. Pipet back and forth a few times and transfer cells to a 15 mL sterile Falcon tube or similar. Add another mL of SOC medium to the cuvette and repeat the process.
12. Grow the 2 mL bacterial culture for 120 min shaking at 30 °C. This step allows phenotypic expression of the resistance marker gene prior to exposure to Km selection. Spread 0.2 mL aliquots of the culture on LB Amp (100 μg/mL) Km (25 μg/

mL) plates and incubate at 30 °C overnight. Alternatively, the pKD46 plasmid could be cured by growing the cells on Km (25 μg/mL) plates at 37 °C overnight.

13. Screening of recombinants: Re-streak candidate *AraC* gene knockout colonies on LB Amp (100 μg/mL) Km (25 μg/mL) plates and incubate at 30 °C overnight.
14. Colony PCR can be used to verify the structure of the recombinant. Use a standard Taq polymerase and a primer located upstream or downstream of the sequences used for targeting the gene replacement: AraC2 (5′- CTG GTG GCG ATC TCT TCA CCG GTA GC), and a primer annealing to the drug marker cassette: k2 (5′- CGG TGC CCT GAA TGA ACT GC). Perform standard PCR as follows: (step 1) 95 °C, 4 min; (step 2) 94 °C, 30 s; (step 3) 55 °C, 1 min; (step 4) 72 °C, 90 s. Repeat last three steps 29 times; (step 5) 72 °C, 10 min; (step 6) hold at 4 °C. When completed, load 10 μL of the PCR on a 1 % agarose gel.

3.1.2 Creation of UTP_{BAD}::surA E. coli Strain

1. The zeoRExBAD (zeoR araC-P_{BAD}) cassette was amplified from the chromosome of *E. coli* strain TG1zeoRExBAD using specific oligonucleotide primers hybridizing with the upstream promoter region of the *surA* gene (5′ primer) and the beginning of their coding sequences (3′ primer), named ZEOBAD surA1 (5′-CGC AAG AGA TGC TGC GTT CGA ACA TTC TGC CGT ATC AAA ACA CTT TGT GAa gca atg ctt gca taa tgt gcc tgt c-3′) and ZEOBAD surA2 (5′-CTG GTA TTC GCG ATC ATG GCG ATA CCG AGA AGC AGC GTT TTC CAG TTC TTC **CAT** cgt ttc act cca tcc aaa aaa acg ggt-3′). The uppercase letters correspond to the sequence hybridizing to surA upstream or coding sequences. The lowercase letters correspond to the sequence hybridizing to the zeoRExBAD cassette. In bold is the first codon of *surA* coding sequence. Use standard PCR conditions, as described above. The PCR fragment with zeoR araC-P_{BAD} cassette is approximately 1.9 kb.
2. Clean the PCR product with QIAquick PCR purification kit (Qiagen) and elute DNA in 50 μL of elution buffer. Digest chromosomal DNA with DpnI (NEB). Mix 43 μL of the PCR product with 1 μL of DpnI (NEB) and 5 μL of NEB buffer 4 (10×) and incubate at 37 °C for 2 h. Gel-purify the PCR product using the Qiaex II gel extraction kit and elute in 50 μL of sterile deionized water.
3. Preparation of electrocompetent UT5601 *E. coli* cells carrying pKD46: Inoculate a fresh colony of UT5601 *E. coli* cells carrying pKD46 into 25 mL LB containing Amp (100 μg/mL) and Km (25 μg/mL). Grow overnight shaking at 30 °C.

4. In a 125 mL flask, inoculate 20 mL of LB containing Amp (100 μg/mL), Km (25 μg/mL) and 0.2 % (v/v) L-Arabinose with 200 μL of a 5 mL overnight culture. Grow cells shaking (210 rpm) at 30 °C to an OD_{600} of 0.5 (~10^8 cells/mL).
5. Proceed as described in **steps 8–11**, Subheading 3.1.1.
6. After electroporation, grow 2 mL of the bacterial culture for 120 min shaking at 30 °C. Spread 0.2 mL aliquots of the culture on low salt LB medium containing 0.2 % (v/v) arabinose and Zeo (40 μg/mL). The presence of L-arabinose is required for *surA* expression in putative UTP_{BAD}::*surA* cells. Incubate at 37 °C overnight to remove pKD46 plasmid.
7. Screening of recombinants: Restreak candidate UTP_{BAD}::*surA* colonies on low salt LB Zeo (40 μg/mL) plates containing 0.2 % (v/v) arabinose and incubate at 30 °C overnight.
8. The insertion of the zeoRExBAD cassette in the promoter region of surA can be tested by colony PCR with oligonucleotides Zeo1 (5′-CAC TGG TCA ACT TGG CCA TGG TTT AG-3′) and SurA2 (5′-CAT TAA TCC ATC AAC GTC GCT TTC CAG CAC-3′). Use standard PCR conditions, as described above. When completed, load 10 μL of the PCR on a 1 % agarose gel (*see* **Note 8**).
9. Elimination of Km selection marker: if desired, the Km selective marker can be removed after transformation with temperature-sensitive helper plasmid pCP20. This FLP expression plasmid is resistant to ampicillin at 30 °C [81].
10. Grow transformed UTP_{BAD}::*surA* bacteria with pCP20 plasmid on LB Amp (100 μg/mL) plates containing 0.2 % (v/v) arabinose and incubate at 30 °C overnight. Induce the flipase and plasmid loss by shifting the temperature to 42 °C and overnight incubation. The resulting integrants are then spread on nonselective LB medium agar at 40 °C for overnight. Cell colonies appearing on plates are again picked and patched onto LB agar plates containing Amp and Km, respectively. Consequently, the integrants are picked for exhibiting sensitivity to both Amp (indicating loss of the helper plasmid) and Km (indicating removal of the marker).

3.1.3 Functional Analysis of SurA Depletion

As mentioned before the expression of SurA in this conditional mutant can be controlled by the presence in the growth medium of the inducer L-arabinose or the repressor D-glucose. The conditional expression of SurA and its impact in the OM folding of the usher protein FimD of type 1 fimbriae are monitored as follows.

1. Inoculate a single colony of UTP_{BAD}::*surA* in 20 mL of liquid brain heart infusion (BHI) medium containing 0.4 % (w/v) L-arabinose and Kanamycin (25 μg/mL) (*see* **Note 1**). Grow

the culture at 37 °C under static conditions for 16 h. Dilute the preinoculum cultures to an OD_{600} of 0.05 in 10 mL of fresh medium containing either 0.4 % (w/v) L-arabinose (inducing conditions) or 0.4 % (w/v) D-glucose (repressing conditions) and Kanamycin (25 μg/mL). Grow the cultures at 37 °C without shaking for type 1 fimbriae expression (FimD) and after 3 h dilute them in 10 mL of inducing or depletion medium to an OD_{600} of 0.05 and grew for an additional 3 h. Repeat this process one more time. Follow the growth of the bacterial cultures by taking OD_{600} readings each hour. Before culture dilution (indicated by roman numbers I, II, and III in Fig. 2a) remove samples (0.25 OD_{600} units) for Western blotting analysis.

2. Harvest bacteria by centrifugation (3300 × *g*, 3 min) and prepare whole cell extracts as follows. Resuspend the cell pellet in 25 μL of PBS and mix it with the same volume of 2× SDS sample buffer.
3. Sonicate the samples on ice (5–10 s) with a thin needle at maximum power and spin (14,000 × *g*, 5 min) to remove insoluble material before loading onto an SDS-PAGE gel.
4. Load samples in 12 % SDS-PAGE and perform Western blotting to detect the levels of folded and unfolded FimD, OmpA; the periplasmic chaperones SurA, Skp, and DegP; and cytoplasmic GroEL as an internal loading control. The procedure for SDS-PAGE and Western blotting has been precisely described elsewhere [33, 87].

3.1.4 Example of Results

As shown in Fig. 2b, the steady state level of SurA during continuous exponential growth with L-arabinose is clearly sufficient for proper expression and folding of FimD and OmpA. However, depletion of SurA from bacteria grown with D-glucose diminished the levels of FimD. The OmpA expression also decreased in depleting medium, although not as dramatically as FimD, since folded OmpA was clearly detectable when SurA was depleted from bacteria. Since DegP is not significantly upregulated under these conditions, these results indicate that SurA is required for FimD folding and insertion in the OM of *E. coli*.

3.2 Biochemical Assessment of OMP Folding

Mobility Shift and Protease Accessibility Assays for OMPs

Herein, we describe two simple methods traditionally used to evaluate the folding of OMPs, which are based in the compact folding adopted by correctly folded OMPs in the OM that confers distinct biochemical properties.

Usually, natively folded OMPs do not denature in SDS sample buffer if not heated before SDS-PAGE, whereas OMPs displaying folding defects are sensitive to SDS. Thus, independently of their oligomeric state, the folding conformation of OMPs can be easily

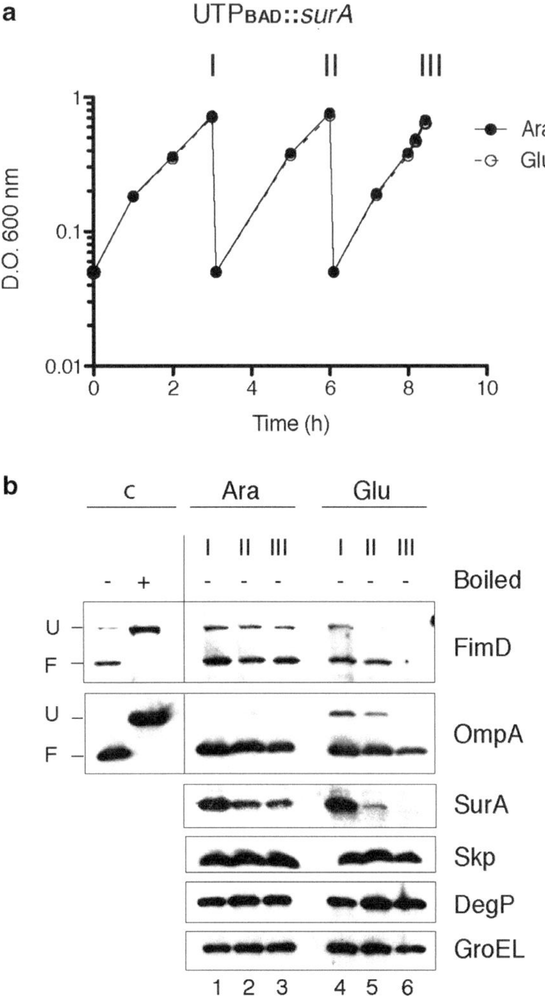

Fig. 2 Expression of FimD in the conditional *E. coli* surA mutant. (**a**) Growth curves of static cultures of UTP_{BAD}::surA strain grown in rich BHI medium containing D-glucose (Glu) or L-arabinose (Ara) and maintained in exponential phase by repeated dilutions with the same medium. Samples from these cultures were taken at the indicated times (I, II, and III) for Western blot analysis. (**b**) Whole-cell protein extracts from bacteria harvested at the indicated times (I, II, and III) from cultures shown in panel A were analyzed by Western blotting with anti-FimD, anti-OmpA, anti-SurA, anti-Skp, anti-DegP, and anti-GroEL antibodies. Control samples were obtained from cultures of UTP_{BAD}::surA grown in medium with L-arabinose. Reproduced from ref. [33] with permission from ASM

distinguished by analyzing their heat-modifiable property in SDS-PAGE [88, 89]. Correctly folded OMPs acquire a compact SDS-resistant structure that migrates faster in SDS-PAGE gels than the unfolded polypeptides.

Also, OMPs are usually highly resistant to proteases when they are correctly assembled in the OM, being either totally protected against the proteolytic treatment or yielding discrete degradation products, such as the membrane embedded β-barrel portion of the OMP. This can be assessed in vivo by incubating bacterial cells with proteases such as trypsin or proteinase K. An increased sensitivity to protease treatment is often seen as an indicative of an alteration in the folding of the protein.

As an example, we will show the heat-modifiable mobility of the OMP intimin and its resistance to the treatment with proteinase K.

Intimins are large polypeptides (ca. 95 kDa) located at the surface of enteropathogenic and enterohemorrhagic *E. coli* strains (EPEC and EHEC, respectively) (Fig. 3a). The N-terminal region of intimins (residues 1–550) anchors the protein in the OM and is highly conserved among intimins from different strains (>95 % identity). This region contains a β-barrel domain embedded in the OM [90]. The extracellular C-terminal region of intimins is less conserved (ca. 50–70 % identity) and forms a rigid rod that binds to the translocated intimin receptor (Tir) [91, 92]. This region contains three immunoglobulin-like domains (named D0, D1, and D2) and a C-type lectin like domain (D3) [92]. $\text{Intimin}_{\text{EPEC}}$ contains two Cys residues (Cys860 and Cys937) in domain D3 that form an intra-domain disulfide bond [93, 94] catalyzed by DsbA the periplasm [27].

3.2.1 Mobility Shift Assay and Proteinase K Digestion of Intimin Expressed in EPEC and E. coli K12 Bacteria

1. For expression of intimin in EPEC bacteria, inoculate a fresh colony of EPEC E2348/69 strain [80] in 25 mL LB medium and grow statically at 37 °C for 16 h.
2. Dilute the bacterial culture in fresh LB medium to a final optical density at 600 nm (OD_{600}) of 0.1 and grow under identical conditions for 2 h in the presence or absence of 10 mM 2-ME. The reducing agent 2-ME is included for reducing the disulphide bond (Cys860 and Cys937) present in intimin that renders this polypeptide sensitive to PK.
3. Measure the optical density at 600 nm of both cultures (+/- 2-ME) and centrifuge 0.5 ODs (3300 × *g*, 3 min). Resuspend the cell pellet in 1 mL of PBS and harvest the cells by centrifugation (3300 × *g*, 3 min).
4. Resuspend the cell pellet in 1 mL of PBS and divide into three aliquots of 330 μL. Dilute two aliquots with 25 μL of PBS and add the same volume of urea–SDS sample buffer (2×). Incubate one of the aliquots at 100 °C for 30 min (boiled sample), and keep the other on ice (unboiled sample).
5. Add 40 μg/mL of PK to the third aliquot on ice (PK treated). Incubate at 37 °C for 20 min. Stop the reaction with 1 μL of PMSF (100 mM). Resuspend it in 25 μL of PBS and add the same volume of urea–SDS sample buffer (2×). Boil it for 30 min.

6. Sonicate the samples on ice (5–10 s) with a thin needle at maximum power and spin (14,000 × *g*, 5 min) to remove insoluble material before loading onto an SDS-PAGE gel.
7. Load samples in 8 % SDS-PAGE and perform Western blotting to detect EPEC intimin with anti-Int280EPEC polyclonal serum (raised against the secreted domains D1, D2, and D3). The procedure for SDS-PAGE and Western blotting has been described in more detail elsewhere [95].
8. For heterologous expression of EPEC intimin in wild type *E. coli* UT5600 or in the isogenic dsbA mutant, transform UT5600 and UT5600*dsbA* bacteria with pCVD438 plasmid (CmR) following a protocol suited for this purpose. Plate transformation at 30 °C on LB plates containing 34 μg/mL Cm and grow overnight.
9. Inoculate a fresh colony of UT5600/pCVD438 or UT5600*dsbA*/pCVD438 bacteria into 25 mL LB medium containing Cm.
10. Grow overnight shaking (160 rpm) at 37 °C for constitutive expression.
11. Dilute the bacterial culture in fresh LB medium to a final optical density at 600 nm (OD_{600}) of 0.1 and grow under identical conditions for 2 h in the presence or absence of 10 mM 2-ME.
12. Repeat from **step 3**.

3.2.2 Example of Results

The intimin band shows a mobility of ~95 kDa in the boiled samples, corresponding to the expected mass of full-length intimin and indicating that the polypeptide was completely denatured after boiling in the SDS–urea buffer (Fig. 3b and c, *lane 2*). In those samples kept at 4 °C intimin showed a faster mobility in SDS-PAGE as expected for a folded β-barrel protein (Fig. 3b and c, *lane 1*). Intimin requires boiling in the presence of 4 M urea and 2 % SDS for complete unfolding indicating the formation of a very stable β-barrel in both EPEC and *E. coli* K-12.

Intimin also shows high resistance to extracellularly added proteases when expressed in EPEC and *E. coli* UT5600/pCVD438. Full-length intimin bands did not exhibit any sign of proteolytic digestion after incubation of intact bacteria with PK (Fig. 3b and c, *lane 3*). Nonetheless, intimin is sensitive to PK when the reducing agent 2-ME was added to the growth medium of EPEC and *E. coli* UT5600/pCVD438. Bacteria grown with 2-ME and incubated with PK showed almost the complete digestion of the full-length intimin band and the simultaneous appearance of a ~66-kDa proteolytic product (Fig. 3b and c, *lane 6*). The growth of EPEC and *E. coli* UT5600/pCVD438 bacteria with 2-ME did not affect folding of the β-barrel of intimin, as demonstrated by its resistance to urea–SDS denaturation and its heat-modifiable mobility (Fig. 3b

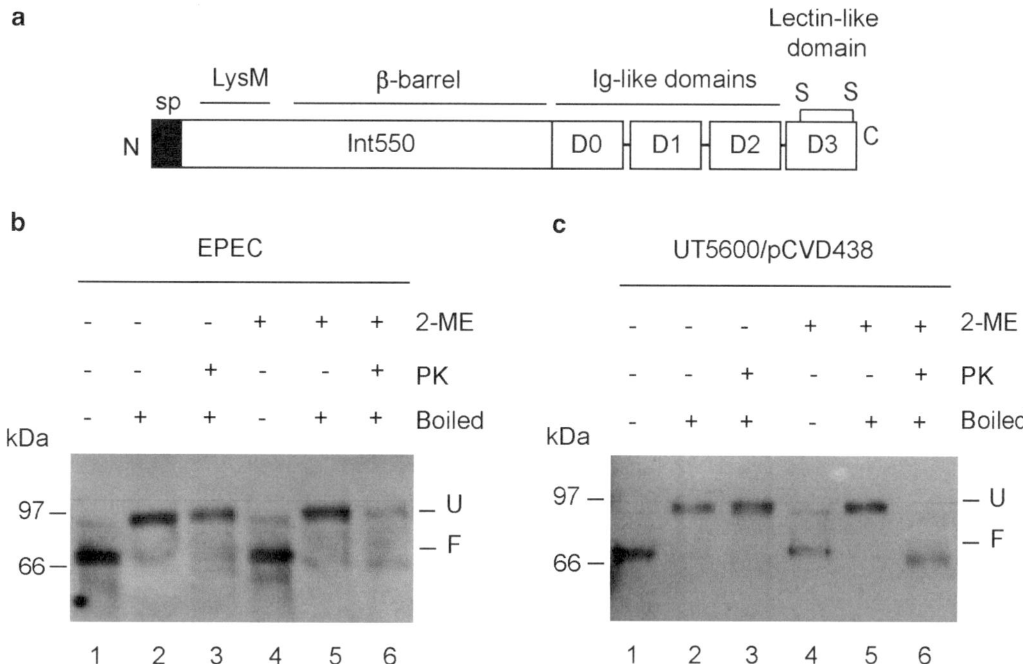

Fig. 3 Intimin domain organization and expression in EPEC (E2348/69) and *E. coli* K-12. (**a**) Schematic drawing of intimin illustrating its domain organization. The first 550 amino acid residues of intimin (Int550) contain a signal peptide (sp) located at the N terminus (N), a putative peptidoglycan-binding domain (LysM), and an OM-embedded domain predicted to fold as a β-barrel. The immunoglobulin-like (D0, D1, and D2) and the lectin-like (D3) domains are displayed on the bacterial surface. Domain D3 contains a disulfide bond indicated with an "S-S". (**b**) Western blotting performed with a rabbit anti-$Int280_{EPEC}$ polyclonal serum (detecting domains D1, D2, and D3) of whole-cell protein extracts from EPEC bacteria grown in LB in the presence (+) or absence (–) of 10 mM 2-ME. Intact bacteria, harvested from this culture, were incubated with (+) or without (–) PK. The whole-cell extracts were prepared in urea–SDS sample buffer (2 % SDS, 4 M urea) and boiled (+) or not (–) as indicated. The mobility of unfolded (U) and folded (F) intimin is labeled on the right, and the masses of protein standards are shown on the left (in kilodaltons). (**c**) Western blot probed with anti-$Int280_{EPEC}$ of whole-cell protein extracts from *E. coli* K-12 strain UT5600 carrying pCVD438. Reproduced from ref. [27] with permission from ASM

and c, *lanes 4 and 5*). These data indicate that such reducing conditions render intimin accessible in the extracellular medium to the action of the protease, and suggest that the increased sensitivity to PK was caused by alterations elsewhere in the protein.

Intimin shows identical heat-modifiable mobility and resistance to urea–SDS denaturation in wild type *E. coli* UT5600/pCVD438 bacteria and *dsbA* mutant UT5600*dsbA*/pCVD438 bacteria (Fig. 4). However, as observed under reducing growth conditions (Fig. 3b and c), intimin produced in *dsbA* mutant bacteria was sensitive to PK digestion (Fig. 4). Therefore, the absence of DsbA makes intimin less stable and susceptible to protease digestion, likely due to the misfolding of the secreted D3 domain that contains the single disulfide bond between (Cys860 and Cys937).

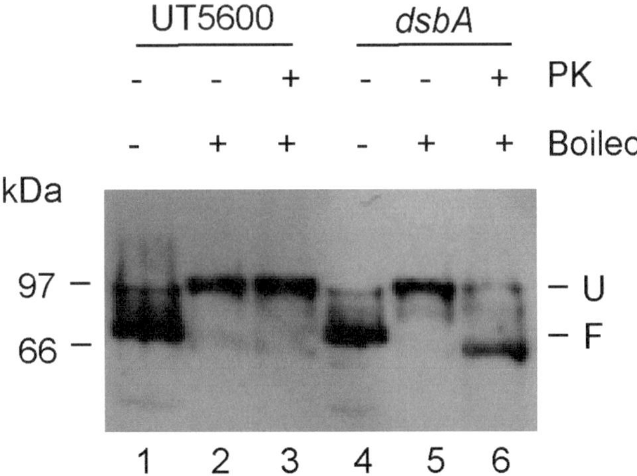

Fig. 4 Intimin expression in a *dsbA* mutant of *E. coli* K-12 and PK sensitivity. Western blot with anti-Int280$_{EPEC}$ of whole-cell protein extracts in urea–SDS sample buffer of *E. coli* UT5600 and its isogenic *dsbA* mutant carrying pCVD438. Samples were treated as in Fig. 3. Reproduced from ref. [27] with permission from ASM

3.3 Disulfide Bond Formation in OMPs

Analysis of the Redox State of Cysteine Residues in OMPs with Alkylating Agents

In this method we determine whether two cysteine residues present in a protein are oxidized by DsbA to form a disulfide bond. This is an important question to address when investigating the molecular mechanisms that take place during the periplasmic transit of OMPs. For instance, DsbA catalyzes the formation of the disulfide bond (Cys860-Cys937) present in the D3 lectin-like domain of EPEC intimin, indicating that this secreted C-terminal domain is at least partially folded prior to its translocation across the OM [27].

For this purpose, an alkylation assay is performed with the polyethylene glycol (PEG)-conjugated maleimide (mPEG-MAL, ca. 5000 Da) or with 4-acetamido-4′-maleimidylstilbene-2,2′-disulfonic acid (AMS, ca. 500 Da) that covalently binds to free sulfhydryl groups in proteins.

Maleimide-activated cross-linkers (mPEG-MAL or AMS) react with the free thiolate groups of reduced cysteine residues (-SH) in proteins at near neutral conditions (pH 6.5–7.5) to form covalent thioether linkages increasing their molecular mass and preventing any further oxidation. Hence, disulfide bonds in protein structures (e.g., between cysteines) must be reduced to free thiols to react with such maleimide reagents. In contrast, cysteine residues involved in a disulfide bond are not modified. Thus, oxidized proteins migrate with the expected size in non-reducing SDS-PAGE, whereas reduced proteins are shifted towards higher molecular weights (*see* **Note 9**).

3.3.1 Alkylation Assay of Intimin Expressed in EPEC and E. coli K12 Bacteria with mPEG-MAL

1. Transform UT5600 and UT5600*dsbA E. coli* bacterial strains with pCVD438 plasmid (CmR) following a protocol suited for this purpose. Plate transformation at 30 °C on LB plates containing 34 μg/mL Cm overnight.
2. For expression of intimin in *E. coli* K-12 strains, inoculate a fresh colony of UT5600 or UT5600*dsbA* bacteria carrying pCVD438 plasmid into 25 mL LB medium containing Cm (34 μg/mL).
3. Grow overnight shaking (160 rpm) at 37 °C for constitutive expression.
4. For expression of intimin in EPEC bacteria, inoculate a fresh colony of EPEC E2348/69 strain [80] in 25 mL LB medium and grow statically at 37 °C for 16 h.
5. Dilute the bacterial culture in fresh LB medium to a final optical density at 600 nm (OD_{600}) of 0.05, and grown under identical conditions for additional 4 h.
6. Harvest UT5600*dsbA*/pCVD438, UT5600/pCVD438, and EPEC bacteria by centrifugation (3300 × *g*, 3 min). Resuspend the cells to an OD_{600} of 1.0 in 2 mL of PBS and pellet again by centrifugation (3300 × *g*, 3 min). Wash twice with 1 mL of PBS by centrifugation (3300 × *g*, 3 min).
7. Resuspend the cell pellet in 0.5 mL of buffer AK and divide the suspension into four aliquots of 100 μL. Incubate two of them with 100 mM dithiothreitol (DTT) on ice for 10 min and then incubate at 60 °C for an additional 10 min (DTT-treated samples) (*see* **Note 10**).
8. Wash DTT-treated and untreated samples with 1 mL of PBS two times by centrifugation (3300 × *g*, 3 min). Resuspend one sample from each group in 50 μL of AK buffer containing 10 mM mPEG-MAL (mPEG-Mal, M_r 5000; Nektar Therapeutics, San Carlos, CA), and resuspend the remaining samples in 50 μL AK buffer. Perform the alkylation reaction for 30 min at room temperature.
9. Precipitate the proteins with trichloroacetic acid (10 % [wt/vol]) for 1 h at 4 °C, and recover the precipitates by centrifugation (14,000 × *g*, 15 min).
10. Wash the protein pellets with 1 mL of ice-cold acetone, followed by centrifugation (14,000 × *g*, 15 min).
11. Dry pellet by placing tube in 95 °C heat block for 5–10 min to drive off acetone and resuspend in 30 μL of AK buffer containing 1 % (w/v) SDS and 5 % (v/v) glycerol. Samples were mixed with the same volume of nonreducing urea–SDS sample buffer (2×) (without 2-ME) before being loaded onto SDS-PAGE gels.

12. The redox state of intimin is analyzed by standard SDS-PAGE and Western blot protocols as described previously. Immunodetection of intimin is performed with rabbit polyclonal anti-Int280$_{EPEC}$ polyclonal serum (1:200; a gift from Gad Frankel). Bound rabbit antibodies were detected with protein A-POD conjugate (1:8000; Zymed).

3.3.2 Example of Results

To gain a direct evidence of the formation disulfide bond (Cys860-Cys937) by DsbA, the in vivo redox state of intimin was compared in wild type and dsbA mutant *E. coli* K-12 strains carrying pCVD438. The experiment shows that intimin reacts to mPEG-MAL when expressed in the dsbA mutant, in which a high-molecular-weight band corresponding to alkylated intimin appears (Fig. 5, *lane 7*). In contrast, intimin expressed in wild type EPEC or UT5600/pCVD438 bacteria was not reactive to the alkylating agent (Fig. 5, *lanes 2 and 5*) unless the disulfide bond in D3 was reduced by the incubation of bacteria with the reducing agent DTT (Fig. 5, *lane 3*).

3.4 Analysis of Quaternary Structure of OMPs

Blue-native PAGE (BN lane-PAGE) and Cross-Linking with DSP to Follow the Quaternary Structure of OMPs

BN-PAGE relies on the solubilization of protein complexes from the membrane with mild non-ionic detergents [96]. These

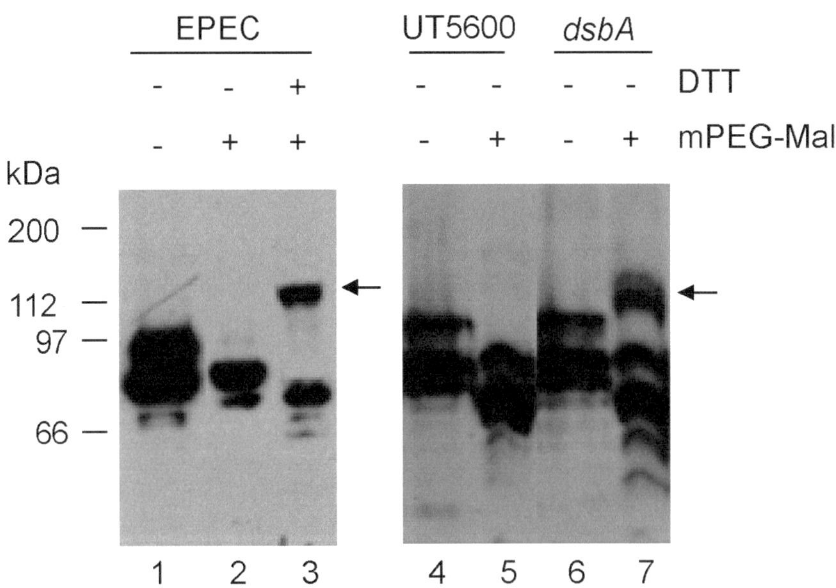

Fig. 5 Alkylation with mPEG-MAL of intimin expressed in EPEC, UT5600/pCVD438 and UTdsbA/pCVD438. Treated samples were subjected to non-reducing SDS-PAGE and Western blotting developed with anti-Int280$_{EPEC}$ serum. Samples incubated with the reducing agent DTT and/or with the alkylating agent mPEG-MAL are indicated (+). The bands corresponding to alkylated intimin polypeptide are labeled with *arrows*. The masses of protein standards are shown on the *left* (in kDa). Reproduced from ref. [27] with permission from ASM

detergents also help to prevent disruption of the protein–protein interactions. The protein complexes are negatively charged with Coomassie brilliant blue G-250 facilitating thus their migration towards the anode and separation according to its size. During electrophoresis, protein complexes separation is obtained with high resolution by the decreasing pore size in the polyacrylamide gradient gel. Always the sample is kept in native condition, the polyacrylamide gels are native (without SDS) and the electrophoresis is under native conditions (without SDS and with Coomassie blue G-250). To this aim, the sample is kept on ice during the preparation steps and running gel. This procedure avoids protein degradation as well as loss of the quaternary structure because separation of protein subunits of the complex.

Intimin is reported to form a homodimer when purified [97]. N-terminal fragments of intimin (Int550, and construct Neae with the D0 Ig like domain) and full length intimin were tested by BN-PAGE to evaluate their dimeric structure [27] (Fig. 6).

To stabilize protein complexes, intact *E. coli* cells can be treated by means of chemical cross-linking in vivo with specific

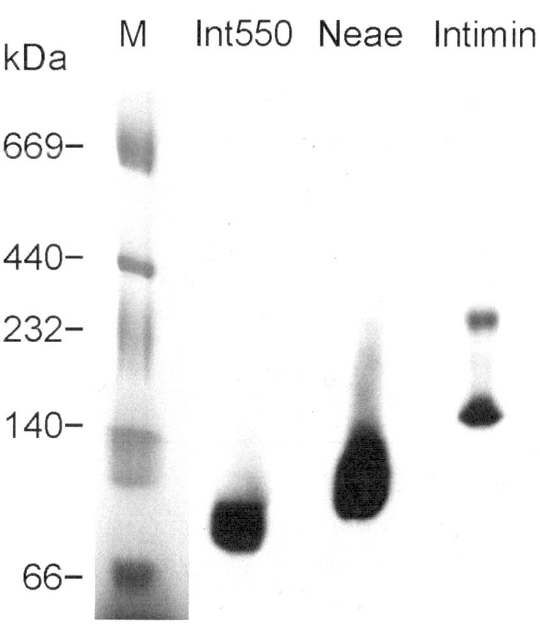

Fig. 6 Outer membrane localization and dimerization of intimin polypeptides. Blue-native PAGE of solubilized OMPs from the outer membrane fractions of *E. coli* UT5600/pInt550, UT5600/pNeae, and EPEC bacteria. Intimin polypeptides were detected by Western blot incubating the membrane with anti-Int280$_{EPEC}$ serum followed by incubation with a mixture of protein A-POD and anti-E-tag mAb-POD. The mass of native protein standards (GE Amersham) is shown on the left (in kDa). Reproduced from ref. [27] with permission from ASM

agent such as dithiobis-succinimidyl propionate (DSP). DSP is a homo-bifunctional amine-reactive N-hydroxysuccinimide (NHS) ester which has a spacer arm of 12 Å and a disulfide bond that can be cleaved with reducing agents (e.g., 2-ME). Reducing agents allow identification of the individual components within an oligomeric complex. The advantage of this technique is that it can be preformed on intact cells without any treatment or purification. The concentration of cross-linker, as well as the length and temperature of the cross-linking reaction, are critical parameters and should not be varied. Few cross-linking results in a ladder of partially cross-linked products whereas over cross-linking can produce an incorrect conclusion because of nonspecific cross-reactions that do not reflect true interactions.

C-terminal β-domains of autotransporters were used as a model to test quaternary structure into the OM. Each construct for of the C-terminal domain is tagged with an epitope (E-tag) to allow the detection with a specific mAb antibody in a Western blot. After induction of these constructs with IPTG, cell cultures were incubated with DSP and bacterial suspensions were mixed with either reducing or non-reducing SDS-PAGE sample buffer, boiled and subjected to Western blotting with anti-E tag mAb-POD [85] (Fig. 7).

3.4.1 BN-PAGE of Outer Membrane Extract

1. Prepare polyacrylamide gradient gels between 4 and 20 % with a gradient forming and based on [96]. This setup allows the separation of protein complexes in the molecular mass range from ~50 kDa to 700 kDa.
2. Prepare separately the higher (20 %) and lower (4 %) percentage of acrylamide, the same volume. Lower percentage of separating gel (4 % acrylamide) and stacking gel (4 % acrylamide) are prepared equal: 242 μL AB solution, 1 mL buffer 3×, 25 μL APS, 2.5 μL TEMED, and 1730 μL water, without glycerol. The higher percentage of separating gel mix (20 % acrylamide): 1212 μL AB solution, 1 mL buffer 3×, 690 μL glycerol, 15 μL APS, 2 μL TEMED, and 81 μL distilled water. Add TEMED and APS at the end of the process, just before opening the valves. The higher percentage of acrylamide solution for the gradient gel must always have 20 % glycerol and be placed in the position closest to the exit in the gradient former. Place the gradient former with a magnetic stir bar inside the higher concentration on a magnetic stirrer and connect with the casting stand. Ensure that all ports are closed, between the collectors of acrylamide solutions and with the casting stand. Use soft agitation to avoid air bubbles inside acrylamide solution. Open the two valves at the same time and let the solutions rise between the glass plates.
3. Add distilled water to the top of separating gel during the polymerization process of the gradient gel that should be done around 90 min. Remove distilled water and dry the area above

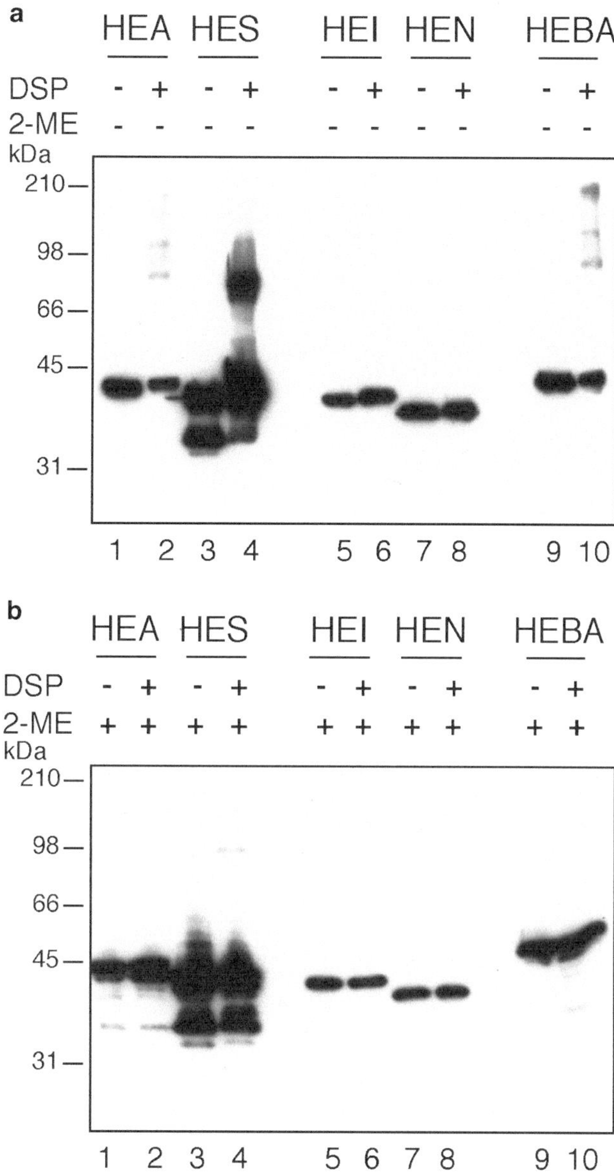

Fig. 7 Analysis of the quaternary structure of the AT C-terminal domains in vivo. (**a**) Cross-linking with DSP of ATs C-terminal domains expressed in *E. coli* to determine the formation of oligomeric complexes in vivo. DSP-treated (+) and untreated (–) samples were subjected to non-reducing SDS-PAGE, and the Western blot was probed with anti-E tag mAb-POD. Samples incubated with the reducing agent 2-ME are indicated (+). (**b**) Samples treated as in A but subjected to reducing SDS-PAGE. Reproduced from [85] with permission from ASM

the separating gel completely with thick blot paper (e.g., Whatman) and add the stacking gel with 12-well comb.

4. To obtain the sample, pellet the membrane fraction of your cell culture by centrifugation after the lysis step (with French Press or sonication). Remove the supernatant containing all

soluble or weakly associated proteins to the membranes. Solubilize the pellet on 20 mM Tris–HCl pH 8.0, 10 mM NaCl, and 1 % (w/v) Zwittergent 3-14 by gentle agitation and keep on ice. Concentrate sample 100× when resuspending the membrane pellet with respect the initial volume culture. Sonicate the sample in a cold sonication bath to improve the solubilization of the samples. It is important to avoid foam formation during the solubilization process. In case of foam it can hinder full sample solubilization. Centrifuge for 30 min at 4 °C and 100,000 × *g* to pellet the unsolubilized material that can affect in a negative way to the subsequent electrophoretic separation of the protein complexes. Before gel loading, 20 μL of this OMP extract is mixed with 2.5 μL glycerol 87 % (v/v) and 2.5 μL Coomassie blue G-250 5 % (w/v) and kept on ice.

5. Protein complex solubilization is the most critical step, and for this it is important to select the appropriate detergent and concentration for solubilization. Protein interactions must be stable in the detergent selected.
6. Load 5–10 μL of OMPs and resuspended protein standards per well, and electrophoresis was run for 45 min at 100 V and for ca. 1 h at 500 V. As soon as the blue Coomassie dye front has reached half of the separating gel, pause the electrophoretic run and remove the blue cathode buffer from the upper buffer chamber. Refilled with colorless blue native cathode buffer and continue the electrophoretic run. The electrophoresis is finished when the blue Coomassie dye front line has reached the bottom of the separating gel.
7. Disassemble the buffer chamber unit and the glass plate sandwich too and remove the stacking gel. Remove blue staining of the gel with distilled water. The blue-gel can transferred directly onto a PVDF membrane using a semidry electrophoresis transfer apparatus to check protein complexes with antibodies.

3.4.2 Treatment with DSP of Intact E. coli Cells

1. 1 mL of a culture of *E. coli* strain selected on late exponential growth phase, OD_{600} around 1–1.5, is centrifuged at 3300 × *g* for 3 min. The pellet is resuspended in 100 μL PBS with or without 2.5 mM DSP, concentrated the sample 10 times. DSP must be prepared freshly, ready to use.
2. Cross-linking is carried out for 30 min at room temperature (RT) with occasional vortex-mixing.
3. Cross-linking reaction must be stopped to avoid undesirable interactions of the oligomeric complex. For this reason, add 5 μL of 1 M Tris–HCl, pH 7.5, to a final concentration of 50 mM. The incubation is done for 15 min at RT to quench the reaction.

4. The cells were washed twice with 10 mM Tris–HCl, pH 7.5, and resuspended in the same buffer at the same volume.
5. One volume of SDS-PAGE with or without 5 % (v/v) 2-ME was added to the samples, and then they were boiled 10 min before gel loading.
6. After chemical cross-linking monomeric and oligomeric species can be immunoprecipitated and/or analyzed by Western blotting using specific antibodies.

3.4.3 Example of Results

The intimin polypeptides run in the BN-PAGE with a mobility corresponding to the expected size of dimers (Fig. 6).

With DSP we evaluated the quaternary structure of C-terminal domains of different ATs from gram-negative bacteria expressed in *E. coli* [85]. DNA fragments encoding the C-terminal domain of the selected ATs, were cloned into the expression vector pAK-Not [78]. The plasmids obtained, named pHEA (C-terminal domain of EhaA), pHES (C-terminal domain of ShdA), pHEI (C-terminal domain of IgAP), pHEN (C-terminal domain of NalP), and pHEBA (C-terminal domain of BruA), encode polypeptides containing the N-terminal signal peptide (sp) of PelB, followed by the His tag (H) and E tag (E) epitopes and their respective C-terminal domains. The expression of these HE-tagged C-terminal domains was induced with IPTG in *E. coli* K-12 strain UT5600. After incubation with DSP and with either non-reducing (Fig. 7a) or reducing SDS-PAGE buffer (Fig. 7b) the samples were analyzed by Western blotting and developed with anti-E-tag antibody. Only HES showed oligomeric state with a strong cross-linking band around 98 kDa corresponding to the predicted size for a dimer. On the contrary, the rest of C-terminal domain ATs were weakly cross-linked (HEA and HEBA) or not cross-linked at all (HEI and HEN), suggesting that they mainly form monomers in the OM in vivo.

4 Notes

1. The BHI medium is used for optimal production of type 1 fimbriae [98] and for expression of endogenous FimD in *E. coli* [33]. LB medium was also used for depletion studies of BamA and SurA [27].
2. Any *E. coli* strain that contains the complete Tn5 transposable element (i.e., DH5αF′IQ, SURE, SURE2, MC1066) encodes the bleomycin (ble) resistance gene. These strains will confer resistance to Zeocin. Plates containing Zeocin are stable for 1 month when stored at 4 °C.
3. SDS-PAGE on polyacrylamide gels of 8 % for OMPs around 90–100 kDa or 10 % for OMPs around 45–65 kDa.

4. We found that intimin requires boiling in the presence of 4 M urea and 2 % SDS for complete unfolding, indicating the formation of a very stable β-barrel in both EPEC and *E. coli* K-12 [27].
5. For POD-developing, a chemiluminescence reaction was prepared using a mixture of 1.25 mM luminol (Sigma) and 42 μM luciferin (Roche) in 100 mM Tris–HCl (pH 8.0). Following a rapid rinse in PBS, the membranes were soaked in this mixture and H_2O_2 added at 0.0075 % (v/v). Alternatively enhanced peroxidase chemiluminescence developer (Roche) was employed. In all cases, after a 1 min incubation in the dark, the PVDF-membranes were exposed to an X-ray film (X-OMAT, Kodak).
6. In this protocol we used proteinase K as the protease of choice. Other proteases can be used (e.g., trypsin). In brief, induced *E. coli* cells were washed and resuspended in PBS, trypsin was added externally (10 μg/mL). Samples were incubated for 20 min at 37 °C and stopped by adding trypsin inhibitor (5 μg/mL; Sigma). Total protein extracts from these cells can be analyzed by Western blotting.
7. mPEG-Mal can be substituted with AMS. 4-acetamido-4′-maleimidylstilbene-2,2′-disulfonic acid (AMS, M_w ~500 Da), as described in [87].
8. In the chromosome of wild type *E. coli*, the surA gene is transcribed with downstream genes pdxA, ksgA, apaG, and apaH. These genes are involved in functions unrelated to surA (e.g., pyridoxal 5′-phosphate biosynthesis and methylation of 16S rRNA). In addition, these genes are transcribed in pdxA-ksgA-apaG-apaH and apaG-apaH transcripts due to the presence of the upstream promoters pdxA and apaG (http://biocyc.org/ecoli).
9. Extraneous thiols (most reducing agents) must be excluded from maleimide reaction buffers, because they will compete for coupling sites.
10. The DTT treatment is performed as a control in order to determine the mobility of the reduced form of the polypeptide.

Acknowledgements

G.B. and E.M. contribute equally to this work. Work in the laboratory of LAF is supported by research grants from the Spanish "Ministerio de Economía y Competitividad" (MINECO) (BIO2011-26689), "Comunidad Autónoma de Madrid" (S2010-BMD-2312), "La Caixa" Foundation, and the European Research Council (ERC-2012-ADG_20120314).

References

1. Silhavy TJ, Kahne D, Walker S (2010) The bacterial cell envelope. Cold Spring Harb Perspect Biol 2(5):a000414
2. Hagan CL, Silhavy TJ, Kahne D (2011) beta-Barrel membrane protein assembly by the Bam complex. Annu Rev Biochem 80:189–210
3. van Ulsen P, Rahman SU, Jong WS et al (2013) Type V secretion: From biogenesis to biotechnology. Biochim et Biophys Acta 1843(8): 1592–1611
4. Rigel NW, Silhavy TJ (2012) Making a beta-barrel: assembly of outer membrane proteins in Gram-negative bacteria. Curr Opin Microbiol 15(2):189–193
5. Du Plessis DJF, Nouwen N, Driessen AJM (2011) The Sec translocase. Biochimica et Biophysica Acta-Biomembranes 1808(3): 851–865
6. Wu T, Malinverni J, Ruiz N et al (2005) Identification of a multicomponent complex required for outer membrane biogenesis in *Escherichia coli*. Cell 121(2):235–245
7. Knowles TJ, Scott-Tucker A, Overduin M et al (2009) Membrane protein architects: the role of the BAM complex in outer membrane protein assembly. Nat Rev Microbiol 7(3):206–214
8. Bardwell JC, McGovern K, Beckwith J (1991) Identification of a protein required for disulfide bond formation *in vivo*. Cell 67(3):581–589
9. Hayano T, Takahashi N, Kato S et al (1991) Two distinct forms of peptidylprolyl-cis-trans-isomerase are expressed separately in periplasmic and cytoplasmic compartments of *Escherichia coli* cells. Biochemistry 30(12): 3041–3048
10. Missiakas D, Betton JM, Raina S (1996) New components of protein folding in extracytoplasmic compartments of *Escherichia coli* SurA, FkpA and Skp/OmpH. Mol Microbiol 21(4): 871–884
11. Lazar SW, Kolter R (1996) SurA assists the folding of *Escherichia coli* outer membrane proteins. J Bacteriol 178(6):1770–1773
12. Chen R, Henning U (1996) A periplasmic protein (Skp) of *Escherichia coli* selectively binds a class of outer membrane proteins. Mol Microbiol 19(6):1287–1294
13. Merdanovic M, Clausen T, Kaiser M et al (2011) Protein quality control in the bacterial periplasm. Annu Rev Microbiol 65(65):149–168
14. Liechti G, Goldberg JB (2012) Outer membrane biogenesis in *Escherichia coli*, *Neisseria meningitidis*, and *Helicobacter pylori*: paradigm deviations in *H. pylori*. Front Cell Infect Microbiol 2:29
15. Geibel S, Procko E, Hultgren SJ et al (2013) Structural and energetic basis of folded-protein transport by the FimD usher. Nature 496(7444):243–246
16. Noinaj N, Kuszak AJ, Gumbart JC et al (2013) Structural insight into the biogenesis of beta-barrel membrane proteins. Nature 501(7467): 385–390
17. Goemans C, Denoncin K, Collet JF (2013) Folding mechanisms of periplasmic proteins. Biochim Biophys Acta 1843(8):1517–1528
18. Bodelón G, Palomino C, Fernández LA (2013) Immunoglobulin domains in *Escherichia coli* and other enterobacteria: from pathogenesis to applications in antibody technologies. FEMS Microbiol Rev 37(2):204–250
19. Geibel S, Waksman G (2014) The molecular dissection of the chaperone-usher pathway. Biochim Biophys Acta 1843(8):1559–1567
20. Ruiz N, Silhavy TJ (2005) Sensing external stress: watchdogs of the *Escherichia coli* cell envelope. Curr Opin Microbiol 8(2):122–126
21. Bitto E, McKay DB (2002) Crystallographic structure of SurA, a molecular chaperone that facilitates folding of outer membrane porins. Structure 10(11):1489–1498
22. Behrens S, Maier R, de Cock H et al (2001) The SurA periplasmic PPIase lacking its parvulin domains functions *in vivo* and has chaperone activity. EMBO J 20(1-2):285–294
23. Xu XH, Wang SY, Hu YX et al (2007) The periplasmic bacterial molecular chaperone SurA adapts its structure to bind peptides in different conformations to assert a sequence preference for aromatic residues. J Mol Biol 373(2):367–381
24. Bitto E, McKay DB (2003) The periplasmic molecular chaperone protein SurA binds a peptide motif that is characteristic of integral outer membrane proteins. J Biol Chem 278(49): 49316–49322
25. Ureta AR, Endres RG, Wingreen NS et al (2007) Kinetic analysis of the assembly of the outer membrane protein LamB in *Escherichia coli* mutants each lacking a secretion or targeting factor in a different cellular compartment. J Bacteriol 189(2):446–454
26. Ieva R, Bernstein HD (2009) Interaction of an autotransporter passenger domain with BamA during its translocation across the bacterial outer membrane. Proc Natl Acad Sci U S A 106(45):19120–19125
27. Bodelón G, Marín E, Fernández LA (2009) Role of Periplasmic Chaperones and BamA (YaeT/Omp85) in Folding and Secretion of

Intimin from Enteropathogenic *Escherichia coli* Strains. J Bacteriol 191(16):5169–5179
28. Sklar JG, Wu T, Kahne D et al (2007) Defining the roles of the periplasmic chaperones SurA, Skp, and DegP in *Escherichia coli*. Genes Dev 21(19):2473–2484
29. Bennion D, Charlson ES, Coon E et al (2010) Dissection of beta-barrel outer membrane protein assembly pathways through characterizing BamA POTRA 1 mutants of *Escherichia coli*. Mol Microbiol 77(5):1153–1171
30. Heuck A, Schleiffer A, Clausen T (2011) Augmenting beta-augmentation: structural basis of how BamB Binds BamA and may support folding of outer membrane proteins. J Mol Biol 406(5):659–666
31. Vertommen D, Ruiz N, Leverrier P et al (2009) Characterization of the role of the *Escherichia coli* periplasmic chaperone SurA using differential proteomics. Proteomics 9(9):2432–2443
32. Hagan CL, Kim S, Kahne D (2010) Reconstitution of outer membrane protein assembly from purified components. Science 328(5980):890–892
33. Palomino C, Marín E, Fernández LA (2011) The Fimbrial Usher FimD Follows the SurA-BamB Pathway for Its Assembly in the Outer Membrane of *Escherichia coli*. J Bacteriol 193(19):5222–5230
34. Ricci DP, Schwalm J, Gonzales-Cope M et al (2013) The activity and specificity of the outer membrane protein chaperone SurA are modulated by a proline isomerase domain. MBio 4(4)
35. Walton TA, Sousa MC (2004) Crystal structure of Skp, a prefoldin-like chaperone that protects soluble and membrane proteins from aggregation. Mol Cell 15(3):367–374
36. Qu J, Mayer C, Behrens S et al (2007) The trimeric periplasmic chaperone skp of Escherichia coli forms 1: 1 complexes with outer membrane proteins via hydrophobic and electrostatic interactions. J Mol Biol 374(1): 91–105
37. Burmann BM, Wang C, Hiller S (2013) Conformation and dynamics of the periplasmic membrane-protein–chaperone complexes OmpX–Skp and tOmpA–Skp. Nat Struct Mol Biol 20(11):1265–1272
38. Schafer U, Beck K, Muller M (1999) Skp, a molecular chaperone of Gram-negative bacteria, is required for the formation of soluble periplasmic intermediates of outer membrane proteins. J Biol Chem 274(35):24567–24574
39. Harms N, Koningstein G, Dontje W et al (2001) The early interaction of the outer membrane protein PhoE with the periplasmic chaperone Skp occurs at the cytoplasmic membrane. J Biol Chem 276(22):18804–18811
40. Ieva R, Tian P, Peterson JH et al (2011) Sequential and spatially restricted interactions of assembly factors with an autotransporter beta domain. Proc Natl Acad Sci U S A 108(31):E383–E391
41. Jarchow S, Luck C, Gorg A et al (2008) Identification of potential substrate proteins for the periplasmic *Escherichia coli* chaperone Skp. Proteomics 8(23-24):4987–4994
42. Ruiz-Perez F, Henderson IR, Leyton DL et al (2009) Roles of periplasmic chaperone proteins in the biogenesis of serine protease autotransporters of Enterobacteriaceae. J Bacteriol 191(21):6571–6583
43. Denoncin K, Schwalm J, Vertommen D et al (2012) Dissecting the *Escherichia coli* periplasmic chaperone network using differential proteomics. Proteomics 12(9):1391–1401
44. Volokhina EB, Grijpstra J, Stork M et al (2011) Role of the periplasmic chaperones Skp, SurA, and DegQ in outer membrane protein biogenesis in *Neisseria meningitidis*. J Bacteriol 193(7):1612–1621
45. Wagner JK, Heindl JE, Gray AN et al (2009) Contribution of the periplasmic chaperone Skp to efficient presentation of the autotransporter IcsA on the surface of *Shigella flexneri*. J Bacteriol 191(3):815–821
46. Spiess C, Beil A, Ehrmann M (1999) A temperature-dependent switch from chaperone to protease in a widely conserved heat shock protein. Cell 97(3):339–347
47. Krojer T, Garrido-Franco M, Huber R et al (2002) Crystal structure of DegP (HtrA) reveals a new protease-chaperone machine. Nature 416(6879):455–459
48. Krojer T, Sawa J, Schafer E et al (2008) Structural basis for the regulated protease and chaperone function of DegP. Nature 453(7197):885–890
49. Jiang JS, Zhang XF, Chen Y et al (2008) Activation of DegP chaperone-protease via formation of large cage-like oligomers upon binding to substrate proteins. Proc Natl Acad Sci U S A 105(33):11939–11944
50. Sawa J, Heuck A, Ehrmann M et al (2010) Molecular transformers in the cell: lessons learned from the DegP protease-chaperone. Curr Opin Struct Biol 20(2):253–258
51. Kim S, Grant RA, Sauer RT (2011) Covalent linkage of distinct substrate degrons controls assembly and disassembly of DegP proteolytic cages. Cell 145(1):67–78
52. Thompson NJ, Merdanovic M, Ehrmann M et al (2014) Substrate occupancy at the onset

of oligomeric transitions of DegP. Structure 22(2):281–290
53. Kim S, Sauer RT (2012) Cage assembly of DegP protease is not required for substrate-dependent regulation of proteolytic activity or high-temperature cell survival. Proc Natl Acad Sci U S A 109(19):7263–7268
54. Ge X, Wang R, Ma J et al (2013) DegP primarily functions as a protease for the biogenesis of beta-barrel outer membrane proteins in the Gram-negative bacterium *Escherichia coli*. FEBS J 281:1226–1240
55. Rizzitello AE, Harper JR, Silhavy TJ (2001) Genetic evidence for parallel pathways of chaperone activity in the periplasm of *Escherichia coli*. J Bacteriol 183(23):6794–6800
56. Bos MP, Robert V, Tommassen J (2007) Biogenesis of the Gram-negative bacterial outer membrane. Annu Rev Microbiol 61: 191–214
57. Schwalm J, Mahoney TF, Soltes GR et al (2013) A role for Skp in LptD assembly in *Escherichia coli*. J Bacteriol 195(16):3734–3742
58. Arie JP, Sassoon N, Betton JM (2001) Chaperone function of FkpA, a heat shock prolyl isomerase, in the periplasm of *Escherichia coli*. Mol Microbiol 39(1):199–210
59. Dartigalongue C, Missiakas D, Raina S (2001) Characterization of the *Escherichia coli* sigma(E) regulon. J Biol Chem 276(24):20866–20875
60. Saul FA, Arie JP, Vulliez-le NB et al (2004) Structural and functional studies of FkpA from *Escherichia coli*, a cis/trans peptidyl-prolyl isomerase with chaperone activity. J Mol Biol 335(2):595–608
61. Ruiz-Perez F, Henderson IR, Nataro JP (2010) Interaction of FkpA, a peptidyl-prolyl cis/trans isomerase with EspP autotransporter protein. Gut Microbes 1(5):339–344
62. Martin JL, Bardwell JCA, Kuriyan J (1993) Crystal structure of the DsbA protein required for disulfide bond formation *in vivo*. Nature 365(6445):464–468
63. Hiniker A, Bardwell JCA (2004) *In vivo* substrate specificity of periplasmic disulfide oxidoreductases. J Biol Chem 279(13):12967–12973
64. Kadokura H, Tian HP, Zander T et al (2004) Snapshots of DsbA in action: detection of proteins in the process of oxidative folding. Science (New York, NY) 303(5657):534–537
65. Negoda A, Negoda E, Reusch RN (2010) Resolving the native conformation of *Escherichia coli* OmpA. FEBS J 277(21):4427–4437
66. Brandon LD, Goldberg MB (2001) Periplasmic transit and disulfide bond formation of the autotransported Shigella protein IcsA. J Bacteriol 183(3):951–958
67. Chng SS, Xue M, Garner RA et al (2012) Disulfide rearrangement triggered by translocon assembly controls lipopolysaccharide export. Science (New York, NY) 337(6102): 1665–1668
68. Kadokura H, Beckwith J (2009) Detecting folding intermediates of a protein as it passes through the bacterial translocation channel. Cell 138(6):1164–1173
69. Rietsch A, Belin D, Martin N et al (1996) An *in vivo* pathway for disulfide bond isomerization in *Escherichia coli*. Proc Natl Acad Sci U S A 93(23):13048–13053
70. McCarthy AA, Haebel PW, Torronen A et al (2000) Crystal structure of the protein disulfide bond isomerase, DsbC, from *Escherichia coli*. Nat Struct Biol 7(3):196–199
71. Missiakas D, Schwager F, Raina S (1995) Identification and characterization of a new disulfide isomerase-like protein (DsbD) in *Escherichia coli*. EMBO J 14(14):3415–3424
72. Heras B, Shouldice SR, Totsika M et al (2009) DSB proteins and bacterial pathogenicity. Nat Rev Microbiol 7(3):215–225
73. Totsika M, Heras B, Wurpel DJ et al (2009) Characterization of two homologous disulfide bond systems involved in virulence factor biogenesis in uropathogenic *Escherichia coli* CFT073. J Bacteriol 191(12):3901–3908
74. Arts IS, Ball G, Leverrier P et al (2013) Dissecting the machinery that introduces disulfide bonds in *Pseudomonas aeruginosa*. MBio 4(6):e00912–e00913
75. Reusch RN (2012) Insights into the structure and assembly of *Escherichia coli* outer membrane protein A. FEBS J 279(6):894–909
76. Ge X, Lyu ZX, Liu Y et al (2014) Identification of FkpA as a key quality control factor for the biogenesis of outer membrane proteins under heat shock conditions. J Bacteriol 196(3): 672–680
77. Roux A, Beloin C, Ghigo JM (2005) Combined inactivation and expression strategy to study gene function under physiological conditions: Application to identification of new *Escherichia coli* adhesins. J Bacteriol 187(3):1001–1013
78. Veiga E, de Lorenzo V, Fernandez LA (1999) Probing secretion and translocation of a beta-autotransporter using a reporter single-chain Fv as a cognate passenger domain. Mol Microbiol 33(6):1232–1243
79. Veiga E, de Lorenzo V, Fernandez LA (2004) Structural tolerance of bacterial autotransporters for folded passenger protein domains. Mol Microbiol 52(4):1069–1080

80. Garmendia J, Phillips AD, Carlier MF et al (2004) TccP is an enterohaemorrhagic Escherichia coli O157: H7 type III effector protein that couples Tir to the actin-cytoskeleton. Cell Microbiol 6(12):1167–1183
81. Datsenko KA, Wanner BL (2000) One-step inactivation of chromosomal genes in *Escherichia coli* K-12 using PCR products. Proc Natl Acad Sci U S A 97(12):6640–6645
82. Cherepanov PP, Wackernagel W (1995) Gene disruption in *Escherichia coli*: TcR and KmR cassettes with the option of Flp-catalyzed excision of the antibiotic-resistance determinant. Gene 158(1):9–14
83. Donnenberg MS, Kaper JB (1991) Construction of an *eae* deletion mutant of Enteropathogenic *Escherichia coli* by using a positive selection suicide vector. Infect Immun 59(12):4310–4317
84. Frankel G, Phillips AD, Novakova M et al (1996) Intimin from enteropathogenic *Escherichia coli* restores murine virulence to a *Citrobacter rodentium* eaeA mutant: Induction of an immunoglobulin a response to intimin and EspB. Infect Immun 64(12):5315–5325
85. Marín E, Bodelón G, Fernández LA (2010) Comparative analysis of the biochemical and functional properties of C-terminal domains of autotransporters. J Bacteriol 192(21):5588–5602
86. Murphy KC (1998) Use of bacteriophage lambda recombination functions to promote gene replacement in *Escherichia coli*. J Bacteriol 180(8):2063–2071
87. Jurado P, Ritz D, Beckwith J et al (2002) Production of functional single-chain Fv antibodies in the cytoplasm of *Escherichia coli*. J Mol Biol 320(1):1–10
88. Nakamura K, Mizushima S (1976) Effects of heating in dodecyl sulfate solution on the conformation and electrophoretic mobility of isolated major outer membrane proteins from *Escherichia coli* K-12. J Biochem 80(6): 1411–1422
89. Schweizer M, Hindennach I, Garten W et al (1978) Major proteins of the *Escherichia coli* outer cell envelope membrane. Interaction of protein II with lipopolysaccharide. Eur J Biochem 82(1):211–217
90. Fairman JW, Dautin N, Wojtowicz D et al (2012) Crystal structures of the outer membrane domain of intimin and invasin from enterohemorrhagic *E. coli* and enteropathogenic *Y. pseudotuberculosis*. Structure 20(7): 1233–1243
91. Kenny B, DeVinney R, Stein M et al (1997) Enteropathogenic *E. coli* (EPEC) transfers its receptor for intimate adherence into mammalian cells. Cell 91(4):511–520
92. Frankel G, Candy DC, Everest P et al (1994) Characterization of the C-terminal domains of intimin-like proteins of enteropathogenic and enterohemorrhagic *Escherichia coli*, *Citrobacter freundii*, and *Hafnia alvei*. Infect Immun 62(5):1835–1842
93. Batchelor M, Prasannan S, Daniell S et al (2000) Structural basis for recognition of the translocated intimin receptor (Tir) by intimin from enteropathogenic *Escherichia coli*. EMBO J 19(11):2452–2464
94. Yu L, Frey EA, Pfuetzner RA et al (2000) Crystal structure of enteropathogenic *Escherichia coli* intimin-receptor complex. Nature 405(6790):1073–1077
95. Lee C. (2007) Western blotting. In: Rosato E (ed) Circadian Rhythms, vol 362. *Methods in Molecular Biology*, Humana Press, pp 391–399
96. Schagger H, von Jagow G (1991) Blue native electrophoresis for isolation of membrane protein complexes in enzymatically active form. Anal Biochem 199(2):223–231
97. Touze T, Hayward RD, Eswaran J et al (2004) Self-association of EPEC intimin mediated by the β-barrel containing anchor domain: a role in clustering of the Tir receptor. Mol Microbiol 51(1):73–87
98. Munera D, Hultgren S, Fernandez LA (2007) Recognition of the N-terminal lectin domain of FimH adhesin by the usher FimD is required for type 1 pilus biogenesis. Mol Microbiol 64(2):333–346

Chapter 8

An In Vitro Assay for Substrate Translocation by FhaC in Liposomes

Enguo Fan, Derrick Norell, and Matthias Müller

Abstract

The two-partner secretion (TPS) pathway is used by gram-negative bacteria to secrete a large family of virulence exoproteins. Its name is derived from the fact that it involves two proteins, a secreted TpsA protein and a cognate TpsB transporter in the outer membrane. A typical TPS system is represented by the filamentous hemagglutinin FhaB (TpsA protein) and its transporter FhaC (TpsB protein) of *Bordetella pertussis*. Results from mutational analysis and heterologous expression experiments suggested that FhaC is essential for FhaB translocation across the outer membrane of bacteria. We have devised a cell-free biochemical assay to reconstitute in vitro the translocation of FhaB into reconstituted membrane vesicles. Thereby the clearest evidence has been provided that the single β-barrel FhaC protein serves as the sole translocator to transport FhaB across the outer membrane. This is the first in vitro assay for protein secretion across the *Escherichia coli* outer membrane and the detailed protocol described here should be amenable to modifications and application to the analysis of related protein transport events occurring at the outer membranes of gram-negative bacteria.

Key words Protein transport, Two-partner secretion, Spheroplasts, β-barrel membrane proteins, Bacterial outer membrane, Reconstitution, Proteoliposome, *Escherichia coli*

1 Introduction

Gram-negative bacteria have evolved at least eight different pathways, termed type-I–type-VIII secretion pathways, for the integration, translocation, and secretion of cytosolically synthesized proteins into and across their inner and outer membranes, respectively [1–3]. Protein transport events occurring at the inner membrane that are mediated by the Sec and Tat machineries are reasonably well understood [3, 4]. In sharp contrast, events of proteins crossing, or inserting into the outer membrane are still poorly defined, which is largely due to the scarcity of appropriate biochemical assays.

A unique feature of the outer membrane of gram-negative bacteria, as well as those of mitochondria and chloroplasts, is the high

Susan K. Buchanan and Nicholas Noinaj (eds.), *The BAM Complex: Methods and Protocols*, Methods in Molecular Biology, vol. 1329, DOI 10.1007/978-1-4939-2871-2_8,

content of β-barrel outer membrane proteins (OMPs) [5–7], which play a key role in basic physiological functions and virulence, efflux of drugs and other toxins, as well as multidrug resistance [8–12]. Therefore, understanding of the biogenesis of OMPs might offer medical benefits. It is now known that proteins of the Omp85 family play an important role in the assembly of β-barrel OMPs and the exposure of proteins at the bacterial cell surface [13–15]. Known crystal structures of Omp85 members include that of BamA [16], a central component of the β-barrel assembly machinery (BAM) responsible for the biogenesis of β-barrel OMPs, and that of FhaC [17], a transporter of the two-partner secretion (TPS) pathway that is embedded in the outer membrane and is designated to translocate its cognate TpsA protein, the filamentous hemagglutinin FhaB, across the outer membrane of *Bordetella pertussis* [10]. Two-partner secretion represents a branch of the Type V-secretion pathway.

Both BamA and FhaC contain a central cylinder-like membrane embedded C-terminal β-barrel, which is composed of 16 transmembrane β-strands, and an N-terminal soluble domain containing one to five *po*lypeptide *tr*ansport-*a*ssociated (POTRA) domains [16, 17]. However, comparing the structures of BamA and FhaC reveals also differences between the two proteins: (1) the overall barrel shapes and shear numbers (S = 20 for FhaC, S = 22 for BamA) within the first and last β-strands are different; (2) BamA has five POTRA domains while FhaC has only two POTRA domains; (3) the conformations of loop 6 containing a conserved VRGF/Y motif and being embedded within the β-barrels of BamA and FhaC seem to differ significantly [16]. These structural differences might reflect the diversity in function of FhaC and BamA: FhaC functions as a translocase [10] whereas BamA primarily functions as an insertase [13, 15]. Currently, it is still a mystery how β-barrel proteins of the Omp85 family can fulfill such seemingly diverse tasks. In order to enable an in-depth biochemical analysis of the function of Omp85 family proteins, we have developed a cell-free system that allows the functional reconstitution of the two-partner secretion pathway in vitro [18]. Using this system we provided the thus far clearest evidence that a single β-barrel TpsB protein (FhaC) by itself enables the translocation of a TpsA passenger (FhaB) across the outer membrane. This system does not only faithfully reproduce in vitro two-partner secretion, but should also be expandable to related bacterial transport events and can therefore be regarded as a useful tool for an in-depth analysis of the OMP folding and assembly pathways.

Here we describe the detailed protocols for the preparation of isolated outer membranes from cells overexpressing the translocator protein FhaC, preparation of proteoliposomes from purified FhaC protein or from the pool of detergent-solubilized outer membrane proteins containing FhaC, the use of spheroplasts for producing FhaB, as well as the detailed experimental setup of our protein translocation system.

2 Materials

2.1 Preparation of Outer Membranes

1. LB medium: 10.0 g/L tryptone/peptone (pancreatic digested casein), 5 g/L yeast extract, 10 g/L NaCl. Prepare 100 mL of medium in a 0.5-L Erlenmeyer flask as overnight starting culture after autoclave (*see* **Note 1**).
2. For OMV medium (*see* **Note 2**), separately prepare the following: one 5-L Erlenmeyer flask containing 10 g each of yeast extract and tryptone/peptone dissolved in 753 mL H_2O; 1 L of 1 M K_2HPO_4; 1 L of 1 M KH_2PO_4; 1 L of 25 % glucose. After autoclaving and before starting to grow the cells, add 166 mL of 1 M K_2HPO_4, 41 mL of 1 M KH_2PO_4, and 41 mL of 25 % glucose into the Erlenmeyer flask already containing 753 mL of yeast extract and tryptone/peptone to prepare 1 L of complete OMV medium.
3. Autoclave 2 L of Milli-Q H_2O.
4. Kanamycin (100 mg/mL): dissolved in water.
5. Triethanolamine acetate (TeaOAc) (1 M): adjusted to pH 7.5 with acetic acid, filtered (*see* **Note 3**) and stored at 4 °C.
6. EDTA (0.5 M): adjusted to pH 8.0 with KOH, filtered and stored at 4 °C.
7. Ultrapure sucrose (2.5 M): heat slightly for better dissolution, store at room temperature.
8. Dithiothreitol (DTT) (1 M): stored at −20 °C in 1 mL aliquots.
9. Isopropyl-β-D-thiogalactopyranoside (IPTG) (1 M): filtered and stored at −20 °C as 1 mL aliquots.
10. Buffer A (Prepare freshly): 50 mM TeaOAc (pH 7.5), 250 mM sucrose, 1 mM EDTA (pH 8.0), and 1 mM DTT.
11. Buffer B (Prepare freshly): 0.5 M TeaOAc (pH 7.5), 10 mM EDTA (pH 8.0), and 10 mM DTT.
12. Sucrose solutions for sucrose gradient centrifugation (prepare freshly) (*see* Table 1).
13. OMV buffer: 50 mM TeaOAc (pH 7.5), 250 mM sucrose, and 1 mM DTT (*see* **Note 4**). Store at −20 °C in 1 mL aliquots.

2.2 Preparation of FhaC Proteoliposomes

1. Na_2HPO_4 (0.5 M).
2. Solution PE: 20 mM Na_2HPO_4, 1 % Elugent.
3. NaCl (1 M).
4. Imidazole, pH 8.0 (1 M).
5. OMV buffer: 50 mM TeaOAc, pH 7.5, 250 mM sucrose, and 1 mM DTT (*see* **Note 4**). Store at −20 °C in 1 mL aliquots.

Table 1
Sucrose solutions for sucrose gradient centrifugation

Sucrose solutions (prepare 100 mL each) (M)	2.5 M Sucrose (mL)	Buffer B (mL)	H_2O (mL)
0.77	30.8	10	59.2
1.44	57.6	10	32.4
2.02	80.8	10	9.2

2.3 Preparation of E. coli Spheroplasts and Translocation Assay

1. LB medium: 10.0 g/L tryptone/peptone (pancreatic digested casein, Carl Roth, Karlsruhe, Germany), 5 g/L yeast extract, 10 g/L NaCl. Store 500 mL autoclaved LB medium at 4 °C.
2. Glucose solution (20 %), autoclaved.
3. Ampicillin (100 mg/mL).
4. E-salts (5×) preparation (recipe for 100 mL): 1 g citric acid monohydrate, 0.1 g $MgSO_4 \cdot 7H_2O$, 6.55 g $K_2HPO_4 \cdot 3H_2O$, 1.75 g $NaNH_4HPO_4 \cdot 4H_2O$. The mixture is filtered and stored at 4 °C.
5. A mix of 18 amino acids (without methionine and cysteine) in water at a concentration of 1 mM each.
6. Tris–HCl, pH 7.5 (1 M).
7. Sucrose (2.5 M).
8. Lysozyme (50 mg/mL): filtered and stored at −20 °C in 1 mL aliquots.
9. EDTA (0.5 M): adjusted to pH 8.0 with KOH, filtered and stored at 4 °C.
10. Glycerol (50 %).
11. E-medium 1 recipe (for 10 mL): *see* Table 2.
12. T/S buffer recipe (for 10 mL): *see* Table 3.
13. L/E buffer recipe (for 10 mL): *see* Table 4.
14. E-medium 2 recipe (for 10 mL): *see* Table 5.
15. Isopropyl-β-D-thiogalactopyranoside (IPTG) (1 M): filtered and stored at −20 °C in 1 mL aliquots.
16. [^{35}S]-EasyTag™ Express Protein Labeling Mix, 407 MBq (11 mCi)/mL (Perkin Elmer, USA). The mixture contains 73 % [^{35}S]-methionine and 22 % [^{35}S]-cysteine. After arrival, prepare 50 μL aliquots and store at −20 °C. (Radioactive material is hazardous. Avoid ingestion or contact with skin or clothing. Always wear gloves when handling. Monitor hands, equipment, and bench frequently. Please note that special safety regulations for handling radioactive material that might apply to your institution!)

Table 2
Recipe for E-medium 1 (10 mL)

Final concentration	Stock solution	Amount required
1× E-salts	5× E-salts	2 mL
100 μM 18 Amino acids	1 mM	1 mL
0.4 % Glucose	20 % Glucose	200 μL
H_2O		6.8 mL

Table 3
Recipe for T/S buffer (10 mL)

Final concentration	Stock solution	Amount required (mL)
100 mM Tris–HCl, pH 7.5	1 M Tris–HCl	1
0.5 M Sucrose	2.5 M Sucrose	2
H_2O		7

Table 4
Recipe for L/E buffer (10 mL)

Final concentration	Stock solution	Amount required
0.1 mg/mL Lysozyme	50 mg/mL	20 μL
8 mM EDTA, pH 8.0	0.5 M EDTA	160 μL
H_2O		9.82 mL

Table 5
Recipe for E-medium 2 (10 mL)

Final concentration	Stock solution	Amount required
1× E-salts	5× E-salts	2 mL
100 μM 18 Amino acids	1 mM	1 mL
2 % Glycerol	50 % Glycerol	400 μL
0.25 M Sucrose	2.5 M Sucrose	1 mL
H_2O		5.6 mL

17. Trichloroacetic acid (10 % w/v).
18. Proteinase K storage buffer: 2.5 mL glycerol, 50 μL 1 M Tris–HCl (pH 7.5), 0.0145 g of $CaCl_2$, add H_2O to 5 mL.
19. Proteinase K solution (10 mg/mL): add 50 mg of Proteinase K to 5 mL storage buffer and mix well by inverting until crystals are dissolved, aliquot into 1 mL aliquots and store at −20 °C.

2.4 Sodium Dodecyl Sulfate–Polyacrylamide Gel Electrophoresis (SDS-PAGE)

1. Tris–HCl, pH 8.8 (2 M): filter and store at 4 °C.
2. Tris–HCl, pH 6.8 (0.5 M): filter and store at 4 °C.
3. SDS (25 % (w/v)): store at room temperature.
4. Acrylamide–bisacrylamide (30 %/0.8 %) solution. TOXIC: always wear gloves when handling acrylamide solutions and gels because acrylamide is a neurotoxin.
5. *N,N,N′,N′*-Tetramethylethylenediamine (TEMED, 99 %, p.a., for electrophoresis, Carl Roth).
6. Ammonium peroxodisulfate (10 %): can be stored at 4 °C for several days, but fresh preparation is preferred.
7. SDS-running buffer: dissolve 150 g Tris base and 720 g glycine in 5 L H_2O to prepare a 5× stock. To prepare 2 L of 1× SDS-running buffer, take 400 mL 5× stock, 4 mL 25 % SDS, add H_2O to 2 L.
8. Loading buffer solution-1: 2 mL 1 M Tris base, 1 mL 0.2 M EDTA, pH 8.0, 7 mL H_2O.
9. Loading buffer solution-2: 4 mL 25 % SDS, 1 mL 1 M Tris base, 3.5 mL glycerol, 3.5 mL 0.1 % bromophenol blue.
10. Dithiothreitol (DTT) (1 M): prepare freshly.
11. SDS-loading buffer (for 1 mL): mix 500 μL solution-1, 400 μL solution-2, and 100 μL 1 M DTT. Always prepare freshly.
12. Prestained molecular weight marker: peqGOLD Protein-Marker IV (PEQLAB Biotechnologie GmbH)
13. Fixing solution: 35 % ethanol, 10 % acetic acid.

3 Methods

3.1 Preparation of Outer Membranes

1. The cell of the French press is precooled in a 4 °C fridge.
2. To prepare overnight starter cultures, 100 mL LB medium containing kanamycin are inoculated with cells from plates or glycerol stocks from an *E. coli* BL21(DE3) strain harboring the *fhaC* gene [19] cloned under an inducible promoter. Cells are grown overnight at 37 °C with sufficient aeration in a rotary shaker (the flask is covered with aluminum foil).
3. In the following day, dilute 20 mL of overnight culture into a flask containing 1 L complete OMV medium and grow

cells at 37 °C. The overexpression of FhaC proteins is induced at an optical density of 0.8–1.0 by adding 1 mM isopropyl-β-D-thiogalactopyranoside (IPTG). Continue to grow the cells for another 3–4 h.

4. Prepare 500 mL buffer A and cool on ice.
5. Place the flasks in an ice/water bath to quickly chill the cell cultures, harvest the cells at 4 °C in a precooled SLA3000 rotor (Sorvall) for 10 min at 4225 × *g* (5000 rpm). All subsequent steps should be performed at 4 °C or on ice. The cell pellets are resuspended in buffer A (*see* **Note 5**) and centrifuged again. Cells obtained from a 1 L culture are resuspended in 5 mL of buffer A (containing 1 tablet protease inhibitor cocktail (Roche) per 15 mL) and finally diluted to 15 mL with buffer A.
6. To break the cells, the cell suspension is pressed two to three times in a drop-wise manner through a precooled 40 K FRENCH Pressure Cell at 8000 psi. This corresponds to a gage pressure setting of 500 when a 1-in. piston cell is used at the "high ratio" selection.
7. To remove cell debris, the broken cells from French press treatment are centrifuged for 10 min at 1954 × *g* (5000 rpm) in a precooled SORVALL SS34 rotor at 4 °C.
8. Collect the supernatant from **step 7** and centrifuge at approx. 150,000 × *g* (40,000 rpm) for 2 h at 4 °C in a precooled BECKMAN 50.2Ti rotor to collect a crude membrane pellet containing inner and outer membranes as well as ribosomes (*see* **Note 6**). The sticky pellet is transferred to a glass homogenizer (Fischer Scientific) and homogenized in approx. 4–5 mL buffer A by use of a loosely fitting glass pestle.
9. To separate the outer membrane from inner membrane and ribosomes taking advantage of their different densities on a sucrose gradient, six discontinuous sucrose gradients are prepared. Each gradient is composed of 9 mL 0.77 M, 16 mL 1.44 M and 9 mL 2.02 M sucrose solution in polyallomer centrifuge tubes (38.5 mL, 25 × 89 mm). Start with the 0.77 M sucrose cushion then followed by 1.44 and 2.02 cushions and always underlay the denser ones using a smoothly running syringe equipped with a blunt wide-bore needle. The gradients should be equilibrated at 4 °C for at least 1 h. Finally, load 2–2.5 mL of the homogenized membranes on the top of each gradient and spin at approx 112,000 × *g* (25,000 rpm) for at least 16 h at 4 °C in a Sorvall AH-629 (36 mL) swing-out rotor.
10. The outer membranes, which look like a light yellow ring and are located at the lower interface between 1.44 and 2.02 M cushion, are collected using a syringe (BRAUN, 1.2 × 40 mm, REF4665120) from the top of the gradient. Place the membrane suspension immediately into a Falcon tube cooled on ice.

11. The collected outer membranes are diluted sevenfold in ice-cold 50 mM TeaOAc (pH 7.5) to lower down the sucrose concentration (the solution withdrawn from the interface presumably equals a mixture of both the 1.44 M and 2.02 M sucrose cushions resulting in a theoretical sucrose concentration of about 1.7 M; the dilution aims to maintain the concentration of sucrose at about 250 mM).
12. The outer membranes are collected by centrifugation at approx. 150,000 × *g* (40,000 rpm) for 2 h at 4 °C in a precooled BECKMAN 50.2Ti rotor. The supernatant is decanted and the outer membrane pellets obtained from a 1 L culture are carefully resuspended in a total of about 1 mL OMV buffer and homogenized using a loosely fitting glass homogenizer. The final desired volume of outer membranes is adjusted to give an absorbance at 280 nm of 40 unit/mL (*see* **Note 7**).
13. Finally, the outer membranes are aliquoted in 50 μL, frozen in liquid nitrogen and stored at −80 °C.

3.2 Preparation of FhaC Proteoliposomes

3.2.1 Purification of FhaC

1. 240 μL outer membranes are solubilized by dilution into 5 mL solution PE and incubated at room temperature for 2 h by end-over-end rotation.
2. To remove insoluble material, a centrifugation at 109,000 × *g* (45,000 rpm) for 1 h at 4 °C using a precooled Sorvall S100-AT4 rotor.
3. Supernatants are combined and applied to a Mono S® HR 5/5 column equilibrated with solution PE. Proteins are eluted at 4 °C by use of a 0–1 M NaCl gradient in solution PE, at a flow rate of 0.5 mL/min. FhaC is eluted between 220 and 400 mM NaCl.
4. FhaC-containing fractions are combined and desalted on a Sephadex G-25 column (NAP 5 columns, GE Healthcare) against solution PE.
5. Apply at 4 °C the desalted materials to a gravity flow column prepared from 1.5 mL of a 50 % slurry of Talon beads (Clontech) that is equilibrated with solution PE. The column is washed three times with 1 mL each of solution PE containing 5 mM imidazole, and bound proteins are eluted by applying three times 1 mL each of solution PE containing 200 mM imidazole. Most FhaC is recovered from the first elution fraction.
6. The imidazole-eluted FhaC is dialyzed using Spectra/Por Dialysis Membrane (Spectrum Laboratories, Inc.) (*see* **Note 8**) three times against 1 L of solution PE and stored at 4 °C.

3.2.2 Preparation of Proteoliposomes

1. Preparation of Bio-Beads (Bio-Beads™ SM-2 Adsorbent 20–50 mesh, Bio-Rad): Put dry beads into a 15 mL Falcon tube and wash twice sequentially with methanol, ethanol, demineralized water, and 20 mM Na_2HPO_4.

2. Preparation of phospholipids (Avanti Polar Lipids, Alabaster, AL): The powder of phospholipids is suspended in H_2O to prepare a stock solution of 50 mg/mL, which is sonicated at low power until an opalescent solution is formed (*see* **Note 9**). For each reconstitution, usually 400 μg phospholipids (*see* **Note 10**) are diluted to get a final volume of 200 μL in solution PE.
3. Set-up of the reconstitution mixture for FhaC-proteoliposome preparation: 0.72 μg or desired amounts of purified FhaC (*see* **Note 11**) are diluted into 400 μL of solution PE (*see* **Note 12**), which are combined with 200 μL of phospholipids (*see* **step 2**), 300 μL of 20 mM Na_2HPO_4, and added to a SafeSeal Microcentrifuge tube (BioScience, Inc.) containing 50–65 mg Bio-Beads (*see* **step 1**). The reconstitution mixture is incubated for 2 h at 4 °C by end-over-end rotation.
4. Spin down the Bio-Beads in a tabletop centrifuge at 13,000 rpm for 2 min at room temperature, and treat the supernatant in the same manner for another two times with fresh batches of Bio-Beads with one incubation step being performed overnight.
5. To recover the proteoliposomes from the final supernatant, an ultra-centrifugation at approx. 186,000 × *g* (550,000 rpm) for 30 min at 4 °C in a precooled Beckmann TLA-55 rotor is performed.
6. After centrifugation, discard the supernatant and resuspend the pellet with 100 μL OMV buffer, then sonicate the suspension for 1–2 min, freeze it in liquid nitrogen and store at −80 °C (*see* **Note 13**).

3.2.3 Preparation of E. coli Spheroplasts and Translocation Assay

An overview of the experimental procedure is shown in Fig. 1.

1. Inoculate from plates or glycerol stocks of *E. coli* strain BL21(DE3) carrying plasmid pTrc99a::FhaB45met-His (*see* **Note 14**) that encodes a 37 kDa N-terminal truncate (FhaB*37) of hemagglutinin FhaB from *Bordetella pertussis* [18]. Grow cells overnight at 37 °C in 3 mL LB medium (*see* **Note 15**) supplemented with 0.4 % glucose and the appropriate antibiotic (ampicillin for pTrc99a::FhaB45met-His).
2. The following day, dilute the overnight culture 1:50 in E-medium 1 and grow to an $OD_{580} = 1$.
3. Harvest 4 mL of cells at 4400 × *g* for 6 min with a Hettich UNIVERSAL 16R centrifuge.
4. Resuspend the cell pellet with 1000 μL T/S buffer, then combine with 1000 μL L/E buffer to prepare spheroplasts, incubate on ice for 15 min.
5. Spheroplasts are harvested by spinning at 4400 × *g* at 4 °C for 15 min in a Hettich UNIVERSAl 16R centrifuge and the

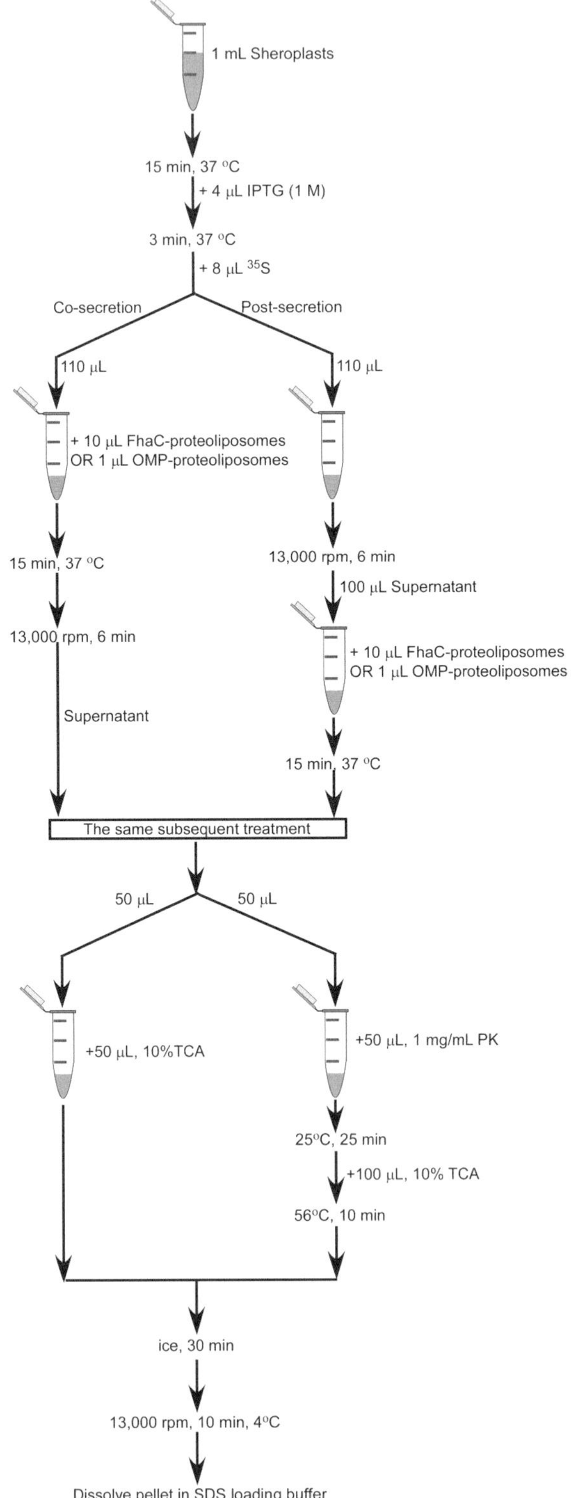

Fig. 1 An overview of the experimental procedure of protein translocation from spheroplasts into proteoliposomes

pellets (*see* **Note 16**) are resuspended in 1000 μL E-medium 2 (*see* **Note 17**).

6. Incubate spheroplasts at 37 °C for 15 min, then add IPTG to a final concentration of 4 mM to induce protein expression followed by another 3 min incubation at 37 °C (*see* **Note 18**).
7. Add [^{35}S]-EasyTag™ Express Protein Labeling Mix (8 μL/mL). For a co-secretional assay immediately mix with proteoliposomes or appropriate controls as shown in Fig. 1 (*see* **Note 19**) then incubate for another 15 min at 37 °C for protein translocation to occur (*see* **Note 20**).
8. Remove spheroplasts by low speed centrifugation at 13,000 rpm for 6 min in a tabletop centrifuge. For a post-secretional assay, follow the modifications of the protocol outlined in Fig. 1.
9. Divide the supernatant into two halves. One half is precipitated directly with TCA (5 % final concentration) on ice for at least 30 min; the other half is treated with 500 μg/mL proteinase K (PK) for 25 min at 25 °C to examine translocation of FhaB*37 into FhaC-proteoliposomes. A successful translocation into the vesicle lumen is indicated by the occurrence of an FhaB*37 band resistant towards PK-treatment (Fig. 2). To inactivate PK, the PK-treated sample is incubated at 56 °C for 10 min immediately after addition of TCA (5 % final), and finally transferred onto ice for precipitation for at least 30 min.
10. Centrifuge both TCA treated sample in a tabletop centrifuge at 14,000 × *g* for 10 min to collect the pellet. After removal of the supernatant, add 15 μL SDS-loading buffer to the pellet (*see* **Note 21**), heat at 95 °C and shake vigorously for 5 min to dissolve the pellet completely. Finally load the sample onto SDS-polyacrylamide gels for electrophoresis.

3.3 SDS-Polyacrylamide Gel Electrophoresis and Autoradiography

1. Large custom-made units are used for gel electrophoresis and dimensions of the gels are: 35 cm × 27.5 cm × 1 cm (W × L × T). These gels are made from about 80 mL of 12 % separating gel and 20 mL of 5 % stacking gel solutions.
2. To prepare 80 mL of 12 % separating gel solution, add 32 mL of 30 % acrylamide–bisacrylamide solution, 16 mL of 2 M Tris–HCl, pH 8.8, 0.8 mL of 10 % SDS to a measuring cylinder and adjust volume to 80 mL with H_2O. Add 0.06 mL TEMED and 0.6 mL 10 % ammonium peroxodisulfate then mix well to start polymerization, immediately pour the mixed solution into gel cassettes and mount them in an upright position, overlay with isobutanol.
3. After polymerization, remove isobutanol from the top of the polymerized gel and rinse with H_2O. To prepare 20 mL of 5 % stacking-gel solution, add 3.33 mL acrylamide–bisacrylamide solution, 2.4 mL 0.5 M Tris–HCl, pH 6.8, 0.2 mL 10 % SDS

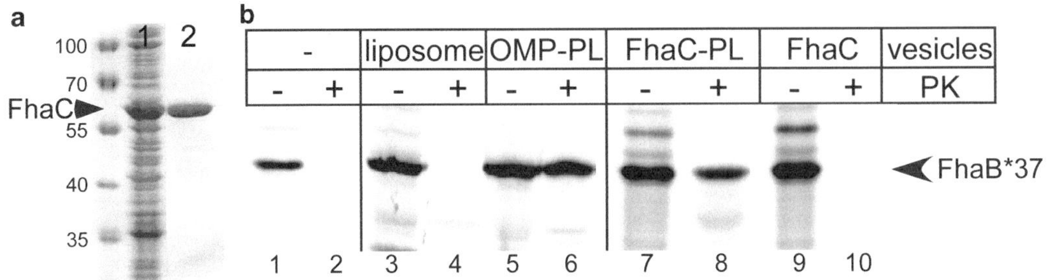

Fig. 2 Spheroplast-secreted FhaB*37 is translocated into proteoliposomes prepared from purified FhaC or the pool of detergent-solubilized outer membrane proteins containing FhaC. (**a**) Purification of FhaC. *Lane 1* shows the protein profile of detergent solubilized outer membrane proteins including FhaC before purification, *lane 2* the purified FhaC. (**b**) Spheroplasts were prepared from *E. coli* BL21(DE3) strain carrying the plasmid pTrc99a::FhaB45met-His, which encodes a truncated version of hemagglutinin FhaB from *Bordetella pertussis*, and induced for the synthesis of FhaB*37 with IPTG (4 mM). 110 μL spheroplasts were mixed with 1 μL of OMV buffer (*lanes 1* and *2*), 1 μL of protein-free liposomes (*lanes 3* and *4*), 1 μL of proteoliposomes prepared from the pool of outer membrane proteins containing FhaC (*lanes 5* and *6*), 10 μL of proteoliposomes prepared from purified FhaC (*lanes 7* and *8*), or equivalent amounts of mock-treated FhaC protein (as in *lanes 7* and *8*), and pulse-labeled with [^{35}S] methionine/cysteine. The spheroplasts were removed by a low speed centrifugation after a 15 min of incubation at 37 °C, and the supernatants containing membrane vesicles were split into two halves; one was precipitated with 5 % TCA and the other one was subjected to proteinase K (500 μg/mL) treatment at 25 °C for 25 min, followed by a precipitation using 5 % TCA and inactivation of proteinase K by incubation at 56 °C for 15 min. Samples were separated by SDS-polyacrylamide gel electrophoresis and visualized by phosphorimaging. Protection of FhaB*37 against proteinase K treatment (i.e., *lanes 6* and *8*) was observed only when the translocation experiments were performed using proteoliposomes containing FhaC, which translocates FhaB*37 into the vesicle lumen. The fact that proteoliposomes prepared from purified FhaC gave rise to the translocation of FhaB*37 indicates that FhaC serves as the sole translocator for its cognate passenger protein FhaB. Lack of PK-protection in the presence of purified FhaC does not occur (*lanes 9* and *10*) because there is no lipid membrane to allow sequestration

to a measuring cylinder and adjust the volume to 20 mL with H_2O. Add 0.016 mL TEMED and 0.16 mL 10 % ammonium peroxodisulfate and mix well to start polymerization, then immediately pour the solution into the gel cassettes, then insert a comb. After polymerization of the stacking-gel, remove the comb and mount the gel in an upright manner in a gel electrophoresis apparatus.

4. Add 1× SDS running buffer to the upper and lower chambers of the electrophoresis apparatus.
5. Load the samples (from **step 10** of Subheading 3.2.3) completely into the wells of the gel and 5 μL of peqGOLD Prestained Protein-Marker IV in one lane. Usually an overnight electrophoresis is carried out at a constant voltage of 110 V until the bromophenol blue dye has run to the bottom of the gel (*see* **Note 22**).
6. After electrophoresis, unload the gel and remove the stacking gel, then incubate the separating gel in a fixing solution for

20 min on a shaking platform. Remove the fixing solution and incubate the gel in H_2O for 10 min, repeat two times.

7. Transfer the gel onto a pre-wetted Whatman 3 MM paper and cover with a plastic wrap, then lay the sandwiched gel on a vacuum dryer (Bio-Rad) with Whatman paper at the bottom. The gel is dried at 70 °C for 2 h with a vacuum pump. After drying, use any radioactive solution to spot on the marker proteins.
8. Autoradiography of the results is obtained by exposing the dried gel to a phosphorimaging screen overnight, which is scanned with a PhosphorImager (e.g., Storm; GE Healthcare) and analyzed using ImageQuant™ software. An autoradiogram of a typical experiment demonstrating translocation of FhaB*37 by FhaC is shown in Fig. 2.

4 Notes

1. All medium and solutions are prepared with deionized water.
2. OMV medium is identical to the medium previously called INV medium (*21*).
3. Stock solutions are routinely filtered through 0.22 μM mixed cellulose ester filters (Millipore) to remove microorganisms and particles.
4. For some of the substrate proteins we tested, translocation into proteoliposomes proved to be unaffected by the presence of DTT.
5. For a fast resuspension, use an automatic pipette equipped with a 5 mL disposable tip.
6. During this 2 h centrifugation, prepare sucrose gradients as indicated in **step 9** (Subheading 3.1).
7. To determine the absorbance of the outer membrane suspension at 280 nm, take 1 μL of membranes and dilute into 100 μL of 2 % SDS.
8. Wear gloves to handle the dialysis tubing. Soak the membrane in distilled water for 30 min at room temperature to remove the preservative (glycerine or sodium azide), then rinse thoroughly in distilled water.
9. Perform sonication overnight in a cold room.
10. Use no more than 400 μg for each reconstitution due to potential contamination of the phospholipids with β-barrel membrane proteins (such as BamA).
11. Titrate the amount of FhaC to be reconstituted with a given amount of phospholipids for obtaining proteoliposomes with highest translocation capacity.

12. For protein-free liposome preparation, simply use 400 μL PE solution; for preparation of OMP-proteoliposomes use 400 μL of solubilized outer membrane proteins as described in Subheading 3.2.2, **step 3**.
13. When proteoliposomes are prepared from an extract of crude outer membrane proteins, they can be stored at −80 °C and reused for two or three times, but remember to shortly sonicate them every time before use. Proteoliposomes prepared from purified proteins, however, use to lose their protein translocation ability after storage at −80 °C.
14. Several other substrates proteins have been tested and all worked in this system. For expression of substrate proteins, we got best results with pTrc99a or pET vectors.
15. The cells can also directly be grown in E-medium 1 overnight.
16. After centrifugation, if the pelleted material is spread out on the walls of a 15 mL Falcon tube rather than forming a centered pellet at the bottom, it is a good indication for a successful spheroplast preparation.
17. The optimal density of the spheroplast suspension has to be determined for each substrate; dilute with E-medium 2 if necessary.
18. After adding IPTG, the incubation time should not exceed 3 min in order to reduce the level of background proteins. Alternatively, one can also try to use rifampicin to see if background can be reduced.
19. The amount of proteoliposomes can be varied and has to be tested to obtain optimal translocation efficiencies.
20. The incubation time can be varied and has to be tested to obtain optimal translocation efficiencies.
21. The color of the loading buffer should remain dark blue. If it changes to yellow, add a few microliters of 1 M Tris base to neutralize residual trichloroacetic acid.
22. Roughly leave 2 cm between the dye front and the end of the gel to avoid low molecular mass radioactive material running off the gel.

Acknowledgements

This work was supported by Sonderforschungsbereich 746 and Forschergruppe 929 and FA 1278/1-1 from the Deutsche Forschungsgemeinschaft.

References

1. Desvaux M, Hebraud M, Talon R et al (2009) Secretion and subcellular localizations of bacterial proteins: a semantic awareness issue. Trends Microbiol 17(4):139–145
2. Economou A, Christie PJ, Fernandez RC et al (2006) Secretion by numbers: protein traffic in prokaryotes. Mol Microbiol 62(2): 308–319
3. Dalbey RE, Kuhn A (2012) Protein traffic in gram-negative bacteria – how exported and secreted proteins find their way. Fems Microbiol Rev 36(6):1023–1045
4. Kudva R, Denks K, Kuhn P et al (2013) Protein translocation across the inner membrane of Gram-negative bacteria: the Sec and Tat dependent protein transport pathways. Res Microbiol 164(6):505–534
5. Webb CT, Heinz E, Lithgow T (2012) Evolution of the beta-barrel assembly machinery. Trends Microbiol 20(12):612–620
6. Schleiff E, Maier UG, Becker T (2011) Omp85 in eukaryotic systems: one protein family with distinct functions. Biol Chem 392(1-2):21–27
7. Walther DM, Rapaport D, Tommassen J (2009) Biogenesis of beta-barrel membrane proteins in bacteria and eukaryotes: evolutionary conservation and divergence. Cell Mol Life Sci 66(17):2789–2804
8. Henderson IR, Nataro JP (2001) Virulence functions of autotransporter proteins. Infect Immun 69(3):1231–1243
9. van Ulsen P, Rahman SU, Jong WS et al (2014) Type V secretion: from biogenesis to biotechnology. Biochim Biophys Acta 1843(8): 1592–1611
10. Jacob-Dubuisson F, Guerin J, Baelen S et al (2013) Two-partner secretion: as simple as it sounds? Res Microbiol 164(6):583–595
11. Dautin N, Bernstein HD (2007) Protein secretion in gram-negative bacteria via the autotransporter pathway. Annu Rev Microbiol 61:89–112
12. Grijpstra J, Arenas J, Rutten L et al (2013) Autotransporter secretion: varying on a theme. Res Microbiol 164(6):562–582
13. Hagan CL, Silhavy TJ, Kahne D (2011) beta-Barrel membrane protein assembly by the BAM complex. Annu Rev Biochem 80:189–210
14. Kim KH, Aulakh S, Paetzel M (2012) The bacterial outer membrane beta-barrel assembly machinery. Protein Sci 21(6):751–768
15. Ricci DP, Silhavy TJ (2012) The Bam machine: a molecular cooper. Biochim Biophys Acta 1818(4):1067–1084
16. Noinaj N, Kuszak AJ, Gumbart JC et al (2013) Structural insight into the biogenesis of beta-barrel membrane proteins. Nature 501(7467):385–390
17. Clantin B, Delattre AS, Rucktooa P et al (2007) Structure of the membrane protein FhaC: a member of the Omp85-TpsB transporter superfamily. Science 317(5840):957–961
18. Fan E, Fiedler S, Jacob-Dubuisson F et al (2012) Two-partner secretion of gram-negative bacteria: a single beta-barrel protein enables transport across the outer membrane. J Biol Chem 287(4):2591–2599
19. Meli AC, Hodak H, Clantin B et al (2006) Channel properties of TpsB transporter FhaC point to two functional domains with a C-terminal protein-conducting pore. J Biol Chem 281(1):158–166

Chapter 9

Measuring Cell–Cell Binding Using Flow-Cytometry

Zachary C. Ruhe, Christopher S. Hayes, and David A. Low

Abstract

Cell–cell adhesion mediates a number of competitive and cooperative microbial interactions. Fluorescence labeling and flow cytometry techniques allow us to observe and measure these interactions rapidly and easily. Here, we describe a method to quantify cell–cell adhesion events between two differentially labeled cell populations.

Key words CdiA, cell–cell binding, Contact-dependent growth inhibition, Flow cytometry, Fluorescent proteins, Self/nonself recognition

1 Introduction

Cell–cell adhesion is an important aspect of microbial life. It is essential for the formation of biofilms and mediates many intercellular interactions. We have developed a method to observe these interactions by combining fluorescent cell labeling with flow cytometry [1–3]. This allows us to observe interactions mediated by the contact-dependent growth inhibition (CDI) system [2, 4, 5]. Additionally, this technique has been useful in the characterization of the CdiA-BamA binding events underpinning cell–cell adhesion [3]. This method can be adapted to analyze other forms of cell–cell adhesion and expanded to study more complex microbial interactions.

The premise of the cell–cell adhesion assay is simple. Two cell populations are differentially labeled with fluorescent proteins, in this case GFP and DsRed. The cells are then mixed and analyzed in a flow cytometer. The cytometer measures the fluorescence of each particle. Single cells display only a single fluorescent signal while mixed cell aggregates emit dual red and green fluorescence. With the aid of analytical software, fluorescent cell populations can be identified and quantified. Since flow cytometry is high-throughput, many thousands of cells and cell aggregates can be analyzed within minutes. This allows statistical assessment of cell–cell interactions

Susan K. Buchanan and Nicholas Noinaj (eds.), *The BAM Complex: Methods and Protocols*, Methods in Molecular Biology, vol. 1329, DOI 10.1007/978-1-4939-2871-2_9, © Springer Science+Business Media New York 2015

and the measurement of rare events that would be virtually undetectable by other methods. Additionally, rapid assessment of binding allows researchers to screen for mutants or clones which influence cell–cell adhesion.

2 Materials

1. Tryptone broth: Weigh 10 g of tryptone and 10 g of NaCl and transfer into a plastic beaker containing a magnetic stir bar. Add 1 L of deionized water and stir until reagents are completely in solution. Transfer to an autoclave-safe container and sterilize by autoclaving. After the solution has cooled, filter the solution through a 0.45 μm nitrocellulose membrane. Store in a sealed container at room temperature.
2. Ampicillin liquid stock: Weigh 1 g of ampicillin powder and transfer to a sterile 15 mL conical tube. Add 10 mL of deionized water and mix until the ampicillin is completely in solution. Filter through a 0.45 μm nitrocellulose membrane. Store at −20 °C and thaw just prior to use.
3. Phosphate-buffered saline (PBS) (1×): Weigh 8 g NaCl, 0.2 g KCl, 1.44 g Na_2HPO_4, and 0.24 g KH_2PO_4 into a plastic beaker containing a magnetic stir bar. Add 800 mL of deionized water and stir until reagents are completely in solution. Adjust the pH to pH 7.4 using concentrated HCl solution. Add deionized water to a final volume of 1 L. Filter through a 0.45 μm nitrocellulose membrane and store in a sealed container at room temperature.
4. Filtered water: Filter 1 L of deionized water through a 0.45 μm nitrocellulose membrane and store in a sealed container at room temperature.
5. Accuri C6 flow cytometer (BD Biosciences) with filter sets FL1 530/30 and FL2 585/40 and the Accuri CFlow analysis software.
6. Fluorescence microscope (Olympus BX51 or equivalent) with filter sets (Omega) XF116-2 (for Ds-Red) and XF103-2 (for EGFP).

3 Methods

3.1 Preparation of Fluorescent Cells

1. Inoculate CDI^+ GFP and $BamA^{Eco}$ DsRed strains (Table 1) (*see* **Note 1**) from a frozen stock into 2 mL TB supplemented with 150 μg/mL ampicillin. Grow the strains overnight at 37 °C in a tube roller.

Table 1
Strains and plasmids

Strain	Genotype
CDI$^+$ GFP (DL4905)	MC4100 *gfpmut3* pDAL660Δ1-39
CDI$^-$ GFP (DL4906)	MC4100 *gfpmut3* pWEB-TNC
BamA^{Eco} DsRed (CH9702)	EPI100 Δ*bamA::cat* pZS21::*bamA* (*E. coli*) pDsRedExpress2
BamA^{ECL} DsRed (CH9703)	EPI100 Δ*bamA::cat* pZS21::*bamA* (*E. cloacae*) pDsRedExpress2

2. The next day, dilute the overnight cultures 1:100 into flasks containing 25 mL TB supplemented with 150 μg/mL ampicillin. Grow cells in a flask shaker at 30 °C until they reach OD_{600} ~0.6–0.8 (*see* **Notes 2** and **3**).
3. Once the cells have reached the appropriate optical density, remove 1 μL of culture and place on a glass microscope slide. Cover the sample with a cover slip and view the cells under 100× magnification on a fluorescent microscope. Fluorescence should be readily detectable without the use of gain. Cells should not appear elongated or otherwise malformed. If fluorescence is faint, it may be necessary to culture the cells for a longer period of time (*see* **Note 4**).

3.2 Analysis of Fluorescent Cells by Flow Cytometry

1. Boot the cytometer and calibrate according to manufacturer's instructions.
2. Open the Accuri CFlow software and open a new file.
3. Select well A1. Name this well "Water Pre-Wash." Set the threshold to SSC-H 6000 (*see* **Note 5**).
4. Decant 1 mL of filtered water into a microfuge tube and load onto the cytometer sipper. Set the cytometer flow speed to "Fast" and run the sample for 2 min (*see* **Note 6**).
5. Dilute CDI$^+$ GFP fluorescent cells ~1:500 into filtered 1× PBS in a microfuge tube. Vortex the cell suspension for 1 min and load onto the cytometer sipper.
6. Select well A2 and name this sample CDI$^+$ GFP. Set the flow speed to "Slow" and run the sample for 50,000 events (*see* **Note** 7).
7. After the run has completed, create a new histogram plot by selecting the "histogram plot" button. Set the *X*-axis to FSC (forward scatter). Create a vertical line gate by clicking the vertical marker tool. Set the gate at 2000 on the FSC scale. This should create a vertical red line that intersects with 2000 on the *X*-axis and divides the plot into two gates, V1-L and V1-R (Fig. 1, left column).

8. Create a second density plot. Set the *Y*-axis to FL1-A (green fluorescence) and the *X*-axis to FL2-A (red fluorescence). Select the "Gate" button and select "exclude V1-L" on all events (*see* **Note 8**). Ensure that the "Show Vertical Markers" box is checked to enable this gate option. Click the "Plot Spec button" and set the *X*- and *Y*-axes to logarithmic. Also check the "Hide first decade" box for both axes. The plots should resemble those shown in Fig. 1 (center-left column).
9. Repeat **steps 5** and **6** with BamA^{Eco} DsRed fluorescent cells in well A4.
10. There may be a noticeable spillover of fluorescence for both the GFP- and DsRed-labeled cells into the opposing channel which will appear as a diagonal trail toward the upper-right of the plot. This can be corrected with color compensation. With the FL1/FL2 plot selected for well A2, create a quadrant gate with the "Quadrant gating" tool that clearly separates GFP cells from nonfluorescent particles (Fig. 1, center-right column). In the plot measurements pane, compare the "Median FL2-A" for quadrants Q1-UL and Q1-LL. This is the median amount of red fluorescence emanating from the GFP+ and nonfluorescent particles, respectively. If the median FL2-A is higher for quadrant Q1-UL, select the "Set Color Compensation" button. Under the "Correct FL2 by subtracting a percentage of:" heading, select the FL1 channel and enter 4 %. Click preview and compare the median FL2-A fluorescence of quadrants Q1-UL and Q1-LL. Adjust this percentage until the median FL2-A fluorescence is identical for both quadrants. In this example, the color compensation is set at 5.76 % (Fig. 1, right column). Select the "Apply to: All samples" button and click "Apply & Close."
11. Repeat the color compensation procedure for the DsRed cells in well A4, this time comparing median FL1-A fluorescence between Q1-LR and Q1-LL. Under the "Correct FL1 by subtracting a percentage of:" heading, set a percentage of FL2 that normalizes the median FL1 fluorescence in quadrants Q1-LR and Q1-LL. In this example, the color compensation is set at 6.15 % (Fig. 1, bottom row). Apply these changes to all samples.
12. Select well A12. Load a fresh filtered water sample into the cytometer sipper. Run the water on "Fast" for 2 min (*see* **Note 9**).
13. Save your work as a CFlow Template file. Power down the cytometer according to manufacturer's instructions.

3.3 Cell–Cell Adhesion Assay

1. Prepare CDI^{+} GFP, CDI^{-} GFP, BamA^{Eco} DsRed and BamA^{ECL} DsRed fluorescent cells as described in Subheading 3.1. Measure the OD_{600} of each culture.

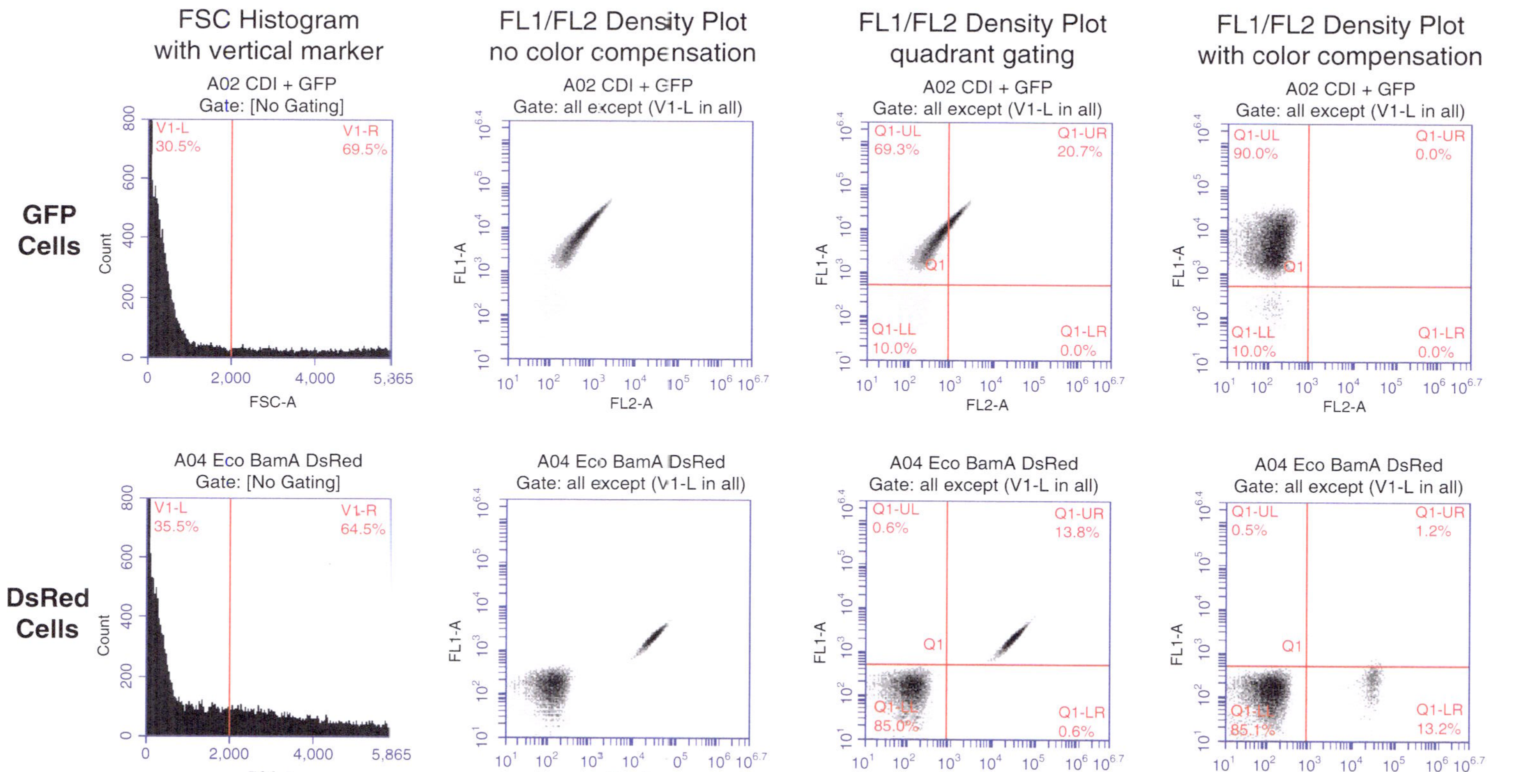

Fig. 1 Flow cytometry. *Left column*: A histogram plot of the FSC channel is used to remove small particulates from the analysis. A vertical line marker (shown in *red*) is placed at the 2000 FSC value. All events in the V1-L gate are ignored in subsequent analyses. *Center-left column*: Fluorescent cells display elevated levels of *green* (FL1) or *red* (FL2) fluorescence in a density plot. Nonfluorescent cells and particles lie near the plot origin. *Center-right column*: A quadrant gate (shown in *red*) is superimposed over the density plot to separate fluorescent and nonfluorescent populations. Measuring the median fluorescence of cells in each quadrant is necessary to properly compensate for fluorescence spillover between channels. *Right column*: Once the color compensation has been adjusted, fluorescent populations appear more circular on the density plot and are completely separated by the quadrant gates

2. Prepare the following samples by diluting fluorescent cells in pre-warmed (30 °C) TB in culture tubes (*see* **Note 10**):

   (a) CDI^{+} GFP at OD_{600} ~0.2.

   (b) CDI^{-} GFP at OD_{600} ~0.2.

   (c) $BamA^{Eco}$ DsRed at OD_{600} ~0.05.

   (d) $BamA^{ECL}$ DsRed at OD_{600} ~0.05.

   (e) CDI^{+} GFP at OD_{600} ~0.2 mixed with $BamA^{Eco}$ DsRed at OD_{600} 0.05.

   (f) CDI^{+} GFP at OD_{600} ~0.2 mixed with $BamA^{ECL}$ DsRed at OD_{600} ~0.05.

   (g) CDI^{-} GFP at OD_{600} ~0.2 mixed with $BamA^{Eco}$ DsRed at OD_{600} ~0.05.

   (h) CDI^{-} GFP at OD_{600} ~0.2 mixed with $BamA^{ECL}$ DsRed at OD_{600} ~0.05.

2. Incubate the cell mixtures at 30 °C in a shaking tube rack for 15 min. Position the tubes at a ~45° angle in the rack to promote mixing (*see* **Note 11**).

3. Boot and calibrate the cytometer according to manufacturer's instructions. Open the template file created in the "Running fluorescent cells in flow cytometer" section above. Decant 1 mL of filtered water into a microfuge tube and load onto the cytometer sipper. Set the cytometer flow speed to "Fast" and run the sample for 2 min (*see* **Note 6**).

4. Dilute the cell mixtures ~1:200 into filtered 1× PBS in a microfuge tube. Vortex each sample for 1 min prior to loading in the cytometer.

5. Run each sample in the cytometer in an appropriately labeled well for 50,000 events on "Slow" (*see* **Note 7**).

6. Select the well containing CDI^{+} GFP cells alone and label this well "CDI^{+} GFP." Using the Polygon gating tool, create a gate that tightly bounds the GFP population of events in the FL1/FL2 density plot, as shown in Fig. 2, gate P1. Select the well containing CDI^{-} GFP cells alone and label this well "CDI^{-} GFP." Ensure that the GFP polygon gate (P1) encompasses the GFP population in this sample (*see* **Note 12**).

7. Select the well containing $BamA^{Eco}$ DsRed cells alone and label this well "Eco BamA DsRed." Create a polygon gate that tightly bounds the DsRed population in the FL1/FL2 density plot, as shown in Fig. 2, gate P2. Select the well containing $BamA^{ECL}$ DsRed cells alone and label this well "ECL BamA DsRed." Ensure that the DsRed polygon gate (P2) encloses the DsRed population in this sample (*see* **Note 12**).

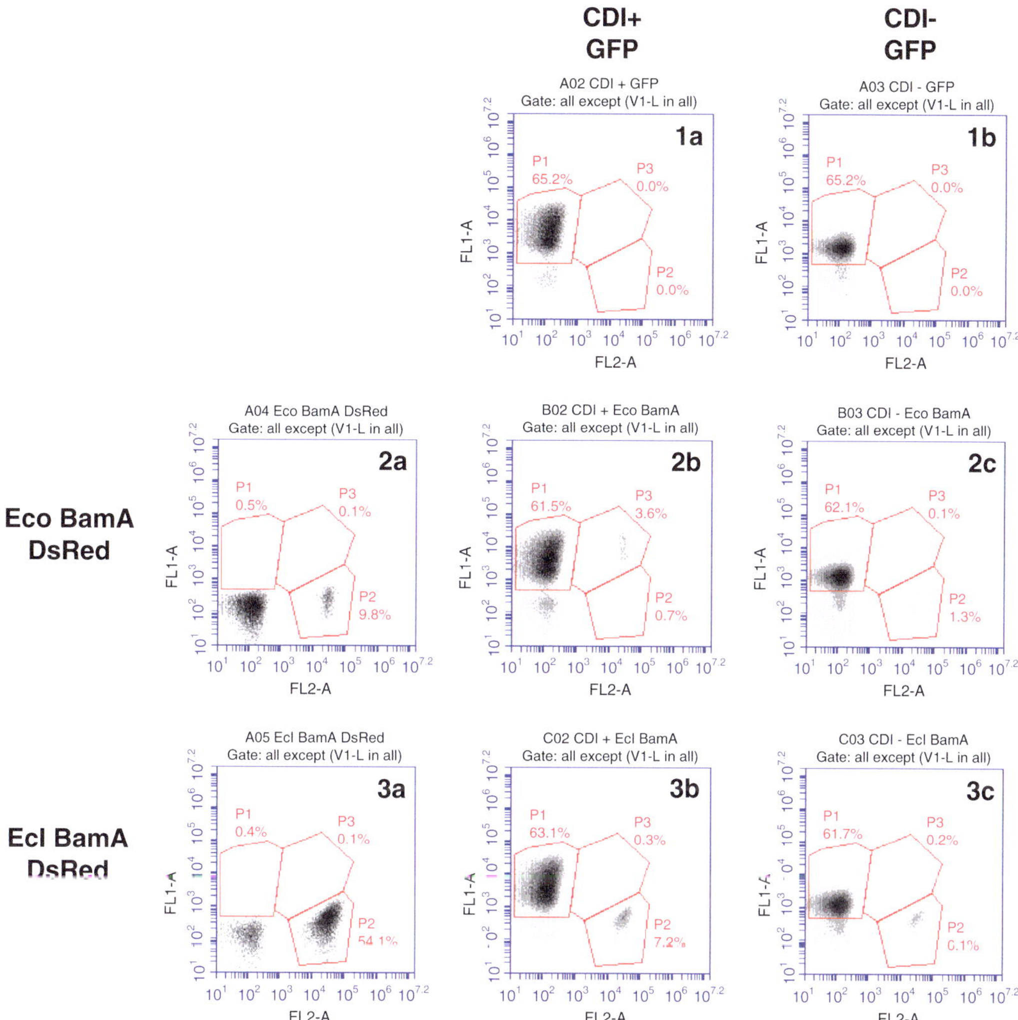

Fig. 2 Cell–cell adhesion. Shown are density plots for CDI$^+$ and CDI$^-$ GFP cells alone (*top row*, 1a and 1b) as well as BamA^{Eco} and BamA^{ECL} DsRed cells alone (*left column*, 2a and 3a). These plots are used to place the unbound GFP and DsRed gates, labeled P1 and P2 respectively. A third gate, P3, is placed to the *right* of the GFP gate and above the DsRed gate to capture dual-fluorescent particles. These particles represent cell–cell adhesion events between the fluorescent populations. CDI$^+$ GFP cells mixed with BamA^{Eco} DsRed cells produce a large number of dual-fluorescent particles (*center panel*, 2b). CDI$^-$ GFP cells and BamA^{ECL} DsRed cells are unable to generate mixed-cell aggregates, yielding a very low number of dual-fluorescent events (3b, 2c, and 3c)

8. Create a third polygon gate that encloses the void adjacent to gates P1 and P2, as shown in Fig. 2, gate P3. This gate will identify the GFP- and DsRed-labeled cells that are adhered to each other (*see* **Note 13**).
9. For each sample run, check that the polygon gates encompass the appropriate populations. In the plot measurements pane, record the event counts for gates P2 and P3 in each mixed-cell sample. Calculate the fraction of bound cells using the following equation:

Table 2
Sample counts

Sample	Unbound DsRed cells	Bound DsRed cells	Fraction bound
CDI$^+$ GFP and BamA^{Eco} DsRed	483	2551	0.841
CDI$^+$ GFP and BamA^{ECL} DsRed	2327	112	0.046
CDI$^-$ GFP and BamA^{Eco} DsRed	5498	220	0.038
CDI$^-$ GFP and BamA^{ECL} DsRed	5249	168	0.031

$$\frac{\text{bound cells}\left(\text{P3}\right)}{\text{unbound cells}\left(\text{P2}\right)+\text{bound cells}\left(\text{P3}\right)}.$$

10. Sample counts and calculation are shown in Table 2 (*see* **Notes 14** and **15**).

4 Notes

1. In this example, we use *E. coli* strains DL4905 and DL4906 that express GFP, and strains CH9702 and CH9703 that express DsRed. The genotypes of these strains are shown in Table 1. These strains have been found to express adequate amounts of fluorescent protein without any overt signs of toxicity under these growth conditions.
2. With GFP and DsRed, better fluorescence is achieved under slow-growth conditions. We routinely culture cells in TB at 30 °C to reduce growth rates.
3. When adapting this assay, you may wish to transfer pDsRedExpress2 to another strain. If this strain harbors *lacI*^{q}, it may be necessary to induce DsRed expression with IPTG. We suggest trying a range of IPTG concentrations to achieve fluorescence while mitigating toxicity from overexpression.
4. Elongated cells are a sign of toxicity from fluorescent protein overexpression. Measures should be taken to reduce expression until the fluorescent cells lack gross symptoms of toxicity.
5. This low SSC threshold is necessary to detect bacterial cells. Particulates in the growth medium and other reagents will easily be detected at this threshold. Therefore, it is necessary to filter all reagents in a 0.45 μM nitrocellulose filter.
6. Inspect the "events per μL" report for this run. This is a good gauge of contaminating particles from the sipper and tubing

within the machine. If there are greater than 100 events per μL, run another sample of filtered water in well B1. Repeat until there are fewer than 100 events detected per μL. If the number of events does not decrease, prepare fresh filtered water and repeat the process.

7. Inspect the "events per μL" report for this run. Check that the cells are dilute enough by ensuring there are fewer than 3000 events per μL. Higher cell densities will interfere with our ability to accurately measure fluorescence and binding in the cytometer. If there are greater than 3000 event per μL, further dilute the cells and repeat the run in a different well.
8. This gate will filter out the majority of small nonfluorescent particles present in the growth medium. If you suspect this gate is excluding cells from your analysis, compare the histogram plots for growth medium with and without cells. Set the vertical line marker to include all events that specifically appear in growth medium with cells.
9. Running water through the cytometer will prevent cells and medium components from adhering to cytometer components. This process should be repeated until there are fewer than 100 events per μL.
10. GFP-labeled cells are in fivefold excess of DsRed-labeled cells. This increases the potential number of binding partners for each DsRed-labeled cell and allows us to accurately measure the ability of DsRed cells to adhere to GFP cells. This ratio can be adjusted depending on the type of assay being performed.
11. Inadequate mixing prevents cell–cell adhesion. Additional steps may be required to ensure samples are mixing properly during this incubation. An easy way to assay this is to place 50 μL of standard DNA loading buffer (with dye) at the bottom of a tube containing 2 mL of TB. Incubate the tube in a shaking tube rack for 1–2 min. If the loading dye has not evenly mixed throughout the solution, adjust the tube angle and shaking speed until you find conditions where the solution mixes.
12. If the fluorescent cells show different levels of fluorescence, expand the gate to include all fluorescent cells from both samples.
13. The bound population of cells may not necessarily lie at the intersection between the GFP- and DsRed-labeled populations. Often the bound population will have increased GFP and/or DsRed fluorescence compared to the unbound population. This is due to fluorescence spillover. Gate P3 should be adjusted to include these events.

14. A high proportion of bound cells in a negative control is often due to high cell density during cytometry. This reflects unbound GFP- and DsRed-labeled cells that pass through the detector coincidentally. Reducing cell density through further dilution into PBS usually resolves this issue.
15. A low proportion of bound cells in a positive control is usually because the adhering protein(s) are not expressed or poorly expressed. Additional assays should be performed to ensure that the adhering proteins are expressed under the growth conditions outlined here.

Acknowledgments

Z.C.R. was supported in part by the Tri-Counties Blood Bank Postdoctoral Fellowship, Santa Barbara. This work was supported by grants from the National Science Foundation (0642052 to D.A.L.) and National Institutes of Health (U01 GM102318 to C.S.H. and D.A.L.).

References

1. Aoki SK, Malinverni JC, Jacoby K et al (2008) Contact-dependent growth inhibition requires the essential outer membrane protein BamA (YaeT) as the receptor and the inner membrane transport protein AcrB. Mol Microbiol 70(2):323–340
2. Aoki SK, Pamma R, Hernday AD et al (2005) Contact-dependent inhibition of growth in *Escherichia coli*. Science 309(5738):1245–1248
3. Ruhe ZC, Wallace AB, Low DA et al (2013) Receptor polymorphism restricts contact-dependent growth inhibition to members of the same species. mBio 4(4), pii: e00480-13
4. Aoki SK, Diner EJ, de Roodenbeke CT et al (2010) A widespread family of polymorphic contact-dependent toxin delivery systems in bacteria. Nature 468(7322):439–442
5. Webb JS, Nikolakakis KC, Willett JL et al (2013) Delivery of CdiA nuclease toxins into target cells during contact-dependent growth inhibition. PloS One 8(2), e57609

Chapter 10

Methods to Characterize Folding and Function of BamA Cross-Link Mutants

Adam J. Kuszak, Nicholas Noinaj, and Susan K. Buchanan

Abstract

The utility of protein engineering, both the mutation and deletion of specific amino acids, to investigate protein structure and function has been demonstrated time and time again, and intermolecular and intramolecular interactions within the BAM complex and its individual components are no exception. Extensive efforts have probed conserved and unique amino acid sequences of the Bam proteins to define their functional roles. This chapter summarizes efforts as applied to the disulfide cross-link mutants of BamA and describes experimental methods used in our studies to determine that lateral opening of the barrel domain is required for function.

Key words BamA, BAM complex, Outer membrane protein, Proteinase K, Disulfide cross-linking, β-barrel membrane protein

1 Introduction

BamA is the central component of the BAM complex [1, 2] and a member of the TspB/Omp85 superfamily of β-barrel membrane proteins that are essential for protein transport across and/or incorporation into the outer membrane of gram-negative bacteria [3]. Prior to assaying mutant BamA function, and in the event of observing a novel phenotype, one must determine whether the mutant BamA is stably folded into the membrane or not and whether any observed loss of function is due to perturbed interactions with the Bam lipoproteins or nascent OMP substrates.

As described in Chapter 4, heat modifiability assays can be used to access whether a β-barrel outer membrane protein is properly folded or not. Here, we use this assay to determine if BamA mutants are being properly folded when expressed and compare the results to wild type BamA.

Limited proteolysis of membrane proteins can be a powerful technique to assess solvent exposed regions, probe conformational

Susan K. Buchanan and Nicholas Noinaj (eds.), *The BAM Complex: Methods and Protocols*, Methods in Molecular Biology, vol. 1329, DOI 10.1007/978-1-4939-2871-2_10,

changes, and monitor protein trafficking. In this application intact bacteria cells, which are heterologously expressing BamA and BamA mutants, are exposed to proteinase K (PK) [4–7]. PK does not readily cross the outer membrane, and therefore only extracellular portions of the BamA are subject to the enzyme, creating conditions for limited proteolysis. When resolved on SDS-PAGE and visualized using Western blot analysis, degradation products of BamA will be observed in PK-treated samples, but full-length BamA will be observed in non-treated samples. If the exogenous BamA was not properly incorporated into the outer membrane, extracellular loops would likely not be exposed. Control experiments utilizing antibodies for soluble proteins like maltose binding protein (MBP) are performed to verify that membrane integrity was maintained during the procedure.

Structure guided disulfide cross-linking is a powerful technique to determine if a conformational change is necessary for function or not in a particular protein. Recently, we used disulfide cross-linking to show that lateral opening of the barrel domain of BamA is required for function [5, 6]. Here, we outline those experiments in more detail.

2 Materials

All solutions are prepared at 25 °C using deionized water with a sensitivity of 18 MΩ cm and are sterile filtered. Use analytical grade reagents for all solutions.

2.1 Cloning and Mutagenesis

1. QuikChange Site-Directed Mutagenesis kit (Agilent Technologies).
2. Miniprep kit (i.e., QIAprep Spin Miniprep Kit (Qiagen)).
3. Template plasmid containing BamA; Here, we use pRSF-1b/BamA from our lab which contains an N-terminal pelB signal sequence followed by a HIS-tag [5, 6].
4. Primer sets (forward and reverse) for each mutation being studied; Here, we present protocols used for our recent work on the lateral gate of BamA as an example [5], however, these methods can be extended to study other mutations as well.
5. LB agar plates containing kanamycin (50 μg/mL) (LB/kan).
6. Inoculation loops and cell spreaders.
7. Sterile culture tubes (15 mL).
8. Plate incubator set at 37 °C.
9. Laboratory incubator shaker set at 220 rpm at 37 °C.
10. Water bath set at 42 °C.
11. UV–Vis spectrometer for measuring DNA concentration.

12. PCR tubes.
13. PCR thermocycler.

2.2 Expression of Wild Type BamA and Mutants

1. Electrocompetent JCM-166 cells (*see* ref. [8]): these cells have endogenous BamA under an arabinose promoter.
2. Plasmids containing BamA and BamA mutants from Subheading 2.1.
3. LB medium: dissolve 5 g NaCl, 10 g peptone, and 5 g yeast extract into 800 mL of water. Adjust the volume to 1 L and autoclave.
4. Arabinose (20 %): dissolve 2 g in 7 mL of water, bring up to 10 mL and sterile filter.
5. UV–Vis spectrophotometer capable of measuring optical density at $\lambda = 600$ nm.
6. Electroporator.
7. SOC medium: 2 % tryptone, 0.5 % yeast extract, 10 mM NaCl, 2.5 mM KCl, 10 mM $MgCl_2$, 10 mM $MgSO_4$, 20 mM glucose. Autoclave and let cool before use.

2.3 Heat Modifiability Assays

1. PBS (10×): 17 mM KH_2PO_4, 50 mM Na_2HPO_4, 1.5 M NaCl, pH 7.4.
2. *N*-dodecyl-β-d-maltoside (DDM) detergent stock solution (10 %).
3. Lysis Buffer: 1× PBS, 1 % DDM, 10 mM EDTA, 10 μg/mL lysozyme, supplemented with 100 μg/mL AEBSF and 10 μg/mL DNase I.
4. Refrigerated benchtop microcentrifuge for 1.5 mL microcentrifuge tubes.
5. Rocking platform at room temperature.

2.4 Proteinase K Digestion Assays

1. Proteinase K stock solution (50 mg/mL) in 20 mM Tris–HCl, pH 7.4, 1 mM $CaCl_2$.
2. Phenylmethanesulfonylfluoride (PMSF) protease inhibitor: 200 mM stock solution in 100 % ethanol.

2.5 Disulfide Cross-Linking

1. Reaction Buffer: 20 mM Tris–HCl, pH 7.5, 100 mM NaCl.
2. Copper sulfate (CuS) (25 mM).
3. Copper-o-phenanthroline (CuP) (25 mM).
4. Dithiothreitol (DTT) (1 M).

2.6 SDS-PAGE

1. SDS-PAGE sample loading buffer (2×): 20 % glycerol, 120 mM Tris–HCl, pH 6.8, 2 % SDS, 0.02 % bromophenol blue (*see* **Note 1**).

2. SDS-PAGE sample loading buffer (2×) with DTT: 20 % glycerol, 120 mM Tris–HCl, pH 6.8, 2 % SDS, 0.02 % bromophenol blue, 100 mM DTT (*see* **Note 1**).
3. SDS-PAGE gels and native gels (*see* **Note 2**). Here, we use 5 % SDS-PAGE cast in-house and precast 4–12 % NuPAGE and 4–12 % NativePAGE gels (Life Technologies).
4. MES-SDS Running Buffer (20×): 1 M MES, 1 M Tris base, 2 % SDS, 20 mM EDTA, pH 7.3.
5. MOPS-SDS Running Buffer (20×): 1 M MOPS, 1 M Tris Base, 2 % SDS, 20 mM EDTA, pH 7.7.
6. Prestained protein markers (e.g., SeeBlue Plus2 Protein Standard (Life Technologies)).
7. Polyacrylamide gel electrophoresis (PAGE) system.
8. Heat block set to 95 °C.
9. Ice bucket and ice.

2.7 Western Blot Analysis

1. Anti-HIS-HRP antibodies (Sigma).
2. Anti-MBP antibodies (NEB).
3. Anti-Mouse IgG HRP secondary antibodies (Sigma).
4. PBST (1×): 1.7 mM KH_2PO_4, 5 mM Na_2HPO_4, 150 mM NaCl, 0.1 % Tween 20, pH 7.4.
5. PBST (1×)+BSA: 1.7 mM KH_2PO_4, 5 mM Na_2HPO_4, 150 mM NaCl, 0.1 % Tween 20, pH 7.4, 1 % bovine serum albumin (BSA).
6. Transfer system for blots (e.g., Life Technologies iBlot® Gel Transfer Device or similar).
7. Western blot box.
8. ECL Western blotting substrate (e.g., Pierce ECL Western Blotting Substrate).
9. Western blot imaging system (e.g., GE ImageQuant LAS 4000 or similar).

3 Methods

3.1 Designing and Cloning BamA Mutants

1. Design DNA oligo primers for the engineered mutation being studied (*see* **Note 3**). Dilute to 1000 ng/μL with water. Make a working solution for each primer at 100 ng/μL.
2. Use the QuikChange Site-Directed Mutagenesis kit following manufacturer's instructions. Briefly, in a PCR tube, add 50 ng DNA template (pRSF1b/BamA in our case), 125 ng of each primer, 1 μL of dNTP mix, 2.5 μL 10× Reaction Buffer, and 1 μL polymerase. Bring to 25 μL final volume with water and mix by pipetting.

3. Place into a thermocycler using a program that matches the manufacturer's instructions (*see* **Note 4**).
4. When finished, add 1 μL of DpnI enzyme to each tube and incubate at 37 °C for 1 h (*see* **Note 5**).
5. Transform each into the provided competent cells using manufacturer's instructions, except only use 200 μL of SOC medium following the heat pulse at 42 °C. Then spread the entire sample onto an LB/kan agar plate and incubate inverted overnight at 37 °C.
6. The next day, prepare at least five 5 mL cultures from 5 different colonies from the plate in LB/kan medium. Allow to grow overnight at 37 °C.
7. The next day, perform DNA plasmid minipreps using manufacturer's instructions for each culture and verify the correct mutation is present using sequencing analysis.
8. Repeat **steps 1–7** for each desired mutation to be analyzed (*see* **Note 6**).

3.2 Expression of BamA Mutants in JCM-166 Cells

1. To assay the mutants for expression and folding, transform plasmids for BamA and each BamA mutant prepared in Subheading 3.1 into JCM-166 electrocompetent cells using an electroporator following manufacturer's instructions.
2. Following electroporation, resuspend the cells in 0.5 mL SOC medium and grow at 37 °C for 1 h.
3. For each sample, (1) pipet 25–50 μL of cells onto a prewarmed LB/kan agar plate and spread across the plate by streaking with an inoculation loop, and (2) pipet 25–50 μL of the cells onto a prewarmed LB/kan agar plate containing 0.1 % arabinose and spread across the plate by streaking with an inoculation loop (*see* **Note 7**).
4. Invert and grow overnight at 37 °C.
5. The next day, for each mutant, compare the growth with and without arabinose. Those mutants showing normal colony growth in the absence of arabinose are likely functioning in a manner similar to wild type BamA. Those mutants showing a growth defect, however, are likely revealing either disrupted BamA function or folding, and can be studied further. Here, we examine if the mutants are being folded properly and making it the outer membrane.
6. For wild type BamA and each mutant showing growth defects, prepare a 5 mL culture using LB/kan medium containing 0.1 % arabinose and grow at 37 °C shaking at 220 rpm to an OD_{600} = ~1.5.
7. For each sample, prepare 1 mL of cells at OD_{600} = 1.0 in a 1.5 mL microcentrifuge tube and spin using a benchtop

microcentrifuge for 1 min at full speed. Remove the supernatant and then resuspend the cells in 1 mL cold 1× PBS by pipetting. Spin again and remove the supernatant.

8. The cells can now be used for further analysis as described in Subheadings 3.4–3.6.

3.3 SDS-PAGE and Western Blot Analyses

Before proceeding with analyzing the BamA mutants, we present the methods required for performing SDS-PAGE and Western blot analyses, since these will be essential for determining the results of the heat modifiability, proteinase K, and disulfide crosslinking assays described next.

3.3.1 Denaturing SDS-PAGE Analysis

Denaturing SDS-PAGE analysis may be performed using standard methods. Briefly, we describe those below.

1. Prior to preparing samples for analysis, assemble the gel apparatus according to manufacturer's instructions. Insert the gel and then fill the inner reservoir completely with 1× MES Running Buffer or 1× MOPS Running Buffer and the outer reservoir at least 1/3 full.
2. Load 20 μL of each sample onto the gel. Be sure to also include a prestained protein standard for when analyzing results and for monitoring the transfer step during Western blot analysis.
3. Run at constant 200 V for 35 min if using 4–12 % SDS-PAGE gradient gels or 70 min if using 5 % SDS-PAGE gels.

3.3.2 Semi-native SDS-PAGE Analysis

1. Prior to preparing the samples for analysis, assemble the gel running apparatus so that the gel may be loaded immediately after the samples are ready. Place the gel apparatus tank into a bed of ice within an ice bucket. Insert the gel and then fill the tank completely with cold 1× MES running buffer. It is important to keep the buffer within the tank cold during the entire experiment (*see* **Note 8**). Once assembled, you can move to the next step.
2. Load 25 μL of each sample onto the native gel assembled in **step 1** (*see* **Note 9**).
3. Run the gel for 60 min at constant 150 V.

3.3.3 Western Blot Analysis

1. Once the gel has finished running from Subheadings 3.3.1 and 3.3.2, transfer to either nitrocellulose or polyvinylidene difluoride (PVDF) membrane using manufacturer's instructions (*see* **Note 10**).
2. Trim the excess membrane and if probing with more than one antibody, use the protein standards to guide cutting the membrane at the appropriate locations. For the proteinase K susceptibility assay, we cut the membrane in half around 55 kDa

and probe the top (higher molecular weight) portion against anti-HIS-HRP antibodies and the bottom (lower molecular weight) portion against anti-MBP antibodies.

3. Place the membrane into a blot box and add 20 mL of 1× PBST + BSA blocking solution and incubate for 1 h on a rotating or rocking platform.
4. Then add either anti-HIS-HRP (1:6000) or anti-MBP (1:5000) antibody and allow to incubate an additional 1 h. Pour off supernatant.
5. Wash each membrane three times for 5 min each using 1× PBST.
6. If using the anti-MBP antibody, proceed to incubate with anti-Mouse HRP secondary antibodies (1:15,000) in 20 mL 1× PBST with 1 % BSA for 30 min. Pour off supernatant and wash membrane three times for 5 min each using 1× PBST.
7. Wash each membrane three times for 5 min each using 1× PBS.
8. Add ECL Western blotting substrate to the membrane and image using a Western blot imaging system, following manufacturer's instructions for both.

3.4 Heat Modifiability Assay to Characterize BamA Folding

Here we present a protocol for heat modifiability assays for BamA using whole cells. For more details about heat modifiability assays for purified protein and in general, please *see* Chapter 4.

1. For each sample from Subheading 3.2, **step 8**, resuspend cells in 500 μL of Lysis Buffer.
2. Rock for 15 min at room temperature.
3. Centrifuge at full speed at 4 °C for 10 min using a refrigerated benchtop microcentrifuge.
4. Pipette 100 μL of the detergent solubilized lysate supernatant into two new microcentrifuge tubes, labeling one "boiled" and the other "unboiled."
5. Add 100 μL of 2× SDS-PAGE sample loading buffer to each tube.
6. Incubate the "boiled" sample tube at 95 °C for 5 min while keeping the "unboiled" sample at room temperature.
7. Proceed with semi-native SDS-PAGE analysis as outlined in Subheading 3.3.2 using a 4–12 % NativePAGE gel in 1× MES Running Buffer and Western blot analysis as outlined in Subheading 3.3.3 using anti-HIS-HRP antibodies.
8. Compare the "boiled" sample to the "unboiled" sample, looking for a shift in the gel to indicate folded and unfolded states (Fig. 1a).

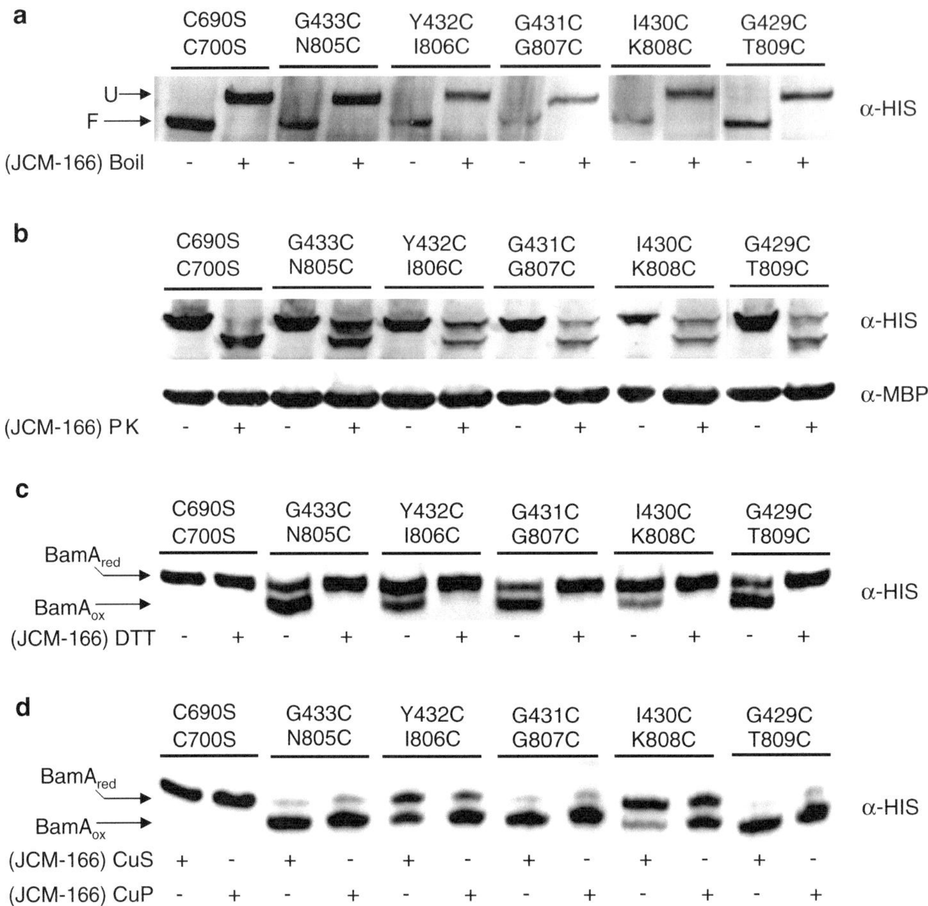

Fig. 1 BamA cross-link mutants are folded and incorporated into the outer membrane. Heat modifiability (panel **a**) and proteinase K (PK) digestion (panel **b**) assays were performed on whole cells expressing the BamA cross-link mutants. All cross-link mutants showed heat modifiability and were susceptible to PK, indicating presentation on the surface and proper folding. (**c**) To analyze disulfide cross-link formation under native conditions, cell lysates were resolved by SDS-PAGE ± DTT. All of the engineered disulfide mutants formed disulfide cross-links within the cell. Additionally, treatment with 100 μM copper sulfate (CuS) or 100 μM copper-o-phenanthroline (CuP) enhanced disulfide formation (panel **d**). BamA was probed using anti-HIS-HRP antibodies while MBP was probed using anti-MBP antibodies. Reproduced from ref. [5] with permission from Elsevier

3.5 Proteinase K Susceptibility Assays to Characterize Surface Exposure

1. For each sample from Subheading 3.2, **step 8**, resuspend cells in 500 μL of 1× PBS and aliquot 250 μL into two new tubes, labeling one as "PK+" and the other as "PK−."
2. Add proteinase K to the "PK+" tube at a final concentration of 500 μg/mL and incubate at 37 °C for 30 min. To stop the reaction, add PMSF to a final concentration of 5 mM and gently invert to mix.
3. Spin both samples at full speed for 1 min using a benchtop microcentrifuge and carefully remove the supernatant.

4. Resuspend the cells in 200 μL 1× SDS-PAGE sample loading buffer and boil for 5 min.
5. Spin at full speed for 10 min using a benchtop microcentrifuge.
6. Proceed with denaturing SDS-PAGE analysis as outlined in Subheading 3.3.1 using a 4–12 % NuPAGE gel in 1× MES Running Buffer and Western blot analysis as outlined in Subheading 3.3.3 using anti-HIS-HRP and anti-MBP antibodies probing for both BamA and MBP, respectively.
7. Compare the "PK+" sample to the "PK–" sample, looking for degradation of the BamA band in the presence of PK. Monitoring MBP verifies that the membrane was not permeabilized during the analysis (Fig. 1b).

3.6 Disulfide Cross-Linking of BamA Barrel Domain

3.6.1 Visualizing the Natively Formed Disulfide Cross-Links

1. For each sample from Subheading 3.2, **step 8**, resuspend cells in 200 μL of 1× PBS and aliquot 100 μL into two new tubes, labeling one as "DTT+" and the other as "DTT–."
2. Add 100 μL of 2× SDS-PAGE sample loading buffer with DTT to the "DTT+" tube and 100 μL of 2× SDS-PAGE sample loading buffer without DTT to the "DTT–" tube.
3. Boil samples at 95 °C for 5 min and then spin for 10 min at full speed using a benchtop microcentrifuge.
4. Proceed with denaturing SDS-PAGE analysis as outlined in Subheading 3.3.1 using a 5 % SDS-PAGE gel in 1× MOPS Running Buffer and Western blot analysis as outlined in Subheading 3.3.3 using anti-HIS-HRP antibodies probing for BamA.
5. Compare the "DTT+" sample to the "DTT–" sample looking for a shift in the BamA bands to indicate the formation of a disulfide bond which can be reduced in presence of DTT (Fig. 1c).

3.6.2 Visualizing the Oxidizer Enhanced Disulfide Cross-Links

To further enhance the disulfide cross-links, incubation with CuS or CuP can be performed prior to processing the cells for SDS-PAGE and Western blot analyses.

1. For each sample from Subheading 3.2, **step 8**, resuspend cells in 200 μL of Reaction Buffer and aliquot 100 μL into two new tubes, labeling one as "CuS" and the other as "CuP."
2. Add CuS and CuP to a final concentration of 100 μM to each tube, respectively, and incubate at 37 °C for 30 min.
3. Spin samples at full speed for 1 min using a benchtop microcentrifuge and carefully remove the supernatant.
4. Resuspend the cells in 200 μL of Reaction Buffer and repeat **step 3**.

5. Resuspend each sample in 200 μL 1× SDS-PAGE sample loading buffer without DTT.
6. Boil samples at 95 °C for 5 min and then spin for 10 min at full speed using a benchtop microcentrifuge.
7. Proceed with denaturing SDS-PAGE analysis as outlined in Subheading 3.3.1 using a 5 % SDS-PAGE gel in 1× MOPS Running Buffer and Western blot analysis as outlined in Subheading 3.3.3 using anti-HIS-HRP antibodies probing for BamA.
8. Compare the "CuS" and "CuP" samples to the "DTT+" and "DTT−" samples from Subheading 3.6.1, **step 4** to determine if disulfide bond formation was enhanced or not (Fig. 1d).

4 Notes

1. For the dye, we advise against using Coomassie G250 here since we have experienced odd results when substituted for bromophenol blue. We hypothesize that the coomassie G250, which is in fact used for blue native PAGE methods, interacts with the samples due to its properties while that is not the case for the bromophenol blue.
2. Most native gels should work here either fixed or gradient. For the examples shown here, we used precast NativePAGE Novex 4–12 % Bis–Tris protein gels (Life Technologies).
3. For designing primers for site-directed mutagenesis, one can use the online QuikChange Primer Design Tool found on the Agilent Technologies website. While convenient and time saving, it does tend to design primers that are often longer than needed. We recommend trimming the primers suggested by the tool to ~35 base pairs long.
4. We routinely do 25–30 cycles rather than the 18–20 suggested by the kit instructions.
5. We routinely do extended incubation times (2–5 h) for the DpnI digestion which reduces the chances of growing colonies which have no mutation.
6. If possible, it is always easier and more efficient to do all the mutations at the same time, rather than doing them one after the other, unless multiple mutations are needed within the same construct as with our disulfide mutants.
7. Depending on the number of mutations, we routinely plate up to six mutations per plate by dividing the plate into six triangular regions, into each of which we streak ~15 μL of cells.
8. Since we keep our electrophoresis equipment at room temperature, it was easier for us to do everything at room tempera-

ture and put the gel apparatus in ice to keep it cool. However, a gel apparatus set up in a cold room would be just as good and would not require the ice bucket or ice.

9. If need, additional sample can be loaded here for better visualization. Alternatively, adjust the volume of Lysis Buffer used to yield a more concentrated sample.
10. We recommend using a blotting system such as the iBlot system (Life Technologies) which can do transfers in 7 min, significantly speeding the time to attaining results.

Acknowledgements

This research was supported by the Intramural Research Program of the NIH, The National Institute of Diabetes and Digestive and Kidney Diseases (NIDDK).

References

1. Ricci DP, Silhavy TJ (2012) The Bam machine: a molecular cooper. Biochim Biophys Acta 1818(4):1067–1084
2. Hagan CL, Silhavy TJ, Kahne D (2011) beta-Barrel membrane protein assembly by the BAM complex. Annu Rev Biochem 80:189–210
3. Heinz E, Lithgow T (2014) A comprehensive analysis of the Omp85/TpsB protein superfamily structural diversity, taxonomic occurrence, and evolution. Front Microbiol 5:370
4. Rigel NW, Ricci DP, Silhavy TJ (2013) Conformation-specific labeling of BamA and suppressor analysis suggest a cyclic mechanism for beta-barrel assembly in *Escherichia coli*. Proc Natl Acad Sci U S A 110(13):5151–5156
5. Noinaj N, Kuszak AJ, Balusek C et al (2014) Lateral opening and exit pore formation are required for BamA function. Structure 22(7):1055–1062
6. Noinaj N, Kuszak AJ, Gumbart JC et al (2013) Structural insight into the biogenesis of beta-barrel membrane proteins. Nature 501(7467):385–390
7. Rigel NW, Schwalm J, Ricci DP et al (2012) BamE modulates the *Escherichia coli* beta-barrel assembly machine component BamA. J Bacteriol 194(5):1002–1008
8. Wu T, Malinverni J, Ruiz N et al (2005) Identification of a multicomponent complex required for outer membrane biogenesis in *Escherichia coli*. Cell 121(2):235–245

Chapter 11

Small Angle X-ray Scattering (SAXS) Characterization of the POTRA Domains of BamA

Pamela Arden Doerner and Marcelo Carlos Sousa

Abstract

BamA is the central component of the BAM complex and contains a C-terminal β-barrel domain embedded in the outer membrane, and a soluble, periplasmic domain, made out of five *po*lypeptide *tr*ansport *a*ssociated (POTRA) motifs. Structural characterization of the POTRA domains was carried out by a combination of crystallographic, NMR and solution Small Angle X-ray Scattering (SAXS) approaches. Despite its limited resolution, SAXS is an excellent complement to NMR and crystallography. It is well suited to validate high-resolution models in solution and is particularly useful to characterize flexible systems such as the POTRA domains of BamA. Here we present a protocol for sample preparation and discuss the considerations of SAXS data collection and quality control, which is applicable to most soluble proteins.

Key words SAXS, BamA, BAM, Conformational change, Structure, OMP

1 Introduction

Small angle X-ray scattering (SAXS) is a powerful technique for the structural characterization of biomolecules in solution. No labeling is necessary and, with bright sources of X-rays at synchrotron facilities and modern detectors, it requires relatively small amounts of sample at concentrations of a few mg/mL for proteins. Data collection is very quick, and simple analysis yields the macromolecular radius of gyration (R_g), which provides an initial indication of molecular size. The maximum dimension (D_{max}) as well as a histogram of interatomic distances, the $P(r)$ function, are also easily derived from the scattering data and provide additional information about molecular shape [1]. The software programs PRIMUS [2] and ScÅtter (https://bl1231.als.lbl.gov/scatter/) offer user-friendly access to this basic data analysis.

While limited in terms of resolution, SAXS is an excellent complement to high-resolution techniques such as X-ray

Susan K. Buchanan and Nicholas Noinaj (eds.), *The BAM Complex: Methods and Protocols*, Methods in Molecular Biology, vol. 1329, DOI 10.1007/978-1-4939-2871-2_11,

crystallography [1]. It is possible to compare experimentally determined and coordinate-derived scattering curves and *P*(*r*) profiles to assess the similarity between crystal and solution structures [3]. In addition, it is often possible to use ab initio methods to calculate molecular envelopes from SAXS data [4–6]. This is very useful not only to compare crystal and solution structures but also to allow "docking" of partial high-resolution structures into the envelopes of complete proteins or multiprotein complexes. Approaches have also been developed to evaluate the structural flexibility of macromolecules and model their conformational changes [7]. The ATSAS software package (http://www.embl-hamburg.de/biosaxs/software.html) provides several tools to carry out these experiments and has been instrumental in popularizing sophisticated SAXS analyses.

Structural characterization of the periplasmic domain of BamA, the central BAM subunit containing five POTRA domains, benefited from several SAXS approaches. Initial crystal structures defined alternate conformations for the first four POTRA domains due to a flexible hinge between POTRAs 2 and 3 [8, 9]. A subsequent crystal structure of a POTRA4-5 fragment allowed construction of spliced models containing all five POTRA domains, which was validated by SAXS [10]. Furthermore, analysis of the scattering curves and *P*(*r*) profiles experimentally determined from the POTRA1-5 fragment and derived from the splice models defined conformational flexibility of the protein, which was further characterized by the Ensemble Optimization Method [7]. Here we present the protocol for sample preparation, data collection and analysis used in these experiments, which can be used for further study of the periplasmic domain of BamA or easily adapted for other soluble macromolecules.

2 Materials

All solutions are prepared with Milli-Q water and reagents of analytical grade or better. All stock solutions are filter-sterilized.

1. *E. coli* Rosetta (DE3) cells (Novagen).
2. Kanamycin stock: 50 mg/mL in water.
3. Luria broth.
4. Isopropyl-β-D-thiogalactopyranoside (IPTG) stock: 0.4 M in water.
5. Sodium chloride stock: 5 M.
6. Imidazole stock: 1 M.
7. Glycerol.
8. cOmplete EDTA-free protease inhibitor (Roche).

9. Lysis Buffer: 25 mM Tris–HCl pH 8.
10. Buffer A: 25 mM Tris–HCl pH 8, 150 mM NaCl.
11. SAXS Buffer: 25 mM Tris–HCl pH 8, 150 mM NaCl, 5 % glycerol.
12. Nickel-NTA resin (Qiagen).
13. His-tagged TEV protease [11].
14. Superdex 200 Size Exclusion Chromatography (SEC) column HiLoad 26/60 (Amersham Pharmacia Biotech).
15. Centrifugal concentrators.
16. Dialysis tubing.
17. Filters (0.1 micron).
18. Sonicator.
19. Centrifuge capable of spinning at 30,000 × *g*.
20. SAXS data collection facility. We assume a synchrotron beamline facility such as Beamline 12.3.1 of the Advanced Light Source (Lawrence Berkeley National Laboratory).

3 Methods

3.1 Protein Expression and Purification

1. Coding sequences for fragments of the periplasmic domain of *E. coli* BamA (amino acids 21–424 for POTRA1-5; 175–424 for POTRA3-5 and 264–424 for POTRA4-5) were cloned into a variant of pET28 vector as described in [10], creating N-terminal fusions to a His-tag followed by a TEV protease cleavage site. These plasmids are used to transform Rosetta DE3 cells (Novagen) for expression.
2. Pick one colony to inoculate 100 mL of LB with 50 μg/mL kanamycin and grow at 37 °C overnight, shaking at 160 rpm.
3. Inoculate 4 × 1 L LB cultures containing 50 μg/mL kanamycin with 10 mL (per 1 L) of the overnight culture and grow at 37 °C until they reach OD_{600} of ~0.7.
4. Induce expression by adding IPTG to the cultures for a final concentration of 0.4 mM and incubate for an additional 1.5 h at 37 °C.
5. Harvest cells by centrifugation. Resuspend cell pellet in Lysis Buffer supplemented with cOmplete EDTA-free protease inhibitor tablets (*see* **Note 1**).
6. Lyse cells by sonication using a macro tip at full power and 2 s pulses followed by 5 s pauses until the cells are broken (*see* **Note 2**).
7. Add sodium chloride to a final concentration of 0.3 M.

8. Remove cell debris by centrifugation at 30,000 × *g* for 30 min at 4 °C, keeping the supernatant.
9. Pack a 10 mL Ni-NTA column and equilibrate with Buffer A.
10. Load the supernatant onto the Ni-NTA column (*see* **Note 3**). Wash with 2 column volumes of Buffer A, followed by 20 column volumes of Buffer A supplemented with 25 mM imidazole. Elute the protein with Buffer A supplemented with 200 mM imidazole.
11. Pool protein-containing elution fractions and measure protein concentration. Add His-tagged TEV protease at a ratio of 1 mg to 20 mg of POTRA domain and DTT at a final concentration of 10 mM. Incubate for 24 h at 4 °C to allow cleavage of the His-tag and dialyze overnight against Buffer A to remove the imidazole (*see* **Note 4**).
12. Pass the preparation through a Ni-NTA column again to remove the tag, the His-tagged TEV protease and any POTRA domain fragment that remains uncleaved by the protease. The purified, untagged POTRA domain protein will be in the flow-through.
13. Concentrate the protein (*see* **Note 5**) and load onto a Superdex 200 SEC column equilibrated with SAXS buffer and collect the elution in 0.4 mL fractions.
14. Set aside the fraction corresponding to the center of the protein peak eluting from the SEC column, which has the highest protein concentration. Pool the other protein containing fractions and concentrate using centrifugal concentrators followed by filtration through 0.1 μm centrifugal filters (*see* **Note 6**).

3.2 Consideration for SAXS Sample Preparation and Data Collection

1. SAXS is a contrast technique where scattering from the macromolecule has to be isolated from that of the solvent. Accurate buffer subtraction is therefore essential so we use the SEC eluted protein and the SEC buffer to carry out the experiments.
2. As the scattering is proportional to the size (and shape) of the particles it is important to avoid even small amounts of aggregated material in the sample. Therefore, whenever possible we use the most concentrated fraction from the SEC elution or dilutions of this fraction to collect data without further manipulation.
3. Glycerol has been shown to protect the sample against radiation damage. Therefore, 5 % glycerol is incorporated in the SAXS Buffer used to develop the SEC column.
4. Concentration-dependent effects are often observed in SAXS data. We thus create concentration series of the proteins with samples at approximately 2, 5, and 10 mg/mL. We use the most concentrated fraction from the SEC elution as one of the

concentrations in the series and, if the protein is concentrated enough, make dilutions from this fraction. If the protein does not display a tendency to aggregate during centrifugal concentration, as is the case for POTRA4-5 fragments, we also prepare higher concentration samples at 20 and 40 mg/mL.

5. Even with glycerol as a radical scavenger is the SAXS buffer, some samples display radiation damage. We thus collect images at increasingly long exposures to X-rays and analyze the data for signs of radiation damage. The exposure times will depend on the X-ray source. At the Advanced Light Source beamline 12.3 (Sibyls beamline) we routinely collect 0.5, 1, 2 and 4 s exposures.
6. To maximize the chances of accurate buffer subtraction, we organize data collection as recommended by the Sibyls beamline protocol [12]: (a) wash sample cuvette with water, (b) wash with SAXS Buffer; (c) collect data on SAXS buffer at all exposures; (d) collect data on protein samples at all exposures in order of increasing concentration; (e) wash sample cuvette with buffer; and (f) collect data on SAXS buffer at all exposures. This allows buffer subtraction to be carried out with two independent buffer measurements.

3.3 SAXS Data Quality Control

The raw SAXS images contain radially symmetric scattering whose intensity decays with scattering angle. Initial image processing depends on the specific detector setup and is typically carried out at the beamline. It involves masking out direct beamstop shadow(s), radial averaging of the integrated intensities and use of detector calibration data to convert the data into a plot of intensity (I) versus q ($q = 4\pi \sin\theta/\lambda$, where 2θ is the scattering angle and λ is the X-ray wavelength). We then perform an evaluation of data quality as described below using the program PRIMUS [2].

1. With the two independent measurements of buffer scattering collected before and after the protein sample, subtract the buffer signal from the total signal to isolate the scattering from the macromolecule. Both subtractions should result in superimposable curves. If this is not the case, additional measurements of buffer scattering can be used for the subtraction to identify the outlier. Buffer subtracted curves are used for all the subsequent analyses.
2. Compare the curves corresponding to each exposure. The curves should have the same shape, with the intensity scaling linearly with the exposure. Radiation damage can be identified by an upward deviation of the curves at low q values.
3. Compare curves corresponding to each concentration in the series. A protein has a characteristic decay of I versus q and the curves should thus have the same shape and scale linearly with

concentration. However, oligomerization and interparticle repulsion are concentration dependent effects observed for some protein samples. Data collected at concentrations displaying these artifacts are not used for further analysis.

4. Inspect the curves collected at the highest protein concentrations for signs of detector saturation. This is easily identified by a flattening of the signal at low q values. Data from the non-saturated region of the curve is nevertheless useful.
5. Curves passing the quality controls described above are scaled together and averaged to generate a consolidated plot of I versus q with error estimates that is used for all further analysis.

4 Notes

1. The resuspended cell pellet can be used immediately or frozen at −70 °C until needed.
2. We resuspend the cells in enough lysis buffer so that a 1/60 dilution in water yields an OD_{600} of approximately 0.8. Lysis is deemed to be complete when the OD_{600} of the 1/60 dilution drops below 0.1 or does not change after additional 1 min sonication periods.
3. Keep the flow-through and analyze by SDS PAGE for expressed protein content. Constructs expressing POTRA domains 3–5 and 4–5 express at high enough levels that can saturate the 10 mL column.
4. Tag cleavage and imidazole removal can be done simultaneously by dialyzing against Buffer A supplemented with 10 mM DTT to keep the protease active during dialysis. This approach is sometimes advantageous as Ni^{2+} released into the eluted fractions from the column can induce aggregation of some proteins.
5. We concentrate as much as possible before the SEC column so that the elution fractions can be concentrated enough to use directly for data collection without additional concentration. For constructs coding POTRA domains we concentrate to 8 mg/mL for POTRA1-5, 10 mg/mL for POTRA3-5, and 40 mg/mL for POTRA4-5.
6. Centrifugal concentrators and filters are often impregnated with preservatives. We pre-run these devices with SAXS Buffer twice before using them with sample to minimize their impact in buffer matching.

Acknowledgements

This work was supported in part by a grant from the National Institutes of Health R01 AI080709 to M.C.S., and Training Grant GM08759 to P.A.D.

References

1. Putnam CD, Hammel M, Hura GL et al (2007) X-ray solution scattering (SAXS) combined with crystallography and computation: defining accurate macromolecular structures, conformations and assemblies in solution. Q Rev Biophys 40(3):191–285
2. Konarev PV, Volkov VV, Sokolova AV et al (2003) PRIMUS: a Windows PC-based system for small-angle scattering data analysis. J Appl Crystallogr 36:1277–1282
3. Svergun D, Barberato C, Koch MHJ (1995) CRYSOL – a program to evaluate x-ray solution scattering of biological macromolecules from atomic coordinates. J Appl Crystallogr 28:768–773
4. Franke D, Svergun DI (2009) DAMMIF, a program for rapid ab-initio shape determination in small-angle scattering. J Appl Crystallogr 42:342–346
5. Svergun DI (1999) Restoring low resolution structure of biological macromolecules from solution scattering using simulated annealing. Biophys J 76(6):2879–2886
6. Svergun DI, Petoukhov MV, Koch MHJ (2001) Determination of domain structure of proteins from X-ray solution scattering. Biophys J 80(6):2946–2953
7. Bernado P, Mylonas E, Petoukhov MV et al (2007) Structural characterization of flexible proteins using small-angle X-ray scattering. J Am Chem Soc 129(17):5656–5664
8. Gatzeva-Topalova PZ, Walton TA, Sousa MC (2008) Crystal structure of YaeT: conformational flexibility and substrate recognition. Structure 16(12):1873–1881
9. Kim KH, Aulakh S, Paetzel M (2011) Crystal structure of beta-barrel assembly machinery BamCD protein complex. J Biol Chem 286(45):39116–39121
10. Gatzeva-Topalova PZ, Warner LR, Pardi A et al (2010) Structure and flexibility of the complete periplasmic domain of BamA: the protein insertion machine of the outer membrane. Structure 18(11):1492–1501
11. Lucast LJ, Batey RT, Doudna JA (2001) Large-scale purification of a stable form of recombinant tobacco etch virus protease. BioTechniques 30(3):544–546, 548, 550 passim
12. Dyer KN, Hammel M, Rambo RP et al (2014) High-throughput SAXS for the characterization of biomolecules in solution: a practical approach. Methods Mol Biol 1091: 245–258

Chapter 12

Assessing the Outer Membrane Insertion and Folding of Multimeric Transmembrane β-Barrel Proteins

Jack C. Leo, Philipp Oberhettinger, and Dirk Linke

Abstract

In addition to the cytoplasmic membrane, Gram-negative bacteria have a second lipid bilayer, the outer membrane, which is the de facto barrier between the cell and the extracellular milieu. Virtually all integral proteins of the outer membrane form β-barrels, which are inserted into the outer membrane by the BAM complex. Some outer membrane proteins, like the porins and trimeric autotransporter adhesins, are multimeric. In the former case, the porin trimer consists of three individual β-barrels, whereas in the latter, the single autotransporter β-barrel domain is formed by three separate polypeptides. This chapter reviews methods to investigate the folding and membrane insertion of multimeric OMPs and further explains the use of a BamA depletion strain to study the effects of the BAM complex on multimeric OMPs in *E. coli*.

Key words BAM complex, BamA depletion strain, β-barrel, Outer membrane, Trimeric autotransporter, Adhesin

1 Introduction

The outer membrane (OM) of Gram-negative bacteria is a remarkable organelle that acts as an additional barrier between the cell and the extracellular milieu, conferring higher tolerance to chemical insult inflicted by such agents as detergents and antibiotics. The OM is a lipid bilayer with an asymmetric structure; the outer leaflet consists exclusively of lipopolysaccharide, whereas the inner leaflet contains glycerophospholipids [1]. The transmembrane proteins of the OM also have a unique structure, with almost all of them belonging to the β-barrel family [2]. Many OM β-barrel proteins (OMPs) are monomeric, but several form higher-order multimers. These multimers fall into two distinct groups: aggregates of single, folded β-barrels such as the porins [3], and OMPs where the single β-barrel consists of several polypeptide chains, such as the transmembrane domains of TolC [4] or trimeric autotransporter adhesins (TAAs) [5] (Fig. 1). The porins, like OmpF, are usually trimeric, and in many cases this trimer is stable enough to resist dissociation

Susan K. Buchanan and Nicholas Noinaj (eds.), *The BAM Complex: Methods and Protocols*, Methods in Molecular Biology, vol. 1329, DOI 10.1007/978-1-4939-2871-2_12,

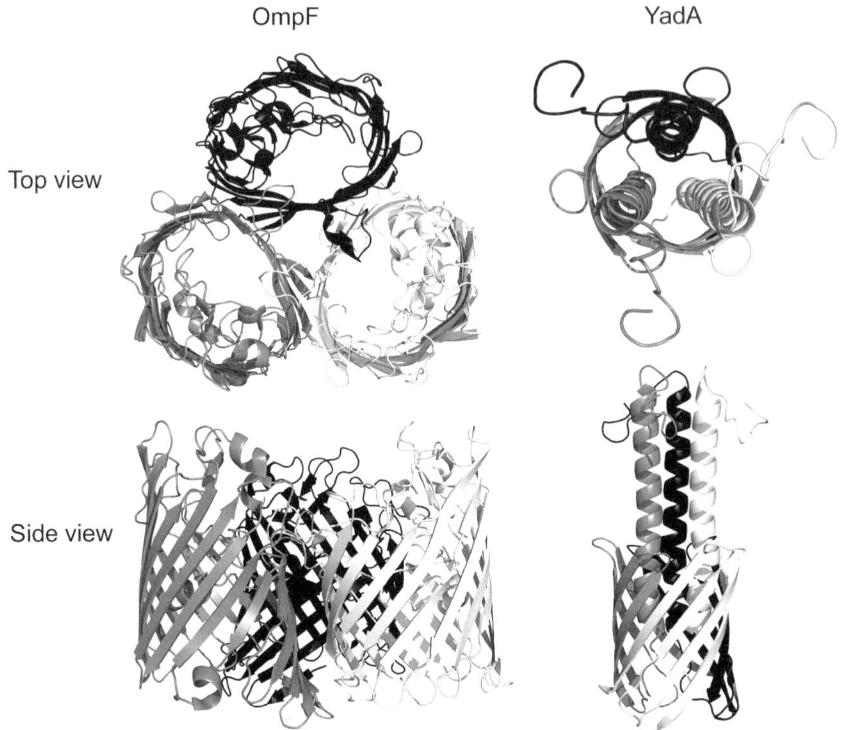

Fig. 1 Examples of multimeric β-barrel OMPs. The porin OmpF (*left*) is a trimer composed of three separate β-barrel subunits, whereas the membrane anchor of the trimeric autotransporter adhesin YadA (*right*) is a single β-barrel comprising three polypeptide chains. Subunits are colored differently. The figures are not to scale. The Protein Data Bank entries used to create the figure are 3K1B and 2LME

in sodium dodecyl sulfate (SDS), but monomeric OMPs can be seen in SDS-polyacrylamide gel electrophoresis (SDS-PAGE) after boiling [6]. In contrast, TAAs form very stable structures that are often resistant to boiling in SDS; additional chaotropic agents such as urea need to be added for full denaturation [7].

All OMPs are inserted into the OM by the β-barrel assembly machinery or BAM complex [8]. The central component of the BAM complex, BamA, is itself a 16-stranded OMP and essential for viability. Interestingly, the C-terminal β-strand of BamA is destabilized, which leads to local perturbation of the hydrophobic lipid environment and possible lateral opening of the BamA β-barrel through which nascent OMPs might enter the OM [9]. In addition to BamA, the complex includes accessory lipoproteins (BamB through BamE), of which only BamD is essential in *E. coli*, though deletion of the others leads to OM biogenesis defects [10]. Because deleting BamA is lethal, creating a knock-out strain to study the effect of BamA on the biogenesis of multimeric OMPs is not feasible. However, a depletion strain where BamA expression can be abrogated under certain conditions has proved useful in assessing the effect of BamA on the biogenesis of OMPs [11, 12].

This chapter reviews methods to study the OM insertion and oligomerization of multimeric OMPs, with emphasis on TAAs as a model system. In addition, we introduce the use of the BamA depletion strain for studying the effect of the BAM complex on multimeric OMP biogenesis.

2 Materials

2.1 Strains and Culture Media

1. Bacterial culture medium, e.g., Lysogeny broth (LB) [13]. 10.0 g/L tryptone/peptone (pancreatic digested casein), 5 g/L yeast extract, 10 g/L NaCl.
2. BamA depletion strain, *E. coli* MC4100A *bamA attB*::(*araC*, P_{BAD}, *bamA*+) Kan^R [11] (*see* **Note 1**).

2.2 OM Isolation

1. A tabletop microcentrifuge.
2. Microcentrifuge tubes (1.5 and 2 mL).
3. 2-(4-(2-hydroxyethyl)-1-piperazinyl)-ethanesulfonic acid (HEPES) stock solution (1 M): weigh 23.8 g of HEPES and dissolve in 90 mL ultrapure H_2O. Adjust the pH to 7.4 with NaOH and, using a measuring beaker, add ultrapure H_2O to 100 mL. This will result in a 1 M HEPES stock solution. Sterile-filter for long-term storage. To make a 10 mM working solution, dilute the stock solution 1:100 in ultrapure H_2O.
4. A bead beater. We have used SpeedMill PLUS (Analytik Jena) with innuSPEED Lysis Tubes B (containing 100 μM glass beads) (catalogue number 845-CS-1030050). Other alternatives include FastPrep®-24 from MP Biomedicals (catalogue number 116004500).
5. *N*-lauroyl sarcosine solution (2 % (w/v)): Weigh out 0.2 g of *N*-lauroyl sarcosine and dissolve in 9 mL 10 mM HEPES, pH 7.4. Fill up to 10 mL with 10 mM HEPES pH 7.4.
6. Sodium dodecyl sulfate–polyacrylamide gel electrophoresis (SDS-PAGE) sample buffer (2×): 100 mM Tris–HCl, pH 7.5, 4 % SDS, 20 % glycerol, 2 mM EDTA, 0.02 % bromophenol blue. The solution is best made from stock solutions, such as 1 M Tris–HCl, pH 7.5, 20 % SDS and 0.5 M EDTA, pH 7.5. To prepare the sample buffer, add 5 mL of the 1 M Tris stock solution, 10 mL of the SDS stock solution, 0.2 mL of the EDTA stock solution, and 10 mL of glycerol (99 % purity). Fill up to 50 mL with ultrapure water and add 10 mg of bromophenol blue. Mix well (*see* **Note 2**).

2.3 Urea Extraction

1. A tabletop ultracentrifuge (e.g., Optima Max ultracentrifuge Beckman Coulter).
2. Ultracentrifuge tubes (1.5 mL) (e.g., Microfuge 1.5 mL Polyallomer tube Beckman).

3. Urea extraction solution: 15 mM Tris–HCl, pH 7.4, 100 mM glycine, 6 M urea. As for the SDS-PAGE sample buffer, the urea extraction solution is best prepared freshly from stock solutions, like 1 M Tris–HCl, pH 7.4 and 1 M glycine. All urea-containing buffers have to be prepared freshly for every application, as urea decomposes over time in aqueous solutions, which can result in objectionable covalent protein modifications. To produce extraction solution, add 0.75 mL of the 1 M Tris–HCl, pH 7.4 stock solution and 5 mL of the 1 M glycine stock solution in 20 mL ultrapure H_2O. Finally dissolve 18.01 g urea into this solution. Fill up to 50 mL with ultrapure water and check the pH after the urea has dissolved completely.
4. SDS-PAGE sample buffer (2×) (*see* Subheading 2.2).

2.4 Experiments with the BamA Depletion Strain

1. Microcentrifuge tubes (1.5 mL).
2. A tabletop centrifuge (e.g., Heraeus Multifuge 3S-R).
3. A spectrophotometer for measuring turbidity (e.g., Eppendorf BioPhotometer).
4. Polystyrene cuvettes.
5. L-arabinose (20 % w/v) and D-glucose (20 % w/v) stock solutions: weigh 10 g of L-arabinose or D-glucose, respectively, and dissolve in 50 mL ultrapure water. Use a syringe linked to a microfilter with a pore size of 0.22 μm for sterilizing the sugar solutions. Prepare aliquots of 1 mL solutions and store at −20 °C until usage.
6. Phosphate buffered saline (PBS).
7. SDS-PAGE sample buffer (2×) (*see* Subheading 2.2).

3 Methods

3.1 Using the BamA Depletion Strain to Assess the Function of BamA in the Biogenesis of Outer Membrane β-Barrel Proteins

The β-barrel assembly machinery of *E.coli* consists of the β-barrel protein BamA and the four accessory lipoproteins BamB-E. To assess the importance of the essential BAM complex component BamA, creating a knock-out of the gene is not feasible as it results in a lethal phenotype. However, by constructing of a conditional depletion strain it is possible to address the function of BamA in the biogenesis of outer membrane β-barrel proteins [11]. In the depletion strain, the wild type *bamA* allele is disrupted by an antibiotic resistance marker. Therefore, an additional copy of the corresponding gene under control of an arabinose-inducible promoter has been introduced upstream. The resulting depletion strain is only able to grow in the presence of L-arabinose in the culture medium. For depletion conditions, arabinose is removed by washing and the bacteria are inoculated into fresh medium containing glucose.

a

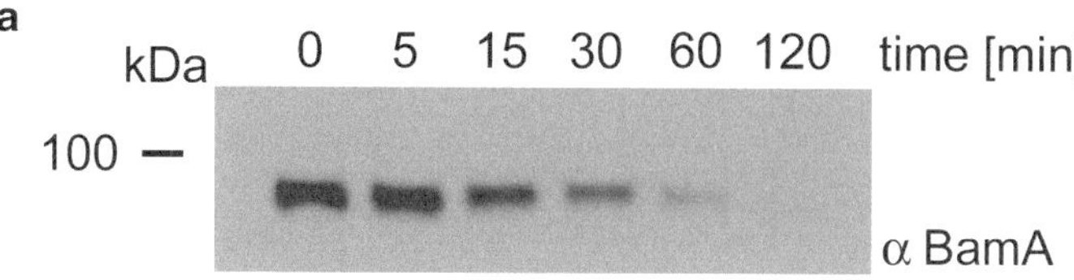

b

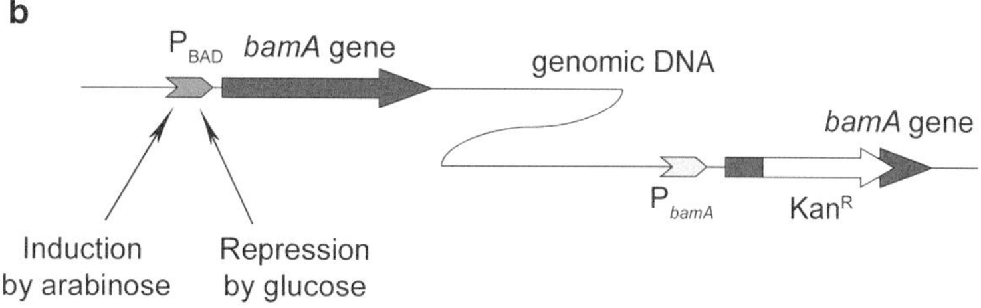

Fig. 2 Depletion of BamA upon growth in medium containing glucose. (**a**) Overnight culture of BamA depletion strain was washed three times with PBS and inoculated into fresh LB medium containing 0.2 % glucose ($t=0$ min). Bacteria were subsequently grown at 37 °C on a shaker and whole-cell lysates were prepared at indicated time points after the start of depletion [11]. (**b**) Genomic organization of the BamA depletion strain. The original *bamA* gene is disrupted by a kanamycin resistance cassette; upstream of this a copy of the *bamA* gene under control of an arabinose-inducible promoter (P_{BAD}) was inserted. Depletion is mediated by supplementing the growth medium with glucose

Depletion of BamA can be tracked by western blotting of whole cell lysates as illustrated in Fig. 2.

1. Create an expression vector containing your gene of interest under control of an inducible promoter (*see* **Note 3**).
2. Transform the *E.coli* BamA depletion strain and plate the bacteria on LB agar containing 0.2 % (w/v) L-arabinose for BamA rescue and appropriate antibiotic for selection of your expression vector.
3. Pick a single colony on the next day and inoculate 5 mL LB medium as an overnight culture with 0.2 % L-arabinose and your antibiotic.
4. Next day, pellet your bacterial cells in a centrifuge at 4500 × *g* for 10 min.
5. Discard the clear supernatant and wash your pellet by resuspending in 5 mL PBS.
6. Centrifuge the bacterial suspension again and repeat **steps 5** and **6**. Efficient washing steps are very important to get rid of residual arabinose.
7. After the washing steps, resuspend your pellet in 5 mL LB medium and check the optical density (OD_{600}).

8. Prepare two Erlenmeyer flasks supplemented with 50 mL pre-warmed LB medium and corresponding antibiotic for the selection of your expression vector. Add L-arabinose to 0.2 % (w/v) from your 20 % stock solution in flask 1 and 0.2 % glucose in flask two. In both cases, dilute the stock solution 1:100 (i.e., add 0.5 mL of the stock solutions).
9. Inoculate these two flasks with bacteria from your overnight culture to an OD_{600} of 0.05.
10. Allow bacteria to grow for 2 h at 37 °C on a shaker until BamA is depleted completely in the arabinose-containing culture (Fig. 2) (*see* **Note 4**).
11. Start expression of your protein of interest by addition of a suitable inducer.
12. Express your protein for an appropriate time period (~1–2 h) and check whether your protein is localized in the outer membrane (*see* Subheading 3.2.2) despite BamA being depleted by glucose. The culture growing with arabinose serves as positive control.

3.2 Testing OM Insertion of Multimeric OMPs

3.2.1 OM Isolation Procedure

There are essentially three ways to extract the OM fraction from Gram-negative bacteria [14]. In all cases, the bacteria are first lysed to produce vesicles from both the outer and inner membranes. These vesicles retain the specific lipid and protein composition of the original membrane. The first method for separating the two membranes is based on the ability of certain detergents to selectively solubilize the inner membrane [15]. The second uses the chaotropic agent sodium carbonate at pH 11 to destabilize the inner membrane vesicles [16]. In the third method, the inner and outer membranes are separated by ultracentrifugation in a sucrose density gradient based on their intrinsic density differences [17]. None of these methods results in complete separation of the inner and outer membranes; rather, each represents a selective enrichment of the OM fraction [14]. For stringent analysis, e.g., determining the subcellular localisation of an uncharacterized protein, using sucrose density gradient centrifugation is recommended. However, this method is laborious and time-consuming; thus, for more rapid analysis of OMs, one of the two other methods can be used. We routinely use a rapid microprocedure for analyzing OM fractions that is based on the insolubility of OM proteins in the detergent *N*-lauroyl sarcosine [18].

1. Prepare an *E. coli* culture. At the desired growth phase, measure the optical density of the culture. Pellet a number of cells equivalent to 20 mL at an OD_{600} value of 1.0 using a centrifuge (10 min at 4000 × *g*) (*see* **Note 5**).
2. Wash the cells once with 2 mL 10 mM HEPES, pH 7.4. Centrifuge as above. The pellet can now be stored at −80 °C for later processing or lysed immediately.

3. Resuspend the cell pellet in 1 mL 10 mM HEPES (*see* **Note 6**).
4. Lyse the cells by sonication, passing through a French pressure cell or using a bead beating apparatus. For small volumes, such as used here, a lysis with 100 μm beads in a bead beater (e.g., SpeedMill PLUS from Analytik Jena) is convenient, fast and sufficient for most applications. Furthermore, several samples can be prepared in parallel. Sonication or French pressing will generally give higher yields of OM samples, but are more laborious to use, especially when handling multiple samples.
5. Move the lysate to a new microcentrifuge tube and centrifuge for 2 min at 15,600 × *g* (approximately 13,000 rpm in a standard microcentrifuge) to pellet unlysed cells and cell debris.
6. Move the supernatant to a fresh microcentrifuge tube (*see* **Note 7**).
7. Centrifuge for 30 min at 15,600 × *g*. After this centrifugation, a slightly brownish membrane pellet should be visible at the bottom of the tube (*see* **Note 8**).
8. Decant the supernatant and resuspend the membrane pellet in 200 μL 10 mM HEPES pH 7.4 (*see* **Note 9**).
9. Add 200 μL of a 2 % (w/v) *N*-lauroyl sarcosine solution in 10 mM HEPES pH 7.4 to the tube. Incubate for 30 min at room temperature, preferably with agitation. This step will solubilize the inner membrane.
10. Centrifuge for 30 min at 15,600 × *g*. After this centrifugation, a clear, gelatinous outer membrane pellet should have formed at the bottom of the tube (*see* **Note 10**).
11. Decant the supernatant. The gelatinous outer membrane pellet should now be more visible, especially against light.
12. Wash the pellet with 200 μL 10 mM HEPES pH 7.4. Do not resuspend the pellet at this stage.
13. Centrifuge for 10 min at 15,600 × *g*.
14. Decant the supernatant and resuspend the pellet in 30–50 μL of 10 mM HEPES, pH 7.4. The sample can now be subjected to SDS-PAGE, urea extraction or testing the heat modifiability of the protein(s) of interest (*see* **Note 11**).

3.2.2 Urea Extraction of OMPs

To analyze the correct insertion of β-barrel proteins in the outer membrane by the BAM complex, isolated membrane fractions can be treated with urea or carbonate. Due to the chaotropic nature of urea, proteins that are only weakly associated with the lipid bilayer can be extracted from the membrane, whereas correctly folded, fully inserted OMPs remain in the membrane.

1. Prepare outer membrane fractions of your bacterial expression strain as described in Subheading 3.2.1 of this chapter.

2. Pellet the membranes by ultracentrifugation (e.g., with an Optima Max ultracentrifuge from Beckman Coulter) at 290,000 × g (Beckman Coulter rotor TLA-55) for 1 h at 4 °C.
3. Resuspend the membrane pellet in 500 μL urea-containing extraction solution.
4. Incubate the sample at 37 °C for 1 h in a water bath.
5. Collect the membranes by ultracentrifugation at 290,000 × g for 90 min at 25 °C. Do not use 4 °C to avoid precipitation of urea.
6. Transfer the clear supernatant into a new tube and store it at −20 °C for further analysis. Resuspend the extracted membrane pellet in 500 μL ultrapure, deionized water and store at −20 °C.
7. The samples can now be analyzed by SDS-PAGE. Correctly inserted β-barrel proteins are detected mainly in the pellet fraction; peripherally attached proteins will have been extracted into the supernatant.

3.3 Assessing Multimerization: Heat Modifiability of Multimeric OMPs

OMPs display a typical property in SDS-PAGE termed heat modifiability: unheated OMPs are resistant to SDS denaturation, and consequently migrate at a slightly different apparent molecular weight than expected. Upon heating, the OMPs denature and run closer to the expected size. Thus, when an unheated and heated OMP sample are run in adjacent lanes in SDS-PAGE, the position of the protein "shifts" to a slightly different apparent molecular weight in the lane with the heated sample [19]. Usually the folded form of the OMP migrates at a lower apparent molecular weight than the denatured protein; however, for some small OMPs such as OmpX, the effect is reversed and the folded protein migrates at an apparent molecular weight that is higher than expected [20, 21].

Multimeric OMPs also display heat modifiability, but this is more pronounced because, when denatured, the OMPs dissociate to monomers and the change in apparent size is large. Porins like OmpF are trimeric in their native state, and this trimeric form (apparent molecular weight ~80 kDa) is resistant do dissociation in SDS at temperatures up to 70 °C [6]. However, upon boiling, the trimer falls apart and the monomer (37 kDa) can be seen in SDS-PAGE. Under certain conditions, the heat modifiability of the dissociated monomer can also be observed [22].

A caveat of examining OMPs in SDS-PAGE is that certain multimeric OMPs, e.g., the membrane anchor of the TAA YadA, do not stain well with Coomassie R-250 [7]. In addition, the β-barrel domains of TAAs are highly stable even when heated and remain largely trimeric even after extended boiling [7]. It is therefore recommended to denature such OMPs by including 4 M urea in the sample buffer and to use colloidal Coomassie G-250 for detection.

1. Resuspend the OMP sample in buffer and add SDS-PAGE sample buffer (*see* **Notes 12** and **13**).
2. Split the sample in half; heat one aliquot for 10 min at 95–100 °C and incubate the other at room temperature (*see* **Note 14**).
3. Apply the samples to a polyacrylamide gel in adjacent lanes. The density of the gel will depend on the size of the OMP of interest, but 12–15 % gels are regularly used.
4. After running, stain the gel to visualize the size differences between the heated and non-heated samples. For monomeric OMPs, the non-heated sample should migrate slightly faster than the heated sample. For multimeric OMPs, the multimers dissociate in the heated sample and consequently the protein should migrate at a significantly slower rate in the unheated sample (*see* **Note 11**).

4 Notes

1. The BamA depletion strain is available from Prof. Tracy Palmer, College of Life Sciences, University of Dundee, Dundee, UK, e-mail: t.palmer@dundee.ac.uk.
2. When adding bromophenol blue, the exact amount is not critical; a "pinch" is sufficient. Most OMPs do not contain disulfides; we therefore do not routinely add reducing agents to our sample buffer. For a reducing sample buffer, add 0.15 g dithiothreitol (to give a concentration of 20 mM) and 2.5 mL β-mercaptoethanol (for a final concentration of 5 %). To make sample buffer with 4 M urea, weigh out 12 g of urea and dissolve this with the other components of the buffer along with a little water. Only fill up to 50 mL with ultrapure water once all the urea has dissolved.
3. Do not use sugar-inducible promoters like P_{BAD} for expressing your protein of interest. Anhydrotetracycline or isopropyl β-d-1-thiogalactoside-inducible promoters are suitable and many different expression vectors are a commercially available.
4. If you use different growth conditions to those described above, it is recommended to perform a preliminary test to ensure the complete depletion of BamA before you start expression of your protein of interest. To ascertain BamA depletion, whole-cell lysates can be analyzed by Western blotting with a BamA antibody [11].

5. *Example*: For testing expression of recombinant outer membrane proteins, we induce expression at mid-log phase (OD_{600} = ~0.5) and grow the cells for a further 2 h. If at this time the OD_{600} = 0.8, we would pellet cells from a volume of 25 mL (20/0.8 = 25).
6. *Optional*: In our experience, lysis can be enhanced by the addition of lysozyme to a final concentration of 0.1 mg/mL, magnesium chloride to 10 mM and a pinch of DNase I. Protease inhibitors may also be included.
7. We have found 2 mL tubes easier to work with, as decanting in particular is more efficient.
8. *Optional*: Centrifugations are best performed at 4 °C, but outer membrane proteins are generally very stable; therefore centrifugations can also be performed at room temperature.
9. Do not use a buffer containing magnesium at this stage, as magnesium will precipitate with *N*-lauroyl sarcosine.
10. The outer membrane pellet can be difficult to see while there is liquid in the tube, so we recommend placing the tubes in the centrifuge in a certain orientation, e.g., with the hinge of the cap always facing outwards, to aid in finding the pellet.
11. For SDS-PAGE samples, we routinely resuspend the pellets in 30 μL HEPES buffer and then add 10 μL 4× (non-reducing) SDS-PAGE sample buffer. The samples are heated for 10 min at 95 °C before separation in a 12 % polyacrylamide gel. For enhanced sensitivity, after electrophoresis the gels can be stained with Coomassie G-250 (colloidal stain).

Example: If you have resuspended your sample in 30 μL of 10 mM HEPES pH 7.4, add 10 μL of 4× SDS-PAGE sample buffer or 30 μL 2× SDS-PAGE sample buffer.

12. For multimeric OMPs, it may be necessary to add 4 M urea to the sample buffer.
13. Heat modifiability can also be assessed from a mixture of OMPs (e.g., from isolated OM preparations), but if the OMP of interest is not abundant, it may be necessary to perform a Western blot to identify the correct bands and visualize the heat modifiability.
14. It is also possible to test several different temperatures to gain an estimate of the stability of the protein/multimer (e.g., 40, 60, and 80 °C).

Acknowledgements

We are grateful to Prof. Tracy Palmer, University of Dundee, for providing us with the BamA depletion strain. For funding, we thank FEMS (an advanced fellowship to J.C.L.), UKT fortüne program (F1433253 to P.O.) and the German Science Foundation (SFB766/B10 to D.L.). We wish to thank Prof. Andrei Lupas (Max Planck Institute for Developmental Biology, Tübingen), and Dr. Monika Schütz and Prof. Ingo B. Autenrieth (University Clinics Tübingen) for continuing support and collaboration.

References

1. Silhavy TJ, Kahne D, Walker S (2010) The bacterial cell envelope. Cold Spring Harb Perspect Biol 2:a000414
2. Remmert M, Biegert A, Linke D et al (2010) Evolution of outer membrane beta-barrels from an ancestral beta beta hairpin. Mol Biol Evol 27:1348–1358
3. Delcour AH (2002) Structure and function of pore-forming β-barrels from bacteria. J Mol Microbiol Biotechnol 4:1–10
4. Koronakis V, Eswaran J, Hughes C (2004) Structure and function of TolC: the bacterial exit duct for proteins and drugs. Ann Rev Biochem 73:467–489
5. Linke D, Riess T, Autenrieth IB et al (2006) Trimeric autotransporter adhesins: variable structure, common function. Trends Microbiol 14:264–270
6. Reid J, Fung H, Gehring K et al (1988) Targeting of porin to the outer membrane of *Escherichia coli*. Rate of trimer assembly and identification of a dimer intermediate. J Biol Chem 263:7753–7759
7. Wollmann P, Zeth K, Lupas AN et al (2006) Purification of the YadA membrane anchor for secondary structure analysis and crystallization. Int J Biol Macromol 39:3–9
8. Knowles T, Scott-Tucker A, Overduin M et al (2009) Membrane protein architects: the role of the BAM complex in outer membrane protein assembly. Nat Rev Microbiol 7:206–214
9. Noinaj N, Kuszak AJ, Gumbart JC et al (2013) Structural insight into the biogenesis of β-barrel membrane proteins. Nature 501:385–390
10. Hagan CL, Silhavy TJ, Kahne D (2011) β-Barrel membrane protein assembly by the Bam complex. Ann Rev Biochem 80:189–210
11. Lehr U, Schütz M, Oberhettinger P et al (2010) C-terminal amino acid residues of the trimeric autotransporter adhesin YadA of *Yersinia enterocolitica* are decisive for its recognition and assembly by BamA. Mol Microbiol 78:932–946
12. Oberhettinger P, Schütz M, Leo JC et al (2012) Intimin and invasin export their C-terminus to the bacterial cell surface using an inverse mechanism compared to classical autotransport. PloS One 7, e47069
13. Bertani G (1951) Studies on lysogenesis. I. The mode of phage liberation by lysogenic *Escherichia coli*. J Bacteriol 62:293–300
14. Thein M, Sauer G, Paramasivam N et al (2010) Efficient subfractionation of gram-negative bacteria for proteomics studies. J Proteome Res 9:6135–6147
15. Arnold T, Linke D (2008) The use of detergents to purify membrane proteins. Curr Protoc Protein Sci. Chapter 4, Unit 4.8.1–4.8.30
16. Molloy MP, Herbert BR, Slade MB et al (2000) Proteomic analysis of the Escherichia coli outer membrane. Eur J Biochem 267: 2871–2881
17. Osborn MJ, Gander JE, Parisi E et al (1972) Mechanism of assembly of the outer membrane of *Salmonella typhimurium*. Isolation and characterization of cytoplasmic and outer membrane. J Biol Chem 247:3962–3972
18. Carlone GM, Thomas ML, Rumschlag HS et al (1986) Rapid microprocedure for isolating detergent-insoluble outer membrane proteins from *Haemophilus* species. J Clin Microbiol 24:330–332
19. Rosenbusch JP (1974) Characterization of the major envelope protein from *Escherichia coli*. Regular arrangement on the peptidoglycan and unusual dodecyl sulfate binding. J Biol Chem 249:8019–8029
20. Arnold T, Poynor M, Nussberger S et al (2007) Gene duplication of the eight-stranded beta-barrel OmpX produces a functional pore: a scenario for the evolution of transmembrane beta-barrels. J Mol Biol 366:1174–1184
21. Rutten L, Geurtsen J, Lambert W et al (2006) Crystal structure and catalytic mechanism of the LPS 3-O-deacylase PagL from *Pseudomonas aeruginosa*. Proc Natl Acad Sci U S A 103: 7071–7076
22. Watanabe Y (2002) Effect of various mild surfactants on the reassembly of an oligomeric integral membrane protein OmpF porin. J Protein Chem 21:169–175

Chapter 13

The Expression, Purification, and Structure Determination of BamA from *E. coli*

Dongchun Ni and Yihua Huang

Abstract

In gram-negative bacteria, assembly of outer membrane proteins requires the multicomponent β-barrel assembly machinery (BAM) complex, of which BamA is an essential and evolutionarily conserved integral outer membrane protein. To understand how BamA facilitates outer membrane protein biogenesis, it is important to obtain sufficient amounts of purified recombinant BamA protein for in vitro functional analysis and structure determination. In this chapter, we describe the protocol that we used in our laboratory for the cloning, expression, and purification of *E. coli* BamA and its N-terminal deletion variants for in vitro functional studies and for structure determination of the β-barrel domain alone (residues 426–810).

Key words BamA, Protein refolding, Outer membrane protein, Outer membrane protein biogenesis

1 Introduction

In *E. coli*, the BAM complex consists of an integral outer membrane protein BamA, and four accessory lipoproteins BamB, BamC, BamD, and BamE [1–5]. BamA, the central component of the BAM complex, contains five polypeptide transport-associated (POTRA) domains (sequentially denoted by P1, P2, P3, P4, and P5) at the N-terminus, followed by a C-terminal transmembrane β-barrel domain [6] (Fig. 1a). It is believed that the N-terminal POTRA domains of BamA function in OMP substrate recognition and serve as a scaffold for the assembly of the BAM complex [7], but the exact function of the β-barrel domain of BamA is unclear. Thus, it is important to obtain sufficient amounts of recombinant protein for structure determination and functional dissection of BamA in vitro. Here, we will describe the methods by which we were able to determine the crystal structure of the β-barrel domain of *E. coli* BamA and to use the structure to guide functional studies.

Susan K. Buchanan and Nicholas Noinaj (eds.), *The BAM Complex: Methods and Protocols*, Methods in Molecular Biology, vol. 1329, DOI 10.1007/978-1-4939-2871-2_13, © Springer Science+Business Media New York 2015

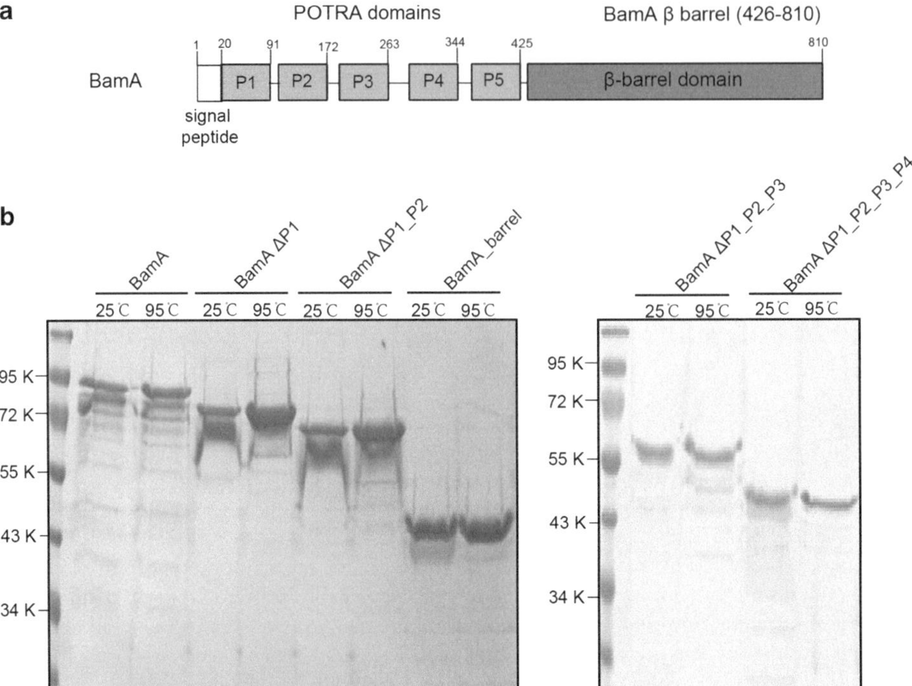

Fig. 1 Expression and purification of BamA and BamA N-terminal deletion variants in native state. (**a**) Domain organization of the full-length BamA from *E. coli*. BamA contains N-terminal five POTRA domains (denoted by P1, P2, P3, P4, and P5) and a C-terminal β-barrel domain (residues 426–810, highlighted in *green*). The signal peptide consists of the first 20 amino acid residues. (**b**) 12 % SDS-PAGE analysis of purified BamA and BamA N-terminal deletion variant proteins (BamA, BamA ΔP1, BamA ΔP1_P2, BamA ΔP1_P2_P3, BamA ΔP1_P2_P3_P4 and BamA_barrel) from bacterial outer membranes (*see* **Note 5**). The protein samples were either left at room temperature or heated for 10 min at 95 °C. None of BamA proteins purified from the outer membrane showed marked mobility difference on 12 % SDS-PAGE under heating and non-heating conditions (Color figure online)

2 Materials

Prepare and store all reagents at room temperature (unless indicated otherwise). Prepare all solutions using ultrapure water (Milli-Q, Millipore) and analytical grade reagents. All buffers are filtered with 0.2 μm membrane filter prior to use. Follow all waste disposal regulations when disposing of waste materials.

2.1 Plasmids and Bacterial Strains

1. pBAD22 (ATCC): Tightly controlled protein expression vector for overexpression of BamA and BamA N-terminal deletion variants in native state (L-arabinose-inducible).
2. pET24b (Novagen): Protein expression vector for overexpression of BamA β-barrel domain as inclusion bodies (IPTG-inducible).

3. DH5α competent cells (Life Technologies): Bacterial strain for cloning.
4. BL21 (DE3) competent cells (NEB): Bacterial strain for protein expression.

2.2 Reagents and Consumables

1. General chemicals: ampicillin, kanamycin, Coomassie brilliant blue R-250, methanol, glacial acetic acid, imidazole and IPTG (Sigma), agarose, L-arabinose, agar, yeast extract, tryptone, glycerol, PBS tablets, NaCl and Tris–HCl (Sinopharm, China), Tacsimate (pH 6.0), Polyethylene Glycol (PEG) 4000 and $HgCl_2$ (Hampton Research), restriction enzymes and DNA ligase (NEB), high fidelity PCR enzyme mix (Thermo Scientific).
2. Crystallization screen kits: Crystal Screen, PEG/ION, Index, membFrac, Additive Screen, Detergent Screen and Heavy Atom kits (Hampton Research).
3. Detergents: Lauryldimethylamine-oxide (LDAO), *n-Dodecyl-β-*D-maltoside (DDM), *n*-decyl-β-D-maltopyranoside (DM), *n*-octyl-β-D-glucoside (OG), *n*-Nonyl β-D-glucopyranoside (NG), Octyl Tetraethylene Glycol Ether (C_8E_4), Octaethylene glycol monododecyl ether ($C_{12}E_8$), *N*-lauroylsarcosine sodium, Foscholine 12(FC12) and Foscholine 10 (FC10) (Anatrace).
4. Consumables: centrifuge tubes (Sigma), 50-kDa cut-off spin concentrators (Amicon), membrane filter 0.2 μm (Millipore), Ni-NTA agarose beads (Qiagen), 5 mL HiTrap Heparin HP column and Superdex 200 10/300 column (GE Healthcare).
5. Crystallization accessories and crystal handling tools: 24-well crystallization plates, 48-well crystallization plates and 20 mm × 20 mm cover slides (Sinopharm, China), mounted CryoLoop (18 mm), CryoTong, CrystalWand, Vial Clamp and CryoCane (Hampton Research), stereomicroscope.

2.3 Solutions

1. SDS-PAGE running buffer (10×): Tris 151 g, Glycine 721 g and SDS 50 g in 5 L water.
2. TAE buffer (50×): Tris 1210 g, glacial acetic acid 285.5 mL and EDTA 145 g in 5 L water.
3. Membrane Protein Extraction Buffer: 1 % (w/v) *N,N-Dimethyldodecylamine N*-oxide (LDAO), 20 mM Tris–HCl, pH 8.0, 20 mM imidazole and 150 mM NaCl.
4. Inner Membrane Solubilization Buffer: 1× PBS, 1 % *N-lauroylsarcosine* sodium.
5. Ni-NTA Agarose Beads Equilibration Buffer: 20 mM Tris–HCl, pH 8.0, 150 mM NaCl, 20 mM imidazole, 1 % LDAO.
6. Ni-NTA Agarose Beads Wash Buffer: 20 mM Tris–HCl, pH 8.0, 150 mM NaCl, 30 mM imidazole, detergent (*see* **Note 1**).

7. Detergent Exchange Buffer: 20 mM Tris–HCl, pH 8.0, 150 mM NaCl, detergent.
8. HiTrap Heparin HP Column Equilibration Buffer: 20 mM Tris–HCl, pH 8.0, 50 mM NaCl, detergent (*see* **Note 2**).
9. HiTrap Heparin HP Column Elution Buffer: 20 mM Tris–HCl, pH 8.0, 1 M NaCl, detergent (*see* **Note 2**).
10. Gel Filtration Buffer: 20 mM Tris–HCl, pH 8.0, 150 mM NaCl, detergent (*see* **Note 3**).
11. Inclusion Body Solubilization Buffer: 20 mM Tris–HCl, pH 8.0, 8 M urea.
12. Refolding Buffer: 20 mM Tris–HCl, pH 8.0, 0.5 % LDAO.
13. Bacteria growth medium: 50 g NaCl, 25 g yeast extract and 50 g tryptone in 5 L water. The medium was sterilized for 15 min at 121 °C.
14. SDS-PAGE gel staining buffer: 0.25 % (w/v) Coomassie brilliant blue R-250, 30 % methanol, 10 % glacial acetic acid.
15. SDS-PAGE gel destaining solution: 300 mL methanol (30 %), 100 mL acetic acid (10 %), 600 mL H_2O.
16. Crystal cryoprotectant: 1.2 % C_8E_4, 30 % PEG 4000, 0.1 M MES (pH 6.0), 6 % Tacsimate, pH 6.0, 15 % glycerol.

3 Methods

3.1 Cloning, Expression, and Purification of BamA and BamA N-Terminal Deletion Variants

1. Clone the full-length BamA gene (including signal peptide) from *E. coli* genomic DNA (*E. coli* strain MG1655) into the pBAD22 vector using the restriction sites KpnI/XbaI, including an additional C-terminal 8× His-tag to facilitate affinity purification. Denote this construct as pBAD-BamA, which encodes the full-length BamA with C-terminal eight consecutive histidines (*see* **Note 4**).
2. Use the pBAD-BamA plasmid as template to construct the deletion mutants: pBAD-BamA ΔP1 (POTRA1 domain deleted), pBAD-BamA ΔP1_P2 (POTRA1 and POTRA2 deleted), pBAD-BamA ΔP1_P2_P3 (POTRA1, POTRA2 and POTRA3 deleted), pBAD-BamA ΔP1_P2_P3_P4 (POTRA1, POTRA2, POTRA3 and POTRA4 deleted), and pBAD-BamA_barrel (all five N-terminal POTRA domains deleted). The N-terminal signal peptide of BamA is kept in all these deletion constructs.
3. Transform each plasmid into BL21(DE3) competent cells and streak-transformed competent cells onto an LB agar plate supplemented with 100 μg/mL ampicillin. Incubate inverted overnight at 37 °C.

4. For expression of full-length BamA, pick a single colony from the agar plate and prepare a 10 mL overnight pre-culture shaking at 200 rpm at 37 °C.
5. Inoculate 1 L of bacterial growth medium supplemented with 100 μg/mL ampicillin with 5 mL of the overnight pre-culture.
6. When the OD_{600} reaches 0.8, induce protein expression by the addition of L-arabinose to a final concentration of 0.5 % and continue to grow at 37 °C for 3 h.
7. Harvest the cells by centrifugation at 4,500 × *g* for 30 min at 4 °C. Pour off medium.
8. Resuspend the cell pellets in 1× PBS buffer and break the cells by a single passage through a French press (JN-3000 PLUS, China) at 16,000 psi.
9. Pellet the cell membranes by centrifugation at 39,000 × *g* for 1 h at 4 °C. To remove inner membranes, resuspend the membranes with Inner Membrane Solubilization Buffer and allow to vigorously stir for 1 h at 4 °C.
10. Isolate the outer membranes by centrifugation at 39,000 × *g* for 1 h at 4 °C. Resuspend the membranes with Membrane Protein Extraction Buffer and stir for 1 h at 4 °C to solubilize the membranes.
11. Centrifuge the solubilized membranes at 39,000 × *g* for 1 h at 4 °C and collect the supernatant which contains the solubilized BamA. Pre-equilibrate Ni-NTA agarose beads with Ni-NTA agarose beads equilibration buffer for about 2 h. Then, incubate the solubilized BamA with the pre-equilibrated Ni-NTA agarose beads for 1 h at 4 °C.
12. Rinse the Ni-NTA agarose beads with Ni-NTA agarose bead equilibration buffer using one column volume.
13. Wash the Ni-NTA agarose beads with Detergent Exchange Buffer using four column volumes.
14. Elute the protein from the Ni-NTA agarose beads with Detergent Exchange Buffer containing 200 mM imidazole.
15. Then apply the sample to a 5 mL HiTrap Heparin HP column, which has been pre-equilibrated with HiTrap Heparin HP Column Equilibration Buffer.
16. Concentrate the peak fractions eluted from the HiTrap Heparin HP column and concentrate the sample to approximately 500 μL. Then apply the sample to a pre-equilibrated (with Gel Filtration Buffer) Superdex 200 10/300 gel filtration column.
17. Pool the peak fractions eluted from the gel filtration column, and run a 12 % SDS-PAGE gel to check the purity of the sample (Fig. 1b).

18. Concentrate the sample to ~10 mg/mL for crystallization screening (*see* **Notes 4** and **5**).
19. Repeat **steps 4–18** for each BamA mutant to be expressed and purified.

3.2 Expression of BamA β-Barrel as Inclusion Bodies for Refolding

1. Clone the BamA DNA sequence that corresponds to residues 426–810 of the full-length BamA protein into vector pET24b using the restriction sites NcoI/XhoI (no signal peptide or affinity tag is included here). Denote the construct as pET24-BamA_barrel.
2. Transform the plasmid into BL21(DE3) competent cells and streak-transformed competent cells onto an LB agar plate supplemented with 50 μg/mL kanamycin. Incubate inverted overnight at 37 °C.
3. For expression of full-length BamA, pick a single colony from the agar plate and prepare a 10 mL overnight pre-culture shaking at 200 rpm at 37 °C.
4. Inoculate 1 L of bacterial growth medium supplemented with 50 μg/mL kanamycin with 5 mL of the overnight pre-culture.
5. When the OD_{600} reaches 1.0, induce protein expression by the addition of IPTG to a final concentration of 0.8 mM and continue to grow at 37 °C for 2 h.
6. Pellet the cells by centrifugation at 4,000 × *g* for 30 min at 4 °C and resuspend in 1× PBS buffer.
7. Lyse the cells by a single passage through a French press (JN-3000 PLUS, China) at 16,000 psi.
8. Remove the unbroken cells and cell debris by centrifugation at 4,000 × *g* for 10 min at 4 °C.
9. Pellet the inclusion bodies by centrifugation at 39,000 × *g* for 30 min at 4 °C.
10. Solubilize the pellet in 1× PBS and 1 % LDAO for 1 h at 4 °C to remove membrane contaminants that may be associated with the inclusion bodies.
11. Pellet the inclusion bodies by centrifugation at 39,000 × *g* for 30 min at 4 °C.

3.3 Refolding BamA β-Barrel from Purified Inclusion Bodies

1. Solubilize the purified inclusion bodies with Inclusion Body Solubilization Buffer, adjusting the final protein concentration to ~30 mg/mL.
2. Add the urea-solubilized protein drop wise into the Refolding Buffer at 4 °C with vigorously stirring to a final protein concentration ~0.5 mg/mL. Continue stirring overnight at 4 °C.
3. Filter the refolded protein sample with 0.2 μm membrane filter to remove the precipitate.

4. Pre-equilibrate a 5 mL HiTrap Heparin HP column with HiTrap Heparin HP Column Equilibration Buffer and apply the refolded protein sample. Detergent exchange is also carried out in this process.
5. Elute refolded BamA β-barrel protein from the HiTrap Heparin HP column using the HiTrap Heparin HP Column Elution Buffer.
6. Concentrate the eluted protein sample using a 50 kDa cut-off spin concentrator, and apply the sample to a pre-equilibrated (with Gel Filtration Buffer) Superdex 200 10/30 gel filtration column (Fig. 2a, b).
7. Pool the eluted peak fractions and concentrate protein to approximately 10 mg/mL using a 50-kDa cut-off spin concentrator for crystallization screening.

3.4 Crystallization of BamA β-Barrel

1. Perform crystallization screening using the sitting drop vapor diffusion method under both 16 °C and 4 °C using the commercially available crystallization screening kits. Monitor crystallization closely using a stereomicroscope.
2. To screen other detergents, repeat the gel filtration column in **step 6** of Subheading 3.3 using the new detergent of interest and repeat **step 1** for crystallization screening. In our studies, the BamA β-barrel protein was screened in nine detergents including LDAO, OG, NG, DDM, DM, $C_{12}E_8$, C_8E_4, FC12, and FC10 (Fig. 2a, b).
3. For all lead conditions, perform optimization of the crystallization conditions using hanging drop vapor diffusion method by mixing protein and precipitants at a ratio of 1:1 using 24-well hanging drop vapor diffusion plates. In our studies, BamA β-barrel crystals (in detergent C_8E_4) appear 1 week after setting up the crystallization trays in a condition that contains 24 % PEG 4000, 0.1 M MES and 7.5 % Tacsimate (pH 6.0) (Fig. 2c).
4. To prepare $HgCl_2$-treated protein crystals, mix the native protein with 2 mM $HgCl_2$ and incubate overnight prior to crystallization trials.

3.5 Data Collection and Structure Determination

1. Harvest the crystals for data collection using standard crystal harvesting tools by transferring the crystals into the cryoprotectant solution for 10 s before flash-freezing in liquid nitrogen.
2. Diffraction data can now be collected at under 100 K with an oscillation angle of 0.5° per image, preferably at a synchrotron source such as the Shanghai Synchrotron Radiation Facility (SSRF, China). Use wavelengths 0.9792 Å and 1.0063 Å for collecting native and $HgCl_2$-treated protein crystals, respectively, and collect 360 images for each crystal. In our studies,

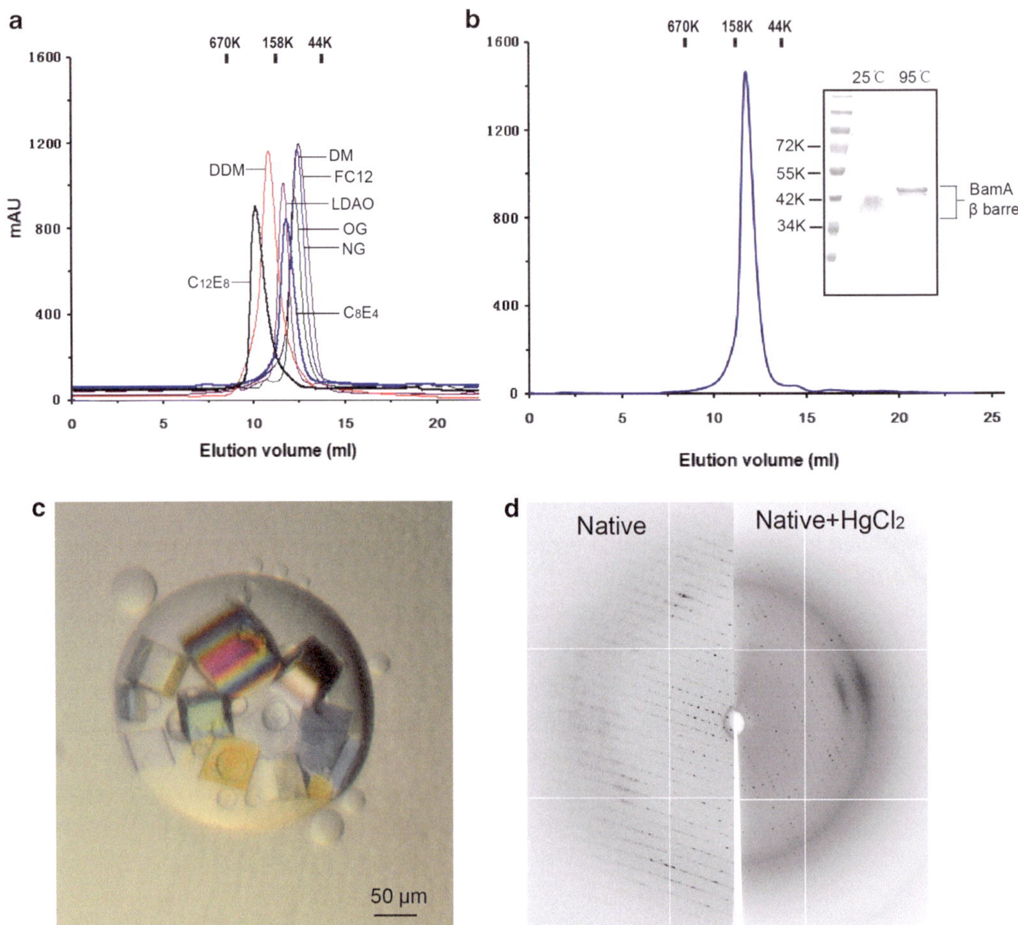

Fig. 2 Characterization of the refolded BamA β-barrel from inclusion bodies. (**a**) Gel filtration profiles of BamA β-barrel (refolded from inclusion bodies) in different detergents on Superdex 200 10/300 column. Detergents have a significant effect on the elution peak position of BamA β-barrel on size exclusion column. (**b**) Gel filtration profile of BamA β-barrel on Superdex 200 10/300 column (gel filtration buffer contained 0.6 % C_8E_4). The insertion showing the sharp mobility difference of the refolded BamA β-barrel (protein from peak fractions, gel filtration buffer contained 0.6 % C_8E_4) on 12 % SDS-PAGE. Samples were boiled at 95 °C or left at room temperature for 10 min. (**c**) Optimized BamA β-barrel crystals. The crystals grew in a condition that contains 24 % PEG4000, 0.1 M MES, and 6 % Tacsimate (pH 6.0). Crystals grew to a final size with a dimension of 60 × 60 × 45 μm within 1 week. (**d**) Representative diffraction patterns of native BamA β-barrel crystals (*left*) and $HgCl_2$-treated BamA β-barrel crystals (*right*). $HgCl_2$-treated protein crystals showing a significant improvement of diffraction spot shape. Both native crystals and $HgCl_2$-treated protein crystals diffracted to approximately 2.6 Å

both native crystals and $HgCl_2$-treated crystals diffracted to ~2.60 Å. The diffraction spots of native crystals were very streaky, which precluded data processing, whereas the shape of diffraction spots of $HgCl_2$-treated crystals were significantly improved (Fig. 2d).

3. Diffraction data can be processed using the HKL2000 package [8]. In our studies, both native and $HgCl_2$-treated crystals belong to space group $P2_12_12$ with unit cell parameters $a = 118.492$ Å, $b = 159.883$ Å, $c = 56.000$ Å, $\alpha = \beta = \gamma = 90°$ with two molecules in the asymmetric unit. Unexpectedly, we found no mercury sites after processing the dataset collected from many $HgCl_2$-treated crystals at the peak wavelength of mercury ($\lambda = 1.0063$ Å) (*see* **Note 6**).
4. To solve the structure using molecular replacement, prepare a search model using the β-barrel domain of BamA from *Neisseria gonorrhoeae* (*Ng*BamA) (PDB ID: 4K3B) [9, 10].
5. Refinement and model building can then be performed to finalize the crystal structure. In our studies, the model was refined using PHENIX [11] to 2.6 Å resolution with $R/R_{free} = 0.20/0.25$.

4 Notes

1. We performed detergent exchange on Ni-NTA agarose column by washing Ni-NTA agarose beads with wash buffer in which the to-be-tested detergent was supplied. Detergent concentrations we used in the wash buffer were 0.2 % LDAO, 0.1 % DDM, 0.1 % DM, 1 % OG, 0.2 % NG, 0.8 % C_8E_4, 0.1 % $C_{12}E_8$, 0.1 % FC12 and 0.2 % FC10.
2. Detergent concentrations in the HiTrap Heparin HP column equilibration buffer and elution buffer are the same as that in the Ni-NTA agarose beads wash buffer.
3. Detergent concentrations in the gel filtration buffers assayed were 0.06 % LDAO, 0.05 % DDM, 0.1 % DM, 0.8 % OG, 0.2 % NG, 0.6 % C_8E_4, 0.05 % $C_{12}E_8$, 0.06 % FC12 and 0.1 % FC10.
4. We initially attempted to express and purify BamA and BamA N-terminal deletion variants in the native state directly from outer membrane for crystallization studies and functional analysis; however, none of these proteins produced crystals with sufficiently high quality for X-ray crystallographic characterization though various detergents had been tested and extensive screening and optimization were pursued.
5. BamA and BamA N-terminal deletion variants purified directly from outer membranes did not show marked mobility shift on 12 % SDS-PAGE under heating and non-heating conditions. By contrast, refolded BamA β-barrel protein exhibited remarkable mobility difference on 12 % SDS-PAGE under heating and non-heating conditions.

6. Although treatment of the crystals with $HgCl_2$ significantly improved the shape of the diffraction spots, we did not find any mercury sites using many complete SAD datasets collected at the peak wavelength of mercury ($\lambda = 1.0063$ Å).

Acknowledgements

The work was supported by grants from the Ministry of Science and Technology and the National Natural Science Foundation of China.

References

1. Knowles TJ, Scott-Tucker A, Overduin M et al (2009) Membrane protein architects: the role of the BAM complex in outer membrane protein assembly. Nat Rev Microbiol 7(3):206–214
2. Wu T, Malinverni J, Ruiz N et al (2005) Identification of a multicomponent complex required for outer membrane biogenesis in *Escherichia coli*. Cell 121(2):235–245
3. Hagan CL, Silhavy TJ, Kahne D (2011) beta-Barrel membrane protein assembly by the Bam complex. Annu Rev Biochem 80:189–210
4. Ricci DP, Silhavy TJ (2012) The Bam machine: a molecular cooper. Biochim Biophys Acta 1818(4):1067–1084
5. Rigel NW, Silhavy TJ (2012) Making a beta-barrel: assembly of outer membrane proteins in Gram-negative bacteria. Curr Opin Microbiol 15(2):189–193
6. Gentle IE, Burri L, Lithgow T (2005) Molecular architecture and function of the Omp85 family of proteins. Mol Microbiol 58(5):1216–1225
7. Hagan CL, Kim S, Kahne D (2010) Reconstitution of outer membrane protein assembly from purified components. Science 328(5980):890–892
8. Otwinowski Z, Minor W (1997) Processing of X-ray diffraction data collected in oscillation mode. Methods Enzymol 276:307–326
9. Collaborative Computational Project N (1994) The CCP4 suite: programs for protein crystallography. Acta Crystallogr D Biol Crystallogr 50(Pt 5):760–763
10. Noinaj N, Kuszak AJ, Gumbart JC et al (2013) Structural insight into the biogenesis of beta-barrel membrane proteins. Nature 501(7467):385–390
11. Adams PD, Grosse-Kunstleve RW, Hung LW et al (2002) PHENIX: building new software for automated crystallographic structure determination. Acta Crystallogr D Biol Crystallogr 58(Pt 11):1948–1954

Chapter 14

Expression and Purification of the Individual Bam Components BamB–E

Suraaj Aulakh, Kelly H. Kim, and Mark Paetzel

Abstract

BamB, BamC, BamD, and BamE are lipoproteins that, along with the integral membrane protein BamA, form the β-barrel assembly machinery (BAM) complex in the outer-membrane of Gram-negative bacteria. Elucidating the roles that these lipoproteins play in the β-barrel assembly process requires both structural and functional studies that rely on milligram quantities of pure protein. Here, we describe a simple protocol for expressing individual BamB–BamE proteins in *Escherichia coli* and purifying them by nickel affinity and size-exclusion chromatography. This protocol yields pure proteins in amounts that are sufficient for crystallization trials, in vitro protein–protein interaction studies, NMR, and other biochemical experiments.

Key words β-barrel assembly machinery, BAM complex, BamB, BamC, BamD, BamE, Lipoprotein, Protein expression, Nickel affinity chromatography, Size-exclusion chromatography

1 Introduction

The lipoproteins BamB, BamC, BamD, and BamE are components of the β-barrel assembly machinery (BAM) complex [1–5]. While the three-dimensional structures of these individual lipoproteins have been determined, how they associate with each other and what exact roles they play within the BAM complex are still not clear. Further structural and functional studies of BamB–BamE are therefore needed to better understand how the BAM complex functions, and such experiments inevitably require obtaining milligram quantities of intact protein sample free of contaminants. Here, we present a simple and quick protocol for expressing and purifying individual BamB–BamE proteins. In addition, Table 1 lists the protein purification strategies for the individual Bam proteins that are currently reported in the literature.

The protocol presented here takes advantage of hexahistidine affinity tags to facilitate protein purification (Fig. 1). For this purpose, constructs for individual BamB–E proteins were created such

Susan K. Buchanan and Nicholas Noinaj (eds.), *The BAM Complex: Methods and Protocols*, Methods in Molecular Biology, vol. 1329, DOI 10.1007/978-1-4939-2871-2_14, © Springer Science+Business Media New York 2015

Table 1
Published purification strategies of Bam proteins

	Amino acid boundaries of cloned construct	Affinity tag	Purification method	References
BamB	21–392	N-terminal hexahistidine tag	Nickel affinity chromatography followed by size-exclusion chromatography	[6]
	21–392	C-terminal hexahistidine tag	Nickel affinity chromatography followed by size-exclusion chromatography	[10]
	21–392	N-terminal glutathione-S-transferase (GST) tag	GST affinity chromatography followed by size-exclusion chromatography	[11]
	21–392	N-terminal hexahistidine tag	Nickel affinity chromatography followed by size-exclusion chromatography	[12]
BamC	26–344 26–344 (E182A, Q183A, K186A)	N-terminal hexahistidine tag	Nickel affinity chromatography followed by size-exclusion chromatography	[7]
	26–344 94–344 220–344 26–217 99–217	N-terminal hexahistidine tag C-terminal hexahistidine tag	(Note: Co-lysed with BamD.) Nickel affinity chromatography followed by size-exclusion chromatography	[8]
	26–344	N-terminal hexahistidine tag	Nickel affinity chromatography followed by size-exclusion chromatography	[13]
	89–344	C-terminal hexahistidine tag	Nickel affinity chromatography followed by size-exclusion chromatography. Treated with subtilisin to yield two domains of BamC (N-terminal 99–217 domain and C-terminal 220–344 domain). Domains separated through second nickel affinity chromatography. Flow-through contained N-terminal domain. C-terminal domain bound to column, and after elution was run again through size-exclusion chromatography	[14]
	26–344 101–344	N-terminal hexahistidine tag N-terminal hexahistidine tag	Nickel affinity chromatography followed by size-exclusion chromatography	[15]

BamD	21–245	N-terminal hexahistidine tag	(Note: Co-lysed with BamC.) Nickel affinity chromatography followed by size-exclusion chromatography	[8]
	33–245	C-terminal hexahistidine tag	Nickel affinity chromatography. The resulting BamD began precipitating. The precipitate was partially redissolved in a buffer containing 50 mM Tris (pH 8.0), 1.5 M urea	[14]
	32–151	C-terminal hexahistidine tag	Nickel affinity chromatography	[10]
	23–280 (C23A) (from *R. marinus*)	N-terminal hexahistidine tag	Nickel affinity chromatography, followed by size-exclusion chromatography	[16]
BamE	21–113	N-terminal hexahistidine tag	Nickel affinity chromatography followed by size-exclusion chromatography	[9]
	40–113	C-terminal hexahistidine tag	Nickel affinity chromatography followed by size-exclusion chromatography	[14]
	1–113	C-terminal hexahistidine tag	(Note: Construct includes signal sequence. Protein was solubilized after lysis using dodecyl-β-d-maltoside.) Nickel affinity chromatography followed by size-exclusion chromatography	[10]
	20–113 (C20S)	C-terminal hexahistidine tag	(Note: periplasmically expressed with a PelB leader sequence in plasmid. Protein was extracted from cells by osmotic shock.) Nickel affinity chromatography followed by size-exclusion chromatography	[17]

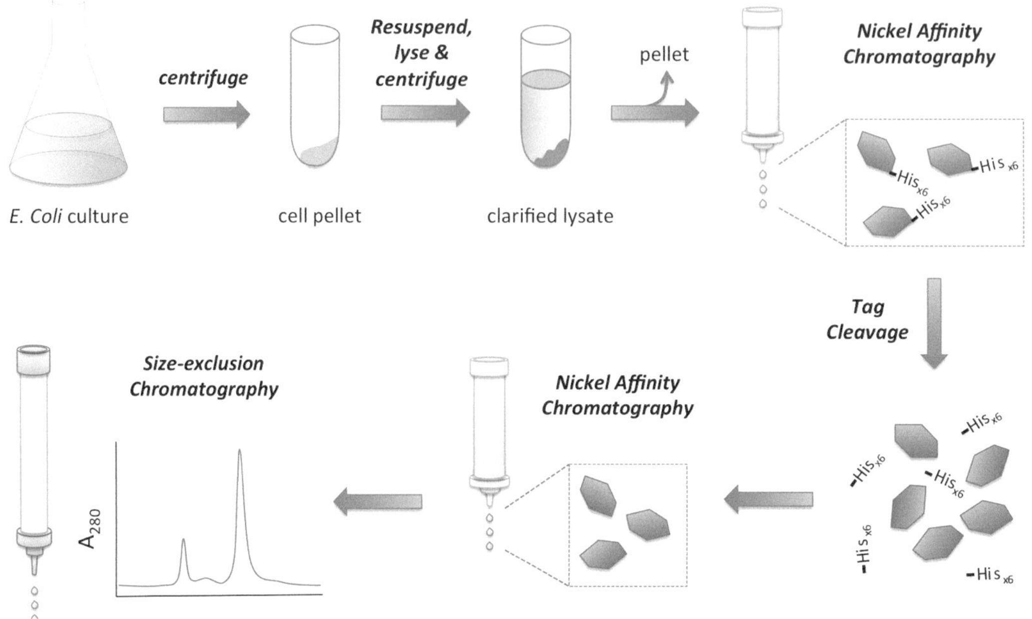

Fig. 1 A schematic overview of the purification protocol for Bam B–E proteins. *E. coli* culture expressing the desired protein is centrifuged to obtain a cell pellet, while the supernatant is discarded. This pellet is then resuspended and lysed to release the expressed protein. Another centrifugation step separates the soluble (containing the Bam lipoprotein) and insoluble fractions into the supernatant and pellet, respectively. Because the proteins are engineered to contain a hexahistidine tag, the soluble fraction is loaded onto a nickel affinity chromatography column and the elution fractions are collected. The hexahistidine tags are then removed via thrombin cleavage and another round of nickel affinity chromatography is performed, where the flow-through is collected to select for the untagged version of the protein. Finally, the untagged proteins are run on a size-exclusion chromatography column where fractions containing proteins of the desired molecular mass are collected

that each possesses an N-terminal hexahistidine tag. It should be noted that all constructs described here do not contain the N-terminal signal sequence and therefore are not secreted from the cytoplasm nor lipidated. Furthermore, the N-terminal cysteine that follows the signal sequence (the site of lipidation) has been removed to prevent potential intermolecular disulfide bond formation during the purification process. Each expression plasmid containing the gene for the individual Bam protein is transformed into and overexpressed in *Escherichia coli* BL21(λDE3) cells. The overexpressed protein is initially purified by nickel affinity chromatography; a subsequent size-exclusion chromatography step separates the different oligomeric states and increases purity and homogeneity.

The protocol is simple to follow and yields approximately 10–20 mg of pure protein from 1 L of bacterial culture, which is sufficient for various downstream experiments such as high-throughput crystallization trials, in vitro protein–protein interaction studies, nuclear magnetic resonance (NMR) spectroscopy, and functional studies. The following protocol is based on protein purification from 1 L of *E. coli* culture, but it can be easily scaled up if needed. The protocol presented here is based on previously published purification procedures for BamB [6], BamC [7, 8], BamD [8], and BamE [9].

2 Materials

2.1 Constructs

1. Agar plates of *E. coli* BL21(λDE3) cells transformed with plasmids coding for kanamycin resistance and one of BamB, BamC, BamD, and BamE with hexahistidine tag (*see* **Note 1**).

2.2 Protein Expression

1. Lysogeny broth (LB): dissolve 10 g NaCl, 10 g tryptone and 5 g yeast extract in 1 L of distilled or deionized water. Autoclave the medium in a large flask.
2. Kanamycin (50 mg/mL).
3. Isopropyl β-D-1-thiogalactopyranoside (IPTG) (1 M).
4. Erlenmeyer flasks.
5. Incubated orbital shaker.

2.3 Cell Lysis

1. Lysis buffer: 20 mM Tris–HCl, pH 8.0, 100 mM NaCl.
2. Protease inhibitor cocktail tablet (EDTA-free).
3. Sonicator.
4. High-pressure cell homogenizer.

2.4 Nickel Affinity Chromatography

1. Nickel-nitrilotriacetic acid (Ni^{2+}–NTA) agarose beads.
2. Gravity-flow chromatography column.
3. Equilibration buffer (Buffer A): 20 mM Tris–HCl, pH 8.0, 100 mM NaCl.
4. Wash buffer 1: Buffer A containing 5 mM imidazole.
5. Wash buffer 2: Buffer A containing 35 mM imidazole.
6. Elution buffer 1: Buffer A containing 100 mM imidazole.
7. Elution buffer 2: Buffer A containing 200 mM imidazole.
8. Elution buffer 3: Buffer A containing 300 mM imidazole.
9. Elution buffer 4: Buffer A containing 400 mM imidazole.
10. Elution buffer 5: Buffer A containing 500 mM imidazole.

2.5 Size-Exclusion Chromatography

1. Running buffer: Buffer A.
2. Size-exclusion column: Sephacryl S-100 HiPrep 26/60 column or Superdex 200 10/300 GL column.
3. FPLC system.

2.6 Miscellaneous

1. Centrifuge.
2. UV–Vis spectrophotometer.
3. Thrombin.
4. Spin-concentrator (or stirred cell).
5. SDS-PAGE apparatus.

3 Methods

3.1 Overexpression in E. coli

1. Pick a single colony from an agar plate and inoculate it into 100 mL LB medium containing kanamycin (50 μg/mL). Grow the starter culture on a shaker (250 rpm) at 37 °C overnight or for 12–16 h.
2. The following morning, use 10 mL of the starter culture to inoculate 1 L LB medium containing kanamycin (50 μg/mL).
3. Grow the culture on a shaker (250 rpm) at 37 °C until OD_{600} reaches 0.6. This will take 3–4 h.
4. Induce protein expression by adding 1 mL of 1 M IPTG to the 1 L culture (*see* **Note 2**).
5. Grow the culture for another 3 h to allow time for protein expression (*see* **Note 3**).
6. Harvest the cells by centrifugation (8000 ×*g*) and store the resulting cell pellet in a 50 mL centrifuge tube at −80 °C until further use.

3.2 Cell Lysis

1. Thaw the frozen cell pellet (*see* **Note 4**) and add 30 mL of lysis buffer containing one dissolved protease inhibitor tablet. Resuspend the cells by pipetting up and down.
2. Sonicate the resuspended cells (three times at 30 % amplitude for 5 s, with 10 s breaks in between) to release the cellular contents (*see* **Note 5**).
3. Pass the sonicated cells through a high-pressure cell homogenizer to yield a homogenous sample of lysed cells.
4. Clarify the lysate by centrifugation (45,000 ×*g*) for 30 min at 4 °C. Save the supernatant and discard the insoluble pellet.

3.3 Nickel Affinity Chromatography

1. While centrifuging the cell lysate, pre-equilibrate a column containing 2 mL of Ni^{2+}-NTA agarose resin by washing it with 20 mL of Buffer A.
2. Carefully pipette the supernatant of the clarified cell lysate into the pre-equilibrated column, so as not to disturb the bed of Ni^{2+}-NTA resin. Collect the flow-through.
3. Wash the column with 20 mL of wash buffer 1 and then with 20 mL of wash buffer 2. Collect the washes.

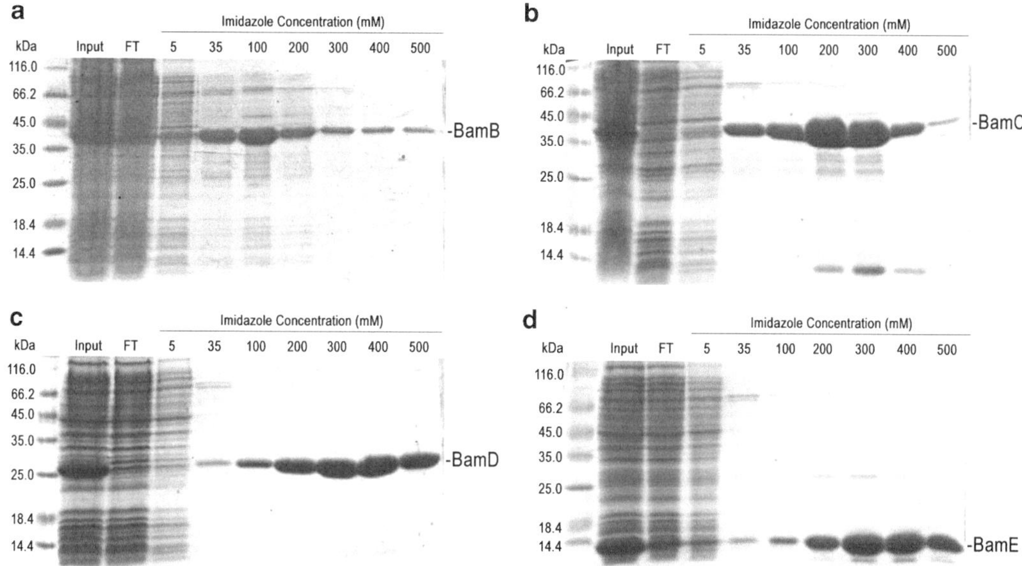

Fig. 2 Nickel affinity chromatography of the individual hexahistidine tagged BamB–E proteins. These SDS-PAGE gels show how the first nickel affinity chromatography step purifies the desired protein from the cell lysate. The gels represent the results for BamB (**a**), BamC (**b**), BamD (**c**), and BamE (**d**). For each gel, "input" indicates the sample that is applied to the chromatography column and "FT" is the flow-through that emerges from the column without binding. As increasing concentrations of imidazole are applied to the column, varying amounts of the desired protein emerge, and the corresponding elution fractions are collected

4. Wash the column subsequently with 2 mL of elution buffer 1, 2, 3, 4, and then 5. Collect the elution fractions in separate tubes.
5. Assess the quality of purification by SDS-PAGE (Fig. 2).
6. Pool fractions that contain the overexpressed protein and determine the protein concentration by measuring absorbance at $\lambda = 280$ nm.

3.4 Affinity Tag Cleavage (Optional)

1. Incubate the purified protein sample with thrombin at 4 °C overnight (10–12 h). Use 10 units of thrombin for every 1 mg of protein (*see* **Note 6**).
2. On the following day, run Ni^{2+}-NTA affinity chromatography as described earlier. The flow-through and the wash fractions should contain the tag-free protein, while the bound fraction contain uncleaved proteins and free tag.
3. Pool the fractions containing the tag-free protein.

3.5 Size Exclusion Chromatography

1. Concentrate the protein sample purified by Ni^{2+}-NTA affinity chromatography to a volume of 1–2 mL (*see* **Note 7**).
2. Centrifuge (14,000 × *g*) the sample for 30 min at 4 °C to pellet down any precipitates and aggregates.

3. Purify the protein further by size-exclusion chromatography (Sephacryl S-100 HiPrep 26/60 column) in Buffer A on an FPLC system.
4. Collect fractions that correspond to the peak of the purified protein (Fig. 3). Typically, the elution volumes for monomeric BamB, BamC, BamD, and BamE are 135 mL, 125 mL, 155 mL, and 200 mL, respectively. Dimeric BamE elutes at an elution volume of 150 mL.

3.6 Protein Storage

1. Concentrate the protein, aliquot, and store at 4 °C (*see* **Note 8**).

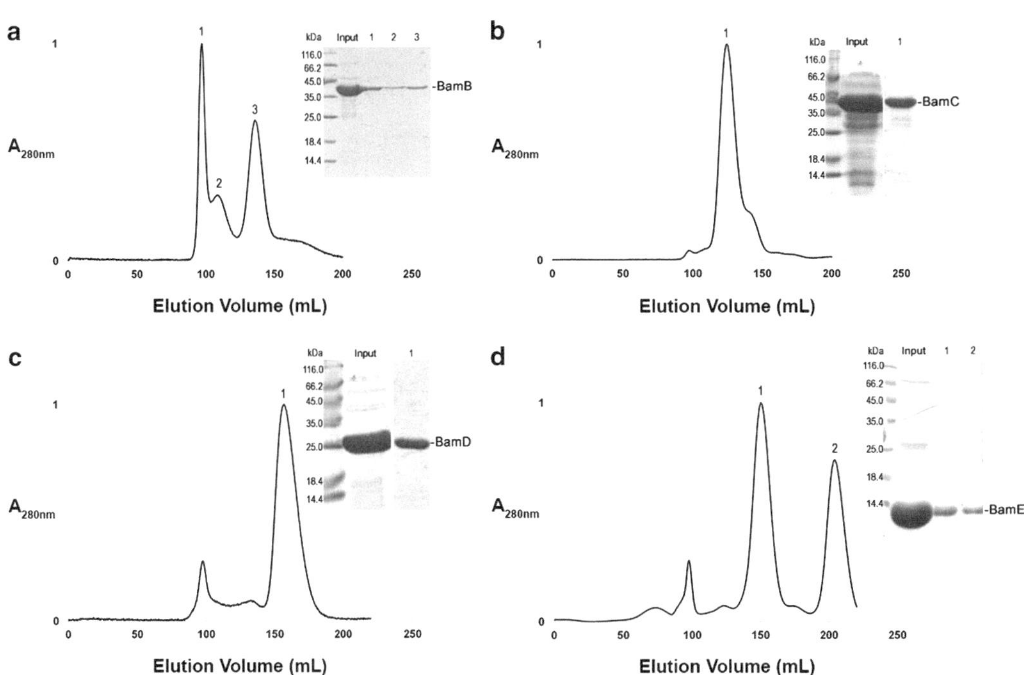

Fig. 3 Size-exclusion chromatography of individual BamB–E proteins. After removal of the hexahistidine tag and a second nickel affinity purification step, size-exclusion chromatography is performed. This figure presents the chromatograms and related SDS-PAGE gels for BamB (**a**), BamC (**b**), BamD (**c**), and BamE (**d**). For each chromatogram, the elution volume is displayed on the *x*-axis with the absolute absorbance at 280 nm displayed on the *y*-axis. The significant peaks observed are labeled numerically, and the corresponding SDS-PAGE gel showing the contents of the peaks is presented. The "input" lane on the gel represents the sample that is applied to the column. For crystallization, fractions corresponding to the monomeric forms of the protein are collected: peak 3 for BamB, peak 1 for BamC, peak 1 for BamD, and peak 2 for BamE. Note that for BamB (**a**), peak 1 corresponds to the aggregated proteins that eluted at the void volume, while peak 2 corresponds to the dimeric form of BamB. Similarly for BamE (**d**), peak 1 corresponds to the dimeric form of BamE

4 Notes

1. There are many commercially available vectors that allow one to create a fusion protein with a hexahistidine tag, either at the N- or the C-terminus. The SDS-PAGE gels and protein yields reported in this article are based on the expression and purification carried out with N-terminally tagged proteins created by standard cloning procedures using the pET28a vector.
2. Before carrying on with the rest of the purification procedures, one may wish to confirm the protein overexpression by SDS-PAGE, to compare cellular protein content before and after the IPTG addition.
3. Alternatively, the protein expression can be induced overnight at 20 °C.
4. To speed up the thawing process, the tube containing the frozen cell pellet can be placed in a beaker filled with warm water on an orbital shaker.
5. Minimize heating of the sample during sonication by placing the tube of resuspended cells in a beaker filled with ice.
6. Alternatively, the hexahistidine tags of BamB, BamD, and BamE can be cleaved off by reaction with thrombin at room temperature for 3 h. BamC is less stable at room temperature and its reaction with thrombin must be carried out at 4 °C.
7. The protein can be concentrated using either the spin-concentrator or a stirred-cell. The cutoff molecular mass of the membrane should be at least twice as small as that of the protein. A membrane with 10 kDa molecular mass cutoff works well for concentration of all Bam lipoproteins.
8. BamB, BamC, and BamE can also be flash frozen in liquid nitrogen and stored in −20 °C or −80 °C. From our experience, BamD seems to be sensitive to cold temperature as it often precipitates irreversibly when frozen and thawed.

References

1. Voulhoux R, Bos MP, Geurtsen J et al (2003) Role of a highly conserved bacterial protein in outer membrane protein assembly. Science 299:262–265
2. Wu T, Malinverni J, Ruiz N et al (2005) Identification of a multicomponent complex required for outer membrane biogenesis in *Escherichia coli*. Cell 121:235–245
3. Bos MP, Robert V, Tommassen J (2007) Biogenesis of the Gram-negative bacterial outer membrane. Annu Rev Microbiol 61:191–214
4. Selkrig J, Leyton DL, Webb CT et al (2014) Assembly of β-barrel proteins into bacterial outer membranes. Biochim Biophys Acta 1848:1542–1550
5. Kim KH, Aulakh S, Paetzel M (2012) The bacterial outer membrane β-barrel assembly machinery. Protein Sci 21:751–768
6. Kim KH, Paetzel M (2011) Crystal structure of *Escherichia coli* BamB, a lipoprotein component of the β-barrel assembly machinery complex. J Mol Biol 406:667–678

7. Kim KH, Aulakh S, Tan W et al (2011) Crystallographic analysis of the C-terminal domain of the *Escherichia coli* lipoprotein BamC. Acta Crystallogr Sect F Struct Biol Cryst Commun 67:1350–1358
8. Kim KH, Aulakh S, Paetzel M (2011) Crystal structure of the β-barrel assembly machinery BamCD complex. J Biol Chem 286: 39116–39121
9. Kim KH, Kang HS, Okon M et al (2011) Structural characterization of *Escherichia coli* BamE, a lipoprotein component of the β-barrel assembly machinery complex. Biochemistry 50:1081–1090
10. Albrecht R, Zeth K (2011) Structural basis of outer membrane protein biogenesis in bacteria. J Biol Chem 286:27792–27803
11. Heuck A, Schleiffer A, Clausen T (2011) Augmenting β-augmentation: structural basis of how BamB binds BamA and may support folding of outer membrane proteins. J Mol Biol 406:659–666
12. Noinaj N, Fairman JW, Buchanan SK (2011) The crystal structure of BamB suggests interactions with BamA and its role within the BAM complex. J Mol Biol 407:248–260
13. Knowles TJ, McClelland DM, Rajesh S et al (2009) Secondary structure and (1)H, (13)C and (15)N backbone resonance assignments of BamC, a component of the outer membrane protein assembly machinery in *Escherichia coli*. Biomol NMR Assign 3:203–206
14. Albrecht R, Zeth K (2010) Crystallization and preliminary X-ray data collection of the *Escherichia coli* lipoproteins BamC, BamD and BamE. Acta Crystallogr Sect F Struct Biol Cryst Commun 66:1586–1590
15. Warner LR, Varga K, Lange OF et al (2011) Structure of the BamC two-domain protein obtained by Rosetta with a limited NMR data set. J Mol Biol 411:83–95
16. Sandoval CM, Baker SL, Jansen K et al (2011) Crystal structure of BamD: an essential component of the β-barrel assembly machinery of Gram-negative bacteria. J Mol Biol 409: 348–357
17. Knowles TJ, Sridhar P, Rajesh S et al (2010) Secondary structure and ^{1}H, ^{13}C and ^{15}N resonance assignments of BamE, a component of the outer membrane protein assembly machinery in *Escherichia coli*. Biomol NMR Assign 4:179–181

Chapter 15

Structure Determination of the BAM Complex Accessory Lipoproteins BamB–E

Kornelius Zeth

Abstract

Outer membrane protein biogenesis is a fundamental and essential process in all Gram-negative bacteria. The key players conducting this process are organized in the β-barrel *a*ssembly *m*achinery (BAM) complex. This complex has recently attracted a lot of attention due to its importance in cell wall generation, maintenance, and the fascinating yet partially unknown mechanism. The currently best studied example is the BAM complex from *E. coli* which comprises five proteins, BamA–BamE, two of which, BamA and BamD, are essential for cell survival. Four of the complex proteins, BamB–BamE, are lipoproteins and are attached to the outer membrane via N-terminal lipid anchors. Two of them, BamB and BamD, comprise protein folds known to mediate protein–protein interactions through WD40 and TPR domains, respectively. Structures of BamB to BamE have been determined using X-ray crystallography, NMR and SAXS techniques. Details on protein preparation, crystallization, data acquisition, and determination of structures are given here along with the brief summary of the currently available structural Bam protein repertoire.

Key words Lipoproteins, TPR domain, WD40 domain, Protein–protein interactions, *E. coli*, X-ray crystallography, NMR, Molecular replacement

1 Introduction

Outer membrane protein (OMP) biogenesis is an energy independent process leading to the maturation of β-barrel structures into the outer membranes of Gram-negative bacteria, mitochondria and chloroplasts. Recent structural and biochemical work is advancing our understanding of the underlying mechanistic processes and necessitates an overview on experimental protocols for the preparation of proteins forming the BAM complex with a focus on the BamB–BamE lipoproteins.

OMP barrel proteins are physically very stable through their continuous hydrogen bond pattern and folding into the lipidic membrane is an exothermic process [1]. Maturation of OMPs through holding or folding processes in the periplasmic space are

Susan K. Buchanan and Nicholas Noinaj (eds.), *The BAM Complex: Methods and Protocols*, Methods in Molecular Biology, vol. 1329, DOI 10.1007/978-1-4939-2871-2_15,

conducted by the chaperones PpiD, SurA, and Skp which maintain the preprotein OMP in a soluble form [2]. This chaperone-OMP complex is finally transferred to the BAM insertase machinery and further transferred into the outer membrane.

In bacteria this complex comprises the integral membrane protein BamA and four associated lipoproteins BamB–BamE. BamA and BamD have proven to be essential factors for cell survival [3]. However, the presence of lipoproteins strongly depends on the bacterial species under investigation. In *Escherichia coli* four lipoproteins BamB–BamE have been identified, while for example *Thermus thermophilus* comprises a simplified version of the bacterial cell wall where only BamA and BamD exist. Recent structures of the bacterial proteins BamB–BamE shed some light on the putative functions of these proteins through fold analysis. Two of these, BamB and BamD, suggest a particular function in this complex through WD40 and TPR domains known to mediate protein–protein interactions in many cellular environments [4, 5]. Furthermore, BamB, BamC, and BamD are all proteins which were derived through gene duplication of an ancient domain motif.

Proteins of the complex are named according to their molecular weight with BamA being the largest protein of ~85 kD (originally also termed Omp85), BamB comprises ~400 residues (42 kD), BamC contains ~350 residues (37 kD), BamD has ~250 residues (28 kD) and the smallest lipoprotein has ~110 residues (12 kD). All proteins possess N-terminal signal sequences which are cleaved after their appearance in the periplasmic space. While BamA–BamD are difficult to access by NMR methodology due to their molecular size, structures of BamE and subdomains of BamC have been solved by X-ray crystallography and NMR. Here we describe our recent studies on BamB–E with a focus on crystallization and structure determination, whilst including any special details about sample expression and preparation prior to crystallization where applicable. Additional protocols for the expression and purification of the Bam lipoproteins are also described in Chapter 14.

2 Materials

2.1 Crystallization Methods

1. Protein samples: prepared either under native conditions or expressed as selenomethionine derived samples for SAD/MAD phasing using the selenium atoms.
2. Tris–HCl, pH 8.0 (1 M).
3. Sitting drop crystallization plates: 96-3 Intelli plate (Art Robbins).
4. Crystallization robot: Honeybee Robot System (Digilab).
5. Cryoprotectant solution: prepared using the reservoir solution supplemented by 20–25 % of glycerol.

6. Subtilisin: used for limited proteolysis.
7. NaCl (5 M) and imidazole (1 M).
8. Buffer X: 10 mM HEPES, 300 mM NaCl, 5 mM $MgCl_2$, 10 mM imidazole, pH 7.5.
9. Ni-NTA column pre-equilibrated with Buffer X.
10. Superdex 75 gel filtration column.
11. Gel filtration buffer (BamC-CT): 10 mM HEPES, pH 7.5, 85 mM NaCl, 20 mM KCl, 3 mM $MgCl_2$, and 1 mM β-mercaptoethanol.
12. Stereomicroscope: for viewing crystallization trays.
13. Tris–HCl, pH 8.0 (50 mM) supplemented with 1.5 M urea.
14. CD spectrometer (BamD): used to verify folded state of BamD in presence of urea.
15. Heavy Atom (Pt) Screen kit (Hampton Research).
16. BamB crystallization conditions: (a) 3.5 M NH_4Cl, 0.1 M Na-acetate, pH 4.6; (b) 2 M LiCl, 0.1 M Na-acetate, pH 4.6.
17. BamC-NT crystallization conditions: 0.1 M citric acid, pH 3.5, 2 M $(NH_4)_2SO_4$.
18. BamC-CT crystallization conditions: 0.1 M MES, pH 6.5, 25 % PEG 1000.
19. BamD crystallization conditions: 0.1 M HEPES, pH 7.5, 20 % PEG 6000.
20. BamE crystallization conditions: 0.1 M HEPES, pH 7.5, 10 % 2-propanol, 20 % PEG 4000.

2.2 Structure Determination Methods

The structure determination methods used in our studies were molecular replacement (MR), single anomalous dispersion (SAD) and multiple isomorphous replacement (MIR). Below we have listed software packages that we utilized for our studies, however, other programs may be used as well to accomplish the same goal.

1. XDS/XSCALE software package or similar.
2. CCP4 suite, PHENIX, or similar.
3. CHAINSAW (CCP4), Phyre (www.sbg.bio.ic.ac.uk/phyre/), Swiss Model (swissmodel.expasy.org/), Sculptor (PHENIX), or similar.
4. CCP4 suite.
5. MOLREP.
6. COOT.
7. SHARP using ShelxD and Solomon options.
8. Arp/Warp.
9. Modern computing system for phasing, modeling building, and refinement.

3 Methods

Structures of BamB, BamC, BamD, and BamE have been independently determined in several laboratories using X-ray crystallography and NMR spectroscopy for the determination at high resolution. SAXS has been used for the determination of the two-domain full-length protein structure of BamC. Our laboratory has determined the structures of BamB–BamE typically truncated by their targeting sequences and the N-terminal Cys residue which couples the protein to the lipid anchor. Additional flexible parts, in particular the elongated and unstructured N-terminus of BamC, have been removed before expression and crystallization. For BamB, residues 17–320 were cloned and crystallized. BamC clones comprised residues 65–320 but crystallized as two fragments (residues 75–193, BamC-NT, and residues 196–320, BamC-CT), for BamD, residues 11–226 were cloned and crystallized and finally for BamE, residues 21–94 were cloned and crystallized (residues given are based on the mature protein—*see* Fig. 1) [4, 6].

3.1 Crystallization of BamB–E

Using standard crystallization methods, broad screen crystallization trials were performed on the five proteins and protein subdomains (BamB, BamC-NT, BamC-CT, BamD, and BamE) and we succeeded in generating and optimizing crystals to the final crystallization conditions summarized in Table 1 [4, 6]. The experimental details to grow these optimized crystals are described in the following protocols.

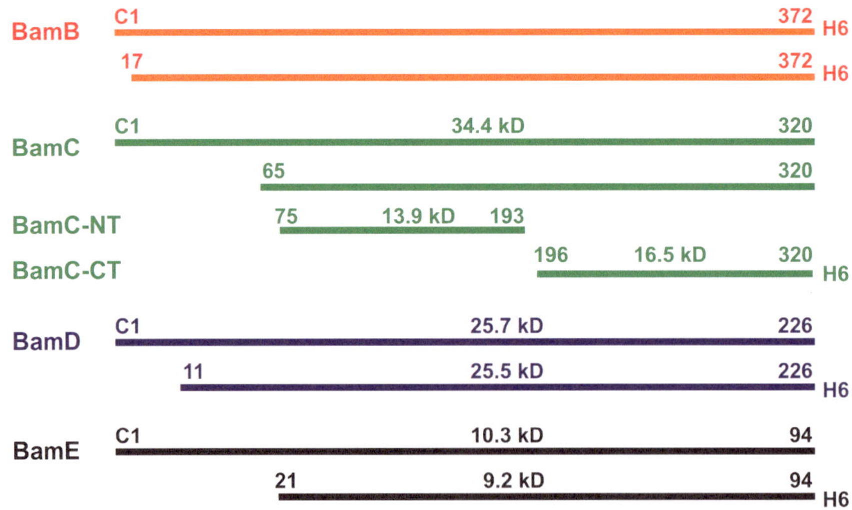

Fig. 1 Scheme showing the size of the mature lipoproteins and the fragments cloned on the basis of secondary-structure predictions. For BamC, the two fragments obtained from subtilisin cleavage of the protein comprising residues 65–320 were purified and crystallized

Table 1
Overview on crystallization conditions of BamB–BamE proteins

Protein	Protein buffer solution and concentration	Crystallization conditions
BamB	6.5 mg/ml in 15 mM Tris–HCl, pH 8	3.5 M NH_4Cl, 0.1 M Na-acetate pH 4.6 or 2 M of LiCl, 0.1 M Na-acetate, pH 4.6
BamC-NT	45 mg ml/ml in 20 mM HEPES pH 7.5, 50 mM NaCl, 1 mM β-ME	0.1 M citric acid pH 3.5, 2 M $(NH_4)_2SO_4$
BamC-CT	6.5 mg/ml in 10 mM HEPES pH 7.5, 100 mM NaCl, 20 mM KCl, 3 mM $MgCl_2$, 1 mM β-ME, 5 % glycerol	0.1 M MES pH 6.5, 25 % PEG 1000
BamD	30 mg/ml in 50 mM Tris pH 8, 1 M urea	0.1 M HEPES pH 7.5, 20 % PEG 6000
BamE	12.8 mg/ml in 10 mM HEPES pH 7.5, 100 mM NaCl, 3 mM MgCl2, 1 mM β-ME	0.1 M HEPES pH 7.5, 10 % 2-propanol, 20 % PEG 4000

3.1.1 Crystallization of BamB

1. For crystallization trials, concentrate the BamB sample to 5.5 mg/mL in 15 mM Tris–HCl, pH 8.0.
2. Prepare the reservoir solutions around the condition listed Table 1 and aliquot 100 μL into each well of a 96-3 Intelli plate (Art Robbins) for sitting drop vapor diffusion method crystallization trials.
3. Using a crystallization robot such as a Honeybee Robot System (Digilab), pipette and mix 0.4 μL of the protein with 0.4 μL of the reservoir solution. Then, store the tray(s) at 20 °C in a vibration free thermal regulated incubator.
4. Using a stereomicroscope, monitor the trays periodically for crystal growth. In our work, crystals appeared after 30–60 days in 3.5 M NH_4Cl, 0.1 M Na-acetate pH 4.6 or 2 M of LiCl, 0.1 M Na-acetate, pH 4.6.
5. Once large enough, harvest the crystals using nylon loops and immediately immerse in a solution of cryoprotectant and flash-freeze in liquid nitrogen.
6. Perform data collection preferably at a synchrotron like the Swiss Light Source (SLS, Villigen) at 100 K. In our work, we typically collected at least 200 images at 1° oscillation using a MARCCD 225 mm detector (MAR Research, Norderstedt, Germany).

3.1.2 Crystallization of BamC Domains (BamC-NT and BamC-CT)

BamC contains 65 N-terminal residues which are predicted to be disordered according to secondary structure prediction server collected at www.lmb.uni-muenchen.de. The protein was cloned, purified and crystallized using a construct containing residues

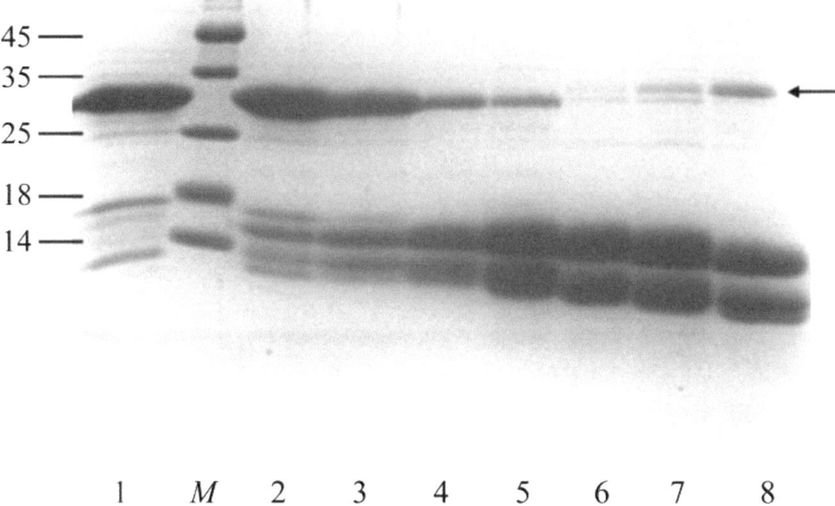

Fig. 2 SDS-PAGE gel of the protein mixture demonstrating the subtilisin treatment of BamC. Lane 1, pure BamC isolated after Ni–NTA affinity chromatography. Lanes 2–8, subtilisin treatment of BamC with increasing subtilisin:BamC ratios (from *left* to *right*) under standard conditions (samples were kept on ice for 1 h followed by addition of PMSF). The subtilisin:BamC ratios were 1:10,000 (lane 2), 1:3000 (lane 3), 1:1000 (lane 4), 1:300 (lane 5), 1:100 (lane 6), 1:30, (lane 7) and 1:10 (lane 8). Lane M contains molecular-weight markers (kDa)

65–320. Extensive crystallization trials in our and other laboratories showed a structural flexibility of the protein domains and hence the crystallization attempts were not successful. Therefore we attempted the crystallization of individual N- and C-terminal domains after the production of the protein and limited proteolysis by the nonspecific protease subtilisin (Figs. 1 and 2). The protein crystallization process of the two fragments can be performed as follows:

1. To prepare for crystallization, following limited proteolysis, treat the sample with NaCl and imidazole to reach final concentrations of 0.3 and 0.02 M, respectively. Filter the sample using a 0.22 μm filter.
2. Pass the sample over a Ni-NTA column pre-equilibrated with 10 mM HEPES, 300 mM NaCl, 5 mM $MgCl_2$, 10 mM imidazole, pH 7.5 (Buffer X).
3. Collect the flow through and elute the Ni-NTA column with Buffer X containing 250 mM imidazole. The flow through contains BamC-NT which was can be concentrated to 6.5 mg/mL, desalted using a concentrator with a 10 kD molecular weight cutoff, and used directly for crystallization.
4. The eluted fractions contain BamC-CT, which can be further purified by size-exclusion chromatography on a Superdex 75 column using 10 mM HEPES, pH 7.5, 85 mM NaCl, 20 mM

KCl, 3 mM $MgCl_2$, and 1 mM β-mercaptoethanol. Collect the peak fractions, concentrate to 45 mg/mL, and proceed with crystallization trials.

5. Perform crystallization screening as described in **steps 2** and **3** in Subheading 3.2.1 around the conditions listed for BamC in Table 1. Using a stereomicroscope, monitor the trays periodically for crystal growth.
6. Once crystals are large enough, harvest the crystals and perform data collection as described in **steps 5** and **6** in Subheading 3.2.1.

3.1.3 Crystallization of BamD

The protein purification and crystallization process was attempted using a slightly truncated expression construct of the *E. coli* BamD sequence. The protein can be prepared for crystallization as follows:

1. BamD will start to precipitate soon after elution from the Ni-NTA column, however, it can be mostly recovered in 50 mM Tris, 1.5 M urea at pH 8.0. In our studies, the addition of urea was essential for the later success of the crystallization trials due its stabilizing effect on the protein. Urea concentrations in the range of 1–2 M allowed us to concentrate the protein to more than 50 mg/mL. Without urea, the protein would begin precipitating at ~2 mg/mL.
2. To confirm urea is not affecting the protein fold, one can perform CD spectroscopy. In our studies, the CD analysis confirmed no difference in secondary structure for urea sample compared to the sample without urea.
3. To prepare for crystallization trials, concentrate the protein 30 mg/mL in 50 mM Tris–HCl, pH 8.0, with 1 M urea.
4. Perform crystallization screening as described in **steps 2** and **3** in Subheading 3.2.1 around the conditions listed for BamC in Table 1. Using a stereomicroscope, monitor the trays periodically for crystal growth.
5. Once crystals are large enough, harvest the crystals and perform data collection as described in **steps 5** and **6** in Subheading 3.2.1.

3.1.4 Crystallization of BamE

BamE, the smallest component of the BAM complex, was truncated by 20 residues on the N-terminus and can be prepared for crystallization as follows:

1. Concentrate BamE to ~13 mg/mL.
2. Perform crystallization screening as described in **steps 2** and **3** in Subheading 3.2.1 around the conditions listed for BamC in Table 1. Using a stereomicroscope, monitor the trays periodically for crystal growth.

3. Once large enough, harvest the crystals and perform data collection as described in **steps 5** and **6** in Subheading 3.2.1. In our studies, initial crystals diffracting to ~3.5 Å resolution proved difficult to reproduce. However, selenomethionine substituted BamE protein, prepared exactly as for the wild type, reproducibly yielded crystals that diffracted to high resolution.

3.2 Structure Determination of BamB–E

Structures of the crystals described in Subheading 3.1 have been solved by the methods of molecular replacement (MR), single anomalous dispersion (SAD) and multiple isomorphous replacement (MIR). The individual approaches used to solve these structures are outlined here. The statistics for the individual datasets are given in Table 2 [4, 6].

3.2.1 Structure Determination of BamB

1. Process the X-ray diffraction data using the XDS/XSCALE software package in combination with programs from the CCP4 suite [7, 8]. In our work, data were collected to 2.6 Å for full length BamB.
2. For molecular replacement, prepare a search model using the coordinates of the eight bladed WD40 structure with PDB code 1W6S (methanol dehydrogenase) [9]. This may be accomplished by using any modeling/threading program such as CHAINSAW (CCP4), Phyre (www.sbg.bio.ic.ac.uk/phyre/), Swiss Model (swissmodel.expasy.org/), or within PHENIX.
3. With the search model, solve the structure of BamB using a molecular replacement program such as the program MOLREP from the CCP4 suite [10].
4. Build the structural model within the electron density maps using a molecular modeling program such as COOT and refine the model using a refinement program such as PHENIX [11, 12]. Final structures were determined using the highest resolution data (*see* **Note 1**).

3.2.2 Structure Determination of the N- and C-Terminal Domains of BamC

1. Process the X-ray diffraction data using the XDS/XSCALE software package in combination with programs from the CCP4 suite for the transformation of intensities to amplitudes [7, 8]. In our studies, data were collected on seleno-methionine substituted samples at the selenium peak to 1.55 Å for the N-terminal domain and to 1.28 Å for the C-terminal domain of BamC.
2. Solve the structure using single anomalous dispersion (SAD) methods. In our work, the heavy atom sites were identified, refined and used for phasing by the program SHARP using ShelxD/Solomon [13–15]. The phased maps were then used for automatic structure modeling in Arp/Warp [16]. These

Table 2
Data collection and refinement statistics [4]

	BamB	BamC$_{ND}$	BamC$_{CD}$	BamD	BamE
Data collection					
Space group	$P4_32_12$	P1	$P2_1$	$P2_1$	C2
Cell dimensions					
a, b, c (Å)	101.84, 101.84, 110.59	46.57, 46.63, 60.20	29.74, 59.12, 31.06	53.13, 33.44, 57.80	69.66, 96.52, 50.49
α, β, γ (°)	90	102.70, 92.86, 118.20	90, 116.37, 90	90, 114.44, 90	90, 134.06, 90
Resolution (Å)	50–2.6 (2.66–2.6)	50–1.55 (1 65–1.55)	50–1.28 (1.33–1.28)	50–1.8 (1.9–1.8)	50–1.8 (1.88–1.8)
R_{sym} or R_{merge}	0.07 (0.91)	0.06 (0.44)	0.06 (0.57)	0.04 (0.51)	0.07 (0.47)
$I/\sigma I$	20.6 (2.2)	11.1 (2.7)	10.3 (1.6)	20.3 (3.9)	25.7 (2.8)
Completeness (%)	99.8 (99.1)	95.1 (93.8)	98.4 (95.6)	99.4 (97.8)	100 (99.8)
Redundancy	8.9	3.4	5	6.2	5.7
Refinement					
Resolution (Å)	50–2.6 (2.66–2.60)	30–1.55 (1 59–1.55)	30–1.28 (1.31–1.28)	30–1.8 (1.85–1.8)	25–1.8 (1.85–1.8)
No. reflections	17,551	56,570	25,297	16,896	21,093
R_{work}/R_{free}	0.21/0.27	0.19/0.23	0.14/0.18	0.19/0.25	0.17/0.22
No. atoms (all)					
Protein	2715	3573	1013	1808	1797
Ligand/ion (sulfate/phosphate/urea)	–/–/–	6/–/–	–/2/–	–/–/1	–/–/–

(continued)

Table 2 (continued)

	BamB	**$BamC_{ND}$**	**$BamC_{CD}$**	**BamD**	**BamE**
Water	67	136	60	90	104
B-factors					
Protein	44.8	15.1	11.5	20.9	17.5
Ligand/ion	–	–	–	–	–
Water	35.5	31.8	35.6	29.5	37.4
R.m.s. deviations					
Bond lengths (Å)	0.019	0.028	0.029	0.022	0.013
Bond angles (°)	2.07	2.03	2.15	1.94	1.34
Ramachandran statistics					
Residues in favored region no (%)	330 (91.7)	395 (95.0)	118 (99.2)	204 (99.5)	192 (98.5)
Residues in allowed region no (%)	26 (7.2)	19 (4.6)	0 (0.0)	1 (0.5)	2 (1.0)
Residues in outlier region no (%)	4 (1.1)	2 (0.4)	1 (0.8)	0 (0)	1 (0.5)
Crystallization conditions	3.5 M NH_4Cl	2 M $(NH_4)_2SO_4$	25 % PEG 1000	20 % PEG 6000	30 % PEG 4000
	0.1 M Na-acetate, pH 4.6	0.1 M citric acid pH 3.5	0.1 M MES, pH 6.5	0.1 M HEPES, pH 7.5	10 % i-PrOH, 0.1 M HEPES pH 7.5
PDB-entry	2YH3	2YH6	2YH5	2YHC	2YH9

Values in parentheses are for highest-resolution shell

initial structures were continuously rebuilt in COOT and refined using the PHENIX and CCP4 program suites. Final structures were determined using the highest resolution data of the two domains (*see* **Note 2**) [10–12].

3.2.3 Structure Determination of BamD

1. For phasing, we prepared protein derivatives with platinum (Pt) compounds. Here, protein crystals were transferred to drops containing the reservoir solution plus 1 mM of an array of Pt-salts, which are commercially available as a kit from Hampton Research.
2. Process the X-ray diffraction data using the XDS/XSCALE software package in combination with programs from the CCP4 suite for the transformation of intensities to amplitudes [7, 8]. In our studies, data were collected to 1.8 Å using protein which had been crystallized in the presence of urea. Interestingly, at this resolution, we were able to identify single urea molecules bound to the protein structure.
3. Using the multiple isomorphous replacement (MIR) method, solve the structure of BamD. In our studies, heavy atom sites were identified, refined and used for phasing by the SHARP program package. Phased maps after solvent flattening were used for automatic structure modeling in ARPWARP. The initial structures were continuously rebuilt in COOT and refined using the PHENIX and CCP4 program suites. Final structures were determined using the highest resolution data (*see* **Note 3**) [10–16].

3.2.4 Structure Determination of BamE from E. coli

1. Process the X-ray diffraction data using the XDS/XSCALE software package in combination with programs from the CCP4 suite for the transformation of intensities to amplitudes [7, 8]. In our studies, data were collected on seleno-methionine substituted sample at the selenium peak to 1.8 Å and used for phasing [7, 8].
2. Solve the structure using single anomalous dispersion (SAD) methods. In our work, the heavy atom sites were identified, refined and used for phasing by the program SHARP using ShelxD/Solomon [13–15]. Phased maps after solvent flattening were used for automatic structure modeling in ARPWARP. The initial structures were continuously rebuilt in COOT and refined using the PHENIX and CCP4 program suites. The final BamE structure was determined using the highest resolution data (*see* **Note 4**) [10–16].

4 Notes

The experimental details described above were collected in our laboratory, however, additional structures of proteins and protein complexes have been reported by other laboratories and the differences in these structures are shortly mentioned in the following notes.

1. Nine crystal structures of BamB homologs have been determined by X-ray crystallography and were submitted for publication and the PDB database (PDB entry codes: 4HDJ, 3Q7M, 3Q7N, 3Q7O, 3Q54, 3PRW, 2YH3, 3P1L, 4IMM), respectively [4, 17–20]. Seven structures were determined from *E. coli*, while two structures were contributed from the species *Pseudomonas aeruginosa* (PDB-entry: 4HDJ—sequence homology of aligned residues 31 %) and *Moraxella catarrhalis* (PDB-entry: 4IMM—sequence homology of aligned residues 25 %). The individual structures deviate by a r.m.s.d of 0.6–1 Å when the *E. coli* structures are compared to PDB-entry 2YH3 (BamB structure determined by our laboratory), while the deviations become more significant for the 4HDJ and 4IMM proteins (r.m.s.d. 2.1 and 1.85 Å, respectively).
2. Several laboratories have attempted the structure determination of BamC using X-ray crystallography, NMR and small angle X-ray scattering for the modeling of the domains. Currently structures of the individual folded N- and C-terminal domains of BamC and a complex comprising the entire N-terminal domain including the disordered portion and full-length BamD are accessible in the PDB database. BamC as full-length protein has only been determined in complex with BamD (PDB-entry: 3TGO). Structures of the N-terminal domain have been determined using X-ray crystallography (PDB-entry: 2YH6) and NMR (PDB-entry: 2LAF) [4, 5, 21, 22]. Structures of the C-terminal domain have been determined by X-ray crystallography (PDB-entry: 3SNS and 2YH5) and NMR (PDB-entry: 2LAE). The structure deviation of the N-terminal domain has been determined to a r.m.s.d of 1.6 Å for the X-ray structures (3TGO/2YH6) and an r.m.s.d of 1.6 Å when the X-ray and NMR structures are compared (2LAF/2YH6). Structure deviation of the C-terminal domain is determined to be r.m.s.d. of 0.4 Å for the crystal structures (3SNS/2YH5) and 0.8 Å for the comparison of X-ray and NMR structure (2YH5/2LAE).
3. Four structures of BamD have been solved by X-ray crystallography, three of which are from *E. coli* (PDB-entries: 3TGO, 3Q5M and 2YHC) while one was derived from *Rhodothermus marinus* (PDB-entry: 3QKY) [4, 5]. The 3TGO structure comprises BamD and the N-terminal extension and first repetitive domain BamC-ND. The r.m.s.d of the proteins relative to 2YHC are 2.0 Å (for 3TGO), 1.0 Å (for 3Q5M) and 2.9 Å (for 3QKY with 25 % sequence similarity; 186 of 251 Cα atoms were superimposed), respectively. While the three *E. coli* structures show essentially the same fold, the structure 3QKY deviates more significantly.

4. Three structures of BamE have been determined, one of which by X-ray crystallography (PDB-entry: 2YH9) and a two structures by NMR techniques (PDB-entry: 2KM7 and 2KXX) [4, 23, 24]. The structures vary significantly in their state of oligomerization and the X-ray structure shows a clear dimer connected by a 3D domain swap which hexamerizes to a symmetric molecule, while the NMR structures are both monomeric. The r.m.s.d. of the NMR structures relative to the X-ray structure are 1.6 (for 2KXX—61 aligned residues) and 1.9 Å (for 2KM7—57 aligned residues), respectively.

Acknowledgements

This work was supported by the Max Planck Society, the Ikerbasque Research Foundation and the Paul Scherrer Institute in Villigen, Switzerland. The author is grateful about the technical support at the beamlines PXI and PXII of the Swiss Light Source (SLS, Villigen, Switzerland).

References

1. McMorran LM, Brockwell DJ, Radford SE (2014) Mechanistic studies of the biogenesis and folding of outer membrane proteins *in vitro* and *in vivo*: what have we learned to date? Arch Biochem Biophys 564:265–280
2. Merdanovic M, Clausen T, Kaiser M et al (2011) Protein quality control in the bacterial periplasm. Annu Rev Microbiol 65:149–168
3. Selkrig J, Leyton DL, Webb CT et al (2014) Assembly of β-barrel proteins into bacterial outer membranes. Biochim Biophys Acta 1843:1542–1550
4. Albrecht R, Zeth K (2011) Structural basis of outer membrane protein biogenesis in bacteria. J Biol Chem 286:27792–27803
5. Kim KH, Aulakh S, Paetzel M (2011) Crystal structure of β-barrel assembly machinery BamCD protein complex. J Biol Chem 286:39116–39121
6. Albrecht R, Zeth K (2010) Crystallization and preliminary X-ray data collection of the *Escherichia coli* lipoproteins BamC, BamD and BamE. Acta Crystallogr Sect F Struct Biol Cryst Commun 66:1586–1590
7. Kabsch W (2010) Integration, scaling, space-group assignment and post-refinement. Acta Crystallogr D66:133–144
8. Winn MD, Ballard CC, Cowtan KD et al (2011) Overview of the CCP4 suite and current developments. Acta Crystallogr D67:235–242
9. Williams PA, Coates L, Mohammed F et al (2005) The atomic resolution structure of methanol dehydrogenase from Methylobacterium extorquens. Acta Crystallogr D61:75–79
10. Vagin A, Teplyakov A (1997) MOLREP: an automated program for molecular replacement. J Appl Cryst 30:1022–1025
11. Emsley P, Lohkamp B, Scott WG et al (2010) Features and development of Coot. Acta Crystallogr D66:486–501
12. Afonine PV, Grosse-Kunstleve RW, Echols N et al (2012) Towards automated crystallographic structure refinement with phenix.refine. Acta Crystallogr D68:352–367
13. Vonrhein C, Blanc E, Roversi P et al (2007) Automated structure solution with autoSHARP. Methods Mol Biol 364:215–230
14. Sheldrick GM (2010) Experimental phasing with SHELXC/D/E: combining chain tracing with density modification. Acta Crystallogr D66:479–485
15. Abrahams JP, Leslie AGW (1996) Methods used in the structure determination of bovine mitochondrial F1 ATPase. Acta Cryst D52:30–42
16. Langer G, Cohen SX, Lamzin VS et al (2008) Automated macromolecular model building for X-ray crystallography using ARP/wARP version 7. Nat Protoc 3:1171–1179

17. Jansen KB, Baker SL, Sousa MC (2011) Crystal structure of BamB from *Pseudomonas aeruginosa* and functional evaluation of its conserved structural features. PLoS One 7, e49749

18. Noinaj N, Fairman JW, Buchanan SK (2011) The crystal structure of BamB suggests interactions with BamA and its role within the BAM complex. J Mol Biol 407:248–260

19. Heuck A, Schleiffer A, Clausen T (2011) Augmenting β-augmentation: structural basis of how BamB binds BamA and may support folding of outer membrane proteins. J Mol Biol 406:659–666

20. Kim KH, Paetzel M (2011) Crystal structure of *Escherichia coli* BamB, a lipoprotein component of the β-barrel assembly machinery complex. J Mol Biol 406:667–678

21. Kim KH, Aulakh S, Tan W et al (2011) Crystallographic analysis of the C-terminal domain of the *Escherichia coli* lipoprotein BamC. Acta Crystallogr F67:1350–1358

22. Warner LR, Varga K, Lange OF et al (2011) Structure of the BamC two-domain protein obtained by Rosetta with a limited NMR data set. J Mol Biol 411:83–95

23. Knowles TJ, Browning DF, Jeeves M et al (2011) Structure and function of BamE within the outer membrane and the β-barrel assembly machine. EMBO Rep 12:123–128

24. Kim KH, Kang HS, Okon M et al (2011) Structural characterization of *Escherichia coli* BamE, a lipoprotein component of the beta-barrel assembly machinery complex. Biochemistry 50:1081–1090

Chapter 16

An In Vitro Assay for Outer Membrane Protein Assembly by the BAM Complex

Giselle Roman-Hernandez and Harris D. Bernstein

Abstract

To elucidate the mechanism of a biochemical process it is often essential to reconstitute the reaction in vitro using the minimal set of factors required to drive the reaction to completion. Here, we describe a method to reconstitute the folding and membrane integration of bacterial outer membrane (OM) proteins that have a characteristic β-barrel structure. In this method the BAM complex, a heteroligomer that catalyzes the membrane integration of β-barrel proteins, is first purified and inserted into small lipid vesicles. Denatured OM proteins are then assembled and integrated into the vesicles in the presence of a molecular chaperone called SurA.

Key words BAM complex, β-barrel proteins, *Escherichia coli*, Membrane proteins, Molecular chaperones, Protein folding, Protein purification, SurA

1 Introduction

The mechanism by which proteins are integrated into the bacterial outer membrane (OM) is a long-standing mystery. While a single hydrophobic α-helix can serve as a membrane-spanning domain for bacterial inner membrane (IM) proteins, the membrane-spanning domains of OM proteins generally fold into a structure known as a β-barrel. A β-barrel is an amphipathic β-sheet wrapped into a closed cylindrical structure that exposes a hydrophobic surface only after it is fully folded. Because of their unique architecture, OM proteins lack extended hydrophobic segments that would presumably cause them to be retained in the IM. Although pioneering genetic and biochemical studies conducted in the 1980s provided strong evidence that OM proteins are transported across the IM through the Sec machinery [1, 2], their subsequent fate was the subject of much debate. Early experiments suggested that OM proteins fold in the periplasm through interactions with lipopolysaccharide (LPS) and then insert, possibly spontaneously, into the OM [3, 4]. Other evidence, however, raised the

Susan K. Buchanan and Nicholas Noinaj (eds.), *The BAM Complex: Methods and Protocols*, Methods in Molecular Biology, vol. 1329, DOI 10.1007/978-1-4939-2871-2_16, © Springer Science+Business Media New York 2015

possibility that OM proteins do not pass through the periplasm but instead remain associated with membranes throughout the localization process [5].

The discovery that the assembly of bacterial β-barrel proteins requires a multiprotein complex known as the *β*-barrel *a*ssembly *m*achine (BAM) complex that is comprised of an integral OM protein (BamA) and four lipoproteins (BamB-E) represented a major breakthrough in our understanding of OM protein biogenesis [6, 7]. More recently, it has been shown that the purified *E. coli* BAM complex and a periplasmic chaperone called SurA that has been implicated in OM protein biogenesis in vivo [8, 9] are sufficient to promote the integration of model OM proteins into proteoliposomes [10–12]. While this observation supports a model in which OM proteins pass through the periplasm and are then handed off to the BAM complex, it is not yet clear how the BAM complex catalyzes the membrane integration of its substrates. The crystal structure of BamA and disulfide bond cross-linking experiments strongly suggest that substrates are released, possibly in a stepwise fashion, through the opening of a lateral gate in the BamA β-barrel domain [13, 14]. Other experiments, however, have suggested that BamA catalyzes the integration of prefolded β-barrels by lowering the kinetic barrier imposed by lipid head groups [15] and that the Bam lipoproteins may also play a role in the integration of at least some OM proteins [11, 16].

In this chapter we describe a method to replicate the membrane integration of OM proteins in vitro using only the purified BAM complex, which is first reconstituted into small lipid vesicles, and SurA. Our method is based on the procedure developed by Kahne and co-workers [10], but differs in that we express all of the Bam subunits together instead of reconstructing the BAM complex from separate BamAB and BamCDE subcomplexes. At least in our hands the BAM complex appears to be more active when it is formed in vivo [12]. We demonstrate the use of the method to replicate the assembly of a model *E. coli* OM protein called EspP (which is a so-called autotransporter protein; *see* Ref. 17), but in principle it can be used to replicate the assembly of many other OM proteins. This method should provide a valuable tool to analyze the function of the BAM complex.

2 Materials

2.1 Protein Purification Buffers and Chromatography Materials

1. Ampicillin (sodium salt) (Sigma Chemical Company, St. Louis, MO, USA): 50 mg/mL stock solution in water. Store at −20 °C.
2. Isoproylthiogalactoside (IPTG) (MP Biomedical, Santa Ana, CA, USA): 100 mM stock solution in water. Sterile filter and store at −20 °C.

3. *N*-dodecyl-β-D-maltoside (DDM) (Anatrace, Maumee, OH, USA): 10 % solution in water. Store at −20 °C.
4. Membrane protein extraction buffer: 50 mM Tris–HCl, pH 8.0, 150 mM NaCl, 1 % DDM.
5. Buffer B: 50 mM Tris–HCl, pH 8.0, 150 mM NaCl, 0.03 % DDM.
6. Ni-NTA agarose (Qiagen, Hilden, Germany).
7. HiLoad Superdex 200 PG column used in conjunction with an AKTA FPLC chromatographic system (GE Healthcare Life Sciences, Piscataway, NJ, USA).
8. Amicon Ultra-15 centrifugal filter units with Ultracel-10 membrane (EMD Millipore, Billerica, MA, USA) (*see* **Note 1**).
9. *DC* Protein Assay Kit (Bio-Rad, Hercules, CA, USA).
10. Kanamycin (Sigma Chemical Company, St. Louis, MO, USA): 50 mg/mL solution in water. Store at −20 °C.
11. TALON metal affinity resin (Clontech, Mountain View, CA, USA).
12. Tris-buffered saline, pH 7.4 (TBS) (KD Medical, Columbia, MD, USA): 25 mM Tris–HCl, 137 mM NaCl, 3 mM KCl.

2.2 Reagent to Prepare Proteoliposomes

1. *E. coli* Polar Lipid Extract (Avanti Polar Lipids, Alabaster, AL, USA).

3 Methods

3.1 Expression and Purification of the BAM Complex

1. Transform an *ompT-* strain such as HDB150 (MC4100 *ompT::spc ΔaraBAD leuD::kan*) with plasmid pJH114 [12] and plate on LB agar containing 100 μg/mL ampicillin (*see* **Note 2**). Inoculate 12 mL LB/100 μg/mL ampicillin with a single colony from the transformation plate and grow overnight in a bacterial shaker at 37 °C (*see* **Note 3**).
2. Centrifuge the overnight culture at 2500 × *g* at room temperature for 10 min. Wash the cells by resuspending them in 12 mL LB and repeating the centrifugation step. Resuspend the cell pellet in 12 mL LB.
3. Add the cells to 1 L LB/100 μg/mL ampicillin in a 2.8 L Fernbach flask and shake at 37 °C.
4. When the culture reaches OD_{600} = 0.5–0.6, add 0.4 mM IPTG to induce synthesis of the BAM complex and shake at 37 °C for an additional 1.5 h (*see* **Note 4**).
5. Centrifuge cultures at 4000 × *g* for 15 min (*see* **Note 5**).
6. Resuspend the cell pellet in 10 mL cold 20 mM Tris–HCl, pH 8.0. Samples should be kept cold in all subsequent steps.

7. Lyse the cells using a French press or other appropriate cell disruption instrument.
8. Remove unbroken cells by centrifuging the lysate at 6000 × *g* at 4 °C for 10 min.
9. Centrifuge the supernatant in a Beckmann 70 Ti rotor (250,000 × *g* , 4 °C, 30 min).
10. Discard the supernatant and resuspend the pellet in membrane protein extraction buffer. Incubate on ice for 1 h.
11. Repeat the ultracentrifugation step (**step 8**).
12. Mix the supernatant containing the soluble membrane proteins with 2 mL Ni-NTA agarose in a 50 mL conical tube and rotate for 1.5 h at 4 °C (*see* **Note 6**).
13. Pour the slurry into a small gravity flow column. After unbound proteins flow through, wash with 1 column volume buffer B containing 50 mM imidazole.
14. Elute the BAM complex in 3.5 mL buffer B containing 500 mM imidazole. Add the elution buffer in 1 mL increments.
15. Inject the eluate onto a Superdex 200 column washed with one column volume (120 mL) water and equilibrated with 1 column volume buffer B.
16. Run the column at 0.5 mL/min and collect 1 mL fractions.
17. The BAM complex should be highly enriched in the fractions that form the major absorbance peak at $\lambda = 280$ nm (A_{280}). Analyze a small aliquot (typically 10 μL) of each peak fraction by SDS-PAGE and Coomassie Blue staining. The relative staining of each of the five BAM complex subunits should be similar in all of the peak fractions (Fig. 1) (*see* **Note 7**).
18. Pool the fractions that contain the purified BAM complex (typically ~5–7 fractions) and concentrate ~tenfold using Amicon Ultra-15 centrifugal filters or equivalent ultrafiltration devices.
19. Determine the concentration of the BAM complex using the Bio-Rad *DC* Protein Assay or equivalent detergent-compatible assay according to the manufacturer's instructions (*see* **Note 8**). Adjust concentration to 20 μM (*see* **Note 8**). The protein can be stored briefly at 4 °C (*see* **Note 9**).

3.2 Insertion of the BAM Complex into Liposomes

1. Suspend *E. coli* phospholipids in water at a concentration of 20 mg/mL and sonicate until well dispersed.
2. Add 40 μL of the phospholipid suspension to 200 μL of the purified BAM complex in a 15 mL tube and incubate on ice for 5 min.
3. Dilute the mixture with 4 mL of 20 mM Tris–HCl, pH 8.0, and incubate on ice for 30 min to reduce the detergent concentration and promote proteoliposome formation.

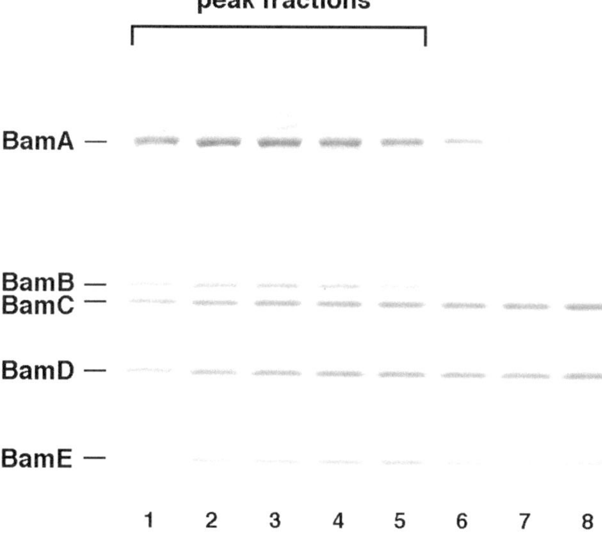

Fig. 1 Analysis of peak fractions eluted from Superdex 200 gel filtration column by SDS-PAGE. Aliquots (10 μL) of peak fractions and side fractions were subjected to SDS-PAGE on an 8–16 % Tris–glycine minigel. The peak fractions (*lanes 1–5*) were highly enriched in the BAM holocomplex. These fractions were pooled and used to generate proteoliposomes. As illustrated here, the side fractions (*lanes 6–8*) often appear to contain incomplete BAM complexes

4. Pellet the proteoliposomes in an ultracentrifuge (135,000 × *g*, 4 °C, 30 min) (*see* **Note 10**).
5. Resuspend the pellet in 200 μL 20 mM Tris–HCl, pH 8.0 (*see* **Note 11**).
6. Flash freeze aliquots of the proteoliposomes in liquid nitrogen and store at −80 °C (*see* **Note 12**).

3.3 Expression and Purification of SurA

1. Transform BL21(DE3) with pSK257 [10], a plasmid that encodes a hexahistidine tagged version of SurA lacking its signal peptide (*see* **Note 13**). Plate cells on LB agar containing 50 μg/mL kanamycin.
2. Inoculate 10–15 mL LB/50 μg/mL kanamycin with a single colony from the transformation plate and grow overnight at 37 °C in a bacterial shaker.
3. Dilute the overnight culture into 1 L LB/50 μg/mL kanamycin in a Fernbach flask and incubate at 37 °C.
4. At OD_{600} = 1.0, add 0.1 mM IPTG. Transfer the culture to a 16 °C shaking water bath and incubate overnight.
5. Harvest and lyse the cells in 10 mL cold 20 mM Tris–HCl, pH 8.0, as described in Subheading 3.1, **steps 5–7**.
6. Centrifuge the lysate at 35,000 × *g* for 20 min at 4 °C.

7. Mix the supernatant with 1 mL pre-equilibrated TALON metal affinity resin. Wash the resin and elute the tagged SurA protein following the manufacturer's instructions.
8. Dialyze the purified SurA overnight against 20 mM Tris–HCl, pH 8.0 (*see* **Note 14**). Determine the protein concentration using the formula $C = A_{280}/\varepsilon \times l$ (Beer–Lambert law) where C is the concentration, ε is the extinction coefficient, and l is the path length. The extinction coefficient can be easily determined using the ProtParam tool (web.expasy.org/protparam/). Flash freeze aliquots in liquid nitrogen and store at −80 °C.

3.4 Expression and Purification of Denatured OM Proteins from Inclusion Bodies

1. Clone the gene that encodes the OM protein of interest without its signal peptide into an appropriate pET vector [18] (*see* **Note 15**). Transform BL21(DE3) with the resulting plasmid and plate cells on LB agar containing 50 μg/mL kanamycin.
2. Inoculate 12 mL LB supplemented with 50 μg/mL kanamycin with a single colony from the transformation plate and grow overnight in a bacterial shaker at 37 °C.
3. Dilute the overnight culture into 1 L LB/50 μg/mL kanamycin in a Fernbach flask and incubate at 37 °C.
4. At $OD_{600} = 0.7$, add 0.5 mM IPTG and continue the incubation at 37 °C for 3 h.
5. Harvest and lyse the cells in 10 mL cold TBS as described in Subheading 3.1, **steps 5–7**.
6. Centrifuge lysate at 5000 × *g* for 10 min at 4 °C.
7. Discard the supernatant and resuspend the pellet in 10 mL TBS (*see* **Note 16**).
8. Repeat **steps 6** and **7** twice, but after the final centrifugation resuspend the pellet in 5 mL 8 M urea.
9. Incubate at room temperature for 1 h.
10. Chill the sample briefly on ice and then centrifuge 37,000 × *g* for 20 min at 4 °C (*see* **Note 10**). Discard the pellet. At this point the protein of interest should be extremely highly enriched in the supernatant.
11. Determine the concentration of the denatured OM protein in the supernatant as described in Subheading 3.3, **step 8**.
12. Flash freeze aliquots of the OM protein in liquid nitrogen and store at −80 °C.

3.5 Assay for OM Protein Assembly

1. Add SurA, the denatured OM protein and the proteoliposomes containing the BAM complex to 20 mM Tris, pH 8.0, in that order (*see* **Note 17**). Mix after the addition of each component. To examine the assembly of EspP(46 + β), a derivative of EspP consisting of a β-barrel plus an embedded α-helical

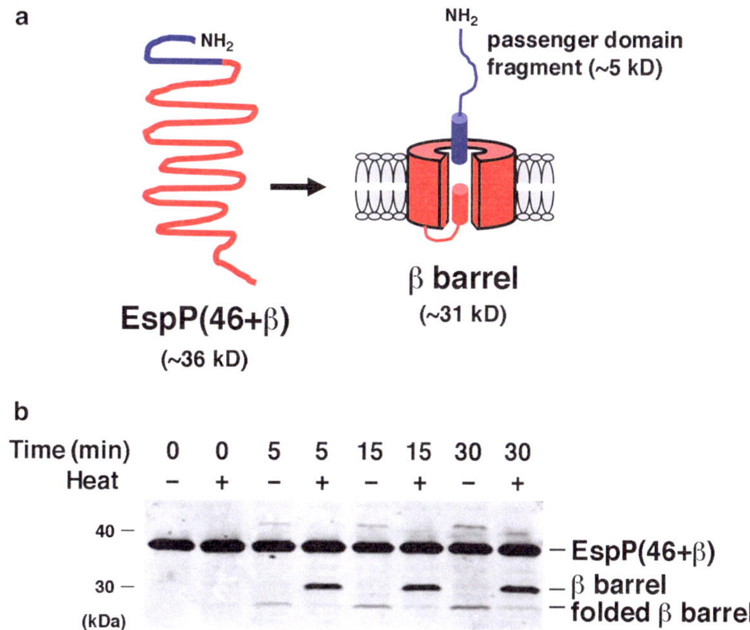

Fig. 2 Assembly of a model OM protein catalyzed by the BAM complex and SurA. (**a**). EspP(46 + β) is a ~36 kD truncated form of the autotransporter EspP. Upon insertion into the OM EspP(46 + β) folds into a β-barrel structure. An α-helix traverses and protrudes through the extracellular side of the β-barrel pore. The protein is then rapidly cleaved into a 46-residue (~5 kD) "passenger domain" fragment and a ~31 kD β-barrel through an autocatayltic intrabarrel reaction. (**b**) EspP(46 + β) was incubated at 30 °C with proteoliposomes containing the BAM complex and SurA. Aliquots were removed at the indicated time points and either heated at 95 °C or maintained at room temperature. Proteins were resolved by SDS-PAGE and transferred to nitrocellulose. Western blotting was then conducted using an antiserum raised against a C-terminal EspP peptide [12]. Because proteolytic maturation requires membrane integration and folding of the protein, the accumulation of the free β-barrel demonstrates that a fraction of the protein was assembled during the incubation. The presence of a rapidly migrating form of the β-barrel in the absence of heat provides further evidence that it was properly folded. Part (**b**) was reproduced from Ref. 12 under a Creative Commons Attribution license

segment that is cleaved in an autocatalytic reaction following membrane integration (Fig. 2a; *see* Ref. 19), we typically use 1–2 μM SurA, 0.1–0.2 μM substrate, and 0.2 μM reconstituted BAM complex. The optimal concentration of each component may vary, however, and should be determined empirically (*see* **Notes 18** and **19**). The volume of the reaction also depends on the nature of the experiment and the sensitivity of the readout assay.

2. Incubate the reaction at 30 °C for 1–30 min. Assembly of EspP(46 + β) can be observed within 1 min and typically peaks

within ~15 min, but the kinetics of assembly of other OM proteins may differ.

3. Monitor the assembly and membrane integration of the OM protein using one or more readout assays. Because many folded β-barrel proteins (including EspP) are resistant to SDS denaturation in the absence of heat [3], the presence of a rapidly migrating form of an OM protein on an SDS gel or a Western blot is indicative of complete assembly (Fig. 2b). Resistance to protease digestion can also often be used to assess integration of an OM protein into proteoliposomes [12]. More specialized assays may also be useful to examine the assembly of specific types of OM proteins. The folding of enzymes like OmpT, for example, can be monitored by activity assays [10, 12]. Likewise, the assembly of EspP (and a subset of other autotransporters) can be assessed by measuring the extent of autoproteolytic processing (Fig. 2b).

4 Notes

1. Units with Ultracel-30, -50 or -100 membranes can presumably also be used.
2. A strain that lacks OmpT is recommended because the BAM complex proteins may be sensitive to the action of this protease. Readily available *rec+* strains (e.g., MG1655, MC4100) grow faster than the *rec-* strains that are commonly used for cloning (e.g., DH5α) and may also be preferable. Plasmid pJH114 is a pBR322-based plasmid that contains the genes encoding all of the *E. coli* BAM complex proteins under the control of the *trc* promoter. To aid in the purification of the BAM complex, an octahistidine tag was attached to the C-terminus of BamE. The addition of IPTG drives the expression of moderately high levels of the BAM complex. We have not used higher level expression vectors, but these vectors may be useful to increase the yield of the BAM complex.
3. We have found that it is often desirable to double the size of the prep described here. For the larger prep we grow two 12 mL overnight cultures and use each overnight culture to inoculate a 1 L culture. Cells are pooled and resuspended in twice the volume of Tris buffer prior to lysis. The volume of the membrane solubilization buffer and Ni-NTA agarose is also doubled.
4. Longer incubation times are not recommended. We have found that the BamCDE subcomplex tends to accumulate if the cultures are incubated for more than 1.5 h in the presence of IPTG.
5. The use of a Beckman JLA-8.1000 rotor or equivalent rotor that accepts 1 L bottles is recommended.

6. Prior to use the Ni-NTA beads should be washed with water and then buffer B.
7. We have found that 8–16 % Tris–glycine minigels (Life Technologies, San Diego, CA, USA) resolve the BAM complex proteins effectively, but other gel systems should also work well. The integrity of the BAM complex in the peak fractions can be confirmed using Blue Native PAGE.
8. At this stage the concentration of the BAM complex should be ≥20 μM if the protein was purified from 2 L of cells and if the yield is optimal. We have found, however, that the BAM complex can insert into liposomes and facilitate OM protein assembly even if the concentration is considerably lower.
9. We have found that the BAM complex tends to aggregate and lose activity if it is frozen in detergent solution prior to reconstitution into proteoliposomes. For that reason we recommend inserting the BAM complex into liposomes immediately after purification, but the purified complex can also be stored overnight at 4 °C.
10. It is convenient to use a tabletop ultracentrifuge with a rotor that holds >2 mL tubes (e.g., Beckman TLA-100.4 rotor) for this purpose.
11. Most of the BAM complex should insert into the lipid vesicles, so the concentration of the protein should not change significantly.
12. Once reconstituted into proteoliposomes the BAM complex can be thawed and refrozen at least a few times without apparent loss of activity.
13. BL21(DE3) competent cells are available commercially (two vendors are New England Biolabs, Ipswich, MA, USA and Life Technologies, San Diego, CA, USA). In pSK257, *E. coli surA* is expressed under the control of a T7 promoter. The gene is expressed upon the addition of IPTG, which activates transcription of a chromosomal copy of the T7 polymerase gene.
14. SurA should be nearly homogeneously pure at this stage.
15. pET vectors can be obtained commercially (two vendors are Clontech, Mountain View, CA, USA and EMD Millipore, Billerica, MA, USA). We recommend using a vector such as pET-28 that facilitates the attachment of an N-terminal hexahistidine tag to the protein of interest. Although the protein is typically very highly enriched in inclusion bodies, in some cases it may be desirable to further achieve a higher degree of purity using a Ni-NTA column.
16. The pellet containing the inclusion bodies should have a chalky appearance and consistency.

17. We and others have found that the assembly of EspP derivatives and other model OM proteins including OmpT, OmpA and BamA does not require the addition of salt [10–12, 20], but the optimal buffer composition may vary.
18. Kahne and co-workers have used different concentrations of the components to examine the assembly of OmpT, OmpA and BamA [10, 11, 20].
19. If the readout assay involves measuring the enzyme activity of the folded OM protein, it may be desirable to add an appropriate substrate molecule before starting the reaction.

Acknowledgment

This work was supported by the Intramural Research Program of the National Institute of Diabetes and Digestive and Kidney Diseases.

References

1. Emr S, Hanley-Way S, Silhavy TJ (1981) Suppressor mutations that restore export of a protein with a defective signal sequence. Cell 23:79–88
2. Brundage L, Hendrick JP, Schiebel E et al (1990) The purified *E. coli* integral membrane protein SecY/E is sufficient for reconstitution of SecA-dependent precursor protein translocation. Cell 62:649–657
3. Freudl R, Schwarz H, Stierhof YD et al (1986) An outer membrane protein (OmpA) of *Escherichia coli* K-12 undergoes a conformational change during export. J Biol Chem 261:11355–11361
4. Sen K, Nikaido H (1990) *In vitro* trimerization of OmpF porin secreted by spheroplasts of *Escherichia coli*. Proc Natl Acad Sci U S A 87: 743–747
5. Danese PN, Silhavy TJ (1998) Targeting and assembly of periplasmic and outer-membrane proteins in *Escherichia coli*. Annu Rev Genet 32:59–94
6. Voulhoux R, Bos MP, Geurtsen J et al (2003) Role of a highly conserved bacterial protein in outer membrane protein assembly. Science 299:262–265
7. Wu T, Malinverni J, Ruiz N et al (2005) Identification of a multicomponent complex required for outer membrane biogenesis in *Escherichia coli*. Cell 121:235–245
8. Lazar SW, Kolter R (1996) SurA assists the folding of *Escherichia coli* outer membrane proteins. J Bacteriol 178:1770–1773
9. Rouviere PE, Gross CA (1996) SurA, a periplasmic protein with peptidyl-prolyl isomerase activity, participates in the assembly of outer membrane porins. Genes Dev 10:3170–3182
10. Hagan CL, Kim S, Kahne D (2010) Reconstitution of outer membrane protein assembly from purified components. Science 328:890–892
11. Hagan CL, Westwood DB, Kahne D (2013) Bam lipoproteins assembly BamA *in vitro*. Biochemistry 52:6108–6113
12. Roman-Hernandez G, Peterson JH, Bernstein HD (2014) Reconstitution of bacterial autotransporter assembly using purified components. ELife 3:04234
13. Noinaj N, Kuszak AJ, Gumbart JC et al (2013) Structural insight into the biogenesis of β-barrel membrane proteins. Nature 501: 385–390
14. Noinaj N, Kuszak AJ, Balusek C et al (2014) Lateral opening and exit pore formation are required for BamA function. Structure 22:1055–1062
15. Gessman D, Chung YH, Danoff EJ et al (2014) Outer membrane β barrel protein folding is physically controlled by periplasmic lipid head groups and BamA. Proc Natl Acad Sci U S A 111:5878–5883
16. Ieva R, Tian P, Peterson JH et al (2011) Sequential and spatially restricted interactions of assembly factors with an autotransporter β domain. Proc Natl Acad Sci U S A 108: E383–E391

17. Dautin N, Bernstein HD (2007) Protein secretion in Gram-negative bacteria via the autotransporter pathway. Annu Rev Microbiol 61:89–112
18. Studier FW, Rosenberg AH, Dunn JJ et al (1990) Use of T7 polymerase to direct expression of cloned genes. Methods Enzymol 185:60–89
19. Dautin N, Barnard TJ, Anderson DE et al (2007) Cleavage of a bacterial autotransporter by an evolutionarily convergent autocatalytic mechanism. EMBO J 26:1942–1952
20. Hagan CL, Kahne D (2011) The reconstituted *Escherichia coli* BAM complex catalyzes multiple rounds of β-barrel assembly. Biochemistry 50:7444–7446

Chapter 17

Identification of BamC on the Surface of *E. coli*

Chaille T. Webb and Trevor Lithgow

Abstract

In order to relate the structural architecture of the BAM complex to its function in outer membrane protein assembly, the arrangement of each component within the complex is vital. This chapter explores the structure and topology of BamC, using a range of biochemical techniques to probe the topology and surface exposure.

Key words Lipoprotein, Immunofluorescence, Surface exposure, Topology, Outer membrane

1 Introduction

Over the past several years, a large amount of structural data have broadened our understanding of the BAM complex. Although these structures provide insights into the individual components, the overall assembly and architecture of the BAM complex remains elusive. The BamA subunit forms a membrane embedded core around which the four remaining lipoprotein subunits (BamB, BamC, BamD, BamE) are arranged. The complex is modular and most likely can rearrange and adjust during substrate binding [1, 2]. Identifying where and how the lipoprotein subunits dock to BamA to build a complete machine is key to understanding the role of the BAM complex in outer membrane protein assembly.

The lipoprotein subunits of the BAM complex were all proposed to be periplasmic due to their known protein–protein interactions. The three-dimensional structures of BamB, BamD and BamE are globular proteins that, apart from their N-terminal lipid anchor, form well-defined and compact structures. BamC however is composed of three distinct domains. The N-terminal 50 amino acids of the mature protein are unstructured yet form the binding interface with BamD required for the assembly of BamC into the BAM complex [2–4]. The remainder of the polypeptide folds into two ~10 kDa globular domains, termed here domains 1 and 2 [5, 6]. Unlike the remaining three lipoproteins, which have defined effects

Susan K. Buchanan and Nicholas Noinaj (eds.), *The BAM Complex: Methods and Protocols*, Methods in Molecular Biology, vol. 1329, DOI 10.1007/978-1-4939-2871-2_17,

on outer membrane protein assembly and membrane permeability defects, there is no distinct phenotype observed in the absence of BamC and consequently its role in outer membrane protein assembly is still unclear.

Proteomic analysis of *Neisseria meningitidis* outer membrane vesicles identified BamC as a putative surface protein and the authors went on to show it could induce an antibody response against *N. meningitidis* strain CU385, suggesting BamC was exposed on the cell surface [7]. We reasoned that perhaps *Ec*BamC may also be on the cell surface and, if so, this requires that BamC serve a function other than to interact with substrate proteins during their assembly into the outer membrane [2].

Bacterial lipoproteins contain an N-terminal lipid modification that attaches the mature protein to the inner leaflet of either the inner or outer membrane, yet the remainder of the protein typically resides within the aqueous environment of the periplasmic space. In *E. coli*, the majority of lipoproteins adopt this topology, but there are a growing number of examples of lipoproteins that traverse the outer membrane presenting some portion of the polypeptide on the cell surface. Examples of experimentally characterized surface-exposed lipoproteins suggest several topologies exist that can achieve surface exposure. The most abundant lipoprotein in *E. coli*, Lpp or Braun's lipoprotein, has two forms which are identified by their distinct subcellular orientations: surface-exposed (free-form) and periplasmic (bound-form) [8]. Wza, the capsule-forming transporter, forms an α-helical channel and the curli transport subunit, CsgG, is also thought to form an oligomeric channel in the outer membrane [9, 10]. Both of these lipoproteins expose specific regions on the cell surface.

While they are "surface-exposed" topologically, these lipoproteins typically display small lengths of polypeptide on the surface making detection by standard means such as protease shaving notoriously difficult. A far more successful approach is to add either a chemical modification or antibody to detect surface exposure. In 1980, the discovery of the lipoprotein TraT on the cell surface was identified using cyanogen bromide-activated dextran coupling to whole cells and lactoperoxidase iodination of surface-exposed proteins that induce a change in protein size that can be detected by SDS-PAGE [11]. More recent imaging techniques used to study CsgG show the protein on the cell surface using intact cell immune-dot blots, indirect immunofluorescence, and immunogold electron microscopy [10]. Surface presentation of the C-terminus of Wza required the attachment of a C-terminal histidine tag that could then be recognized by immunofluorescence microscopy of whole cells [9]. The detection of the "free-form" of Lpp was observed using a more recent technique of labeling surface-exposed proteins with NHS-LC-LC-biotin [8]. Other surface-exposed lipoproteins from a range of gammaproteobacteria have been identified using immunofluorescence microscopy and immunoelectron microscopy [12].

In this chapter, we will present the techniques we used to identify and characterize the presence of BamC on the cell surface and confirm that this topology is adopted when BamC is assembled in the BAM complex.

2 Materials

2.1 General Media and Buffers

1. Luria Broth (LB): combine 10 g bacto-tryptone, 5 g yeast extract and 5 g NaCl with 900 mL water, adjust to pH 7.5 with NaOH and make up to 1 L with water [13]. Autoclave to sterilize.
2. PBS (10×): combine 80 g NaCl, 2 g KCl, 14.4 g Na_2HPO_4 and 2.4 g KH_2PO_4 with 900 mL water, adjust to pH7.5 with NaOH and make up to 1 L with water. Autoclave to sterilize. To make 1× PBS, dilute 10 mL 10× PBS with 90 mL water (sterilized).

2.2 Strains and Plasmids

1. *See* Table 1.
2. *See* Table 2.

2.3 Antibodies

1. Polyclonal Primary Antibodies: recombinant forms of BamC (residues 26–344) and SurA were expressed and purified. BamC antibodies were generated in mice and rabbit and SurA antibodies in rabbit.
2. Detection of Secondary Antibodies: For immunodetection experiments, AlexaFluor 488 goat anti-mouse IgG and AlexaFluor 594 goat anti-rabbit IgG were used (Molecular Probes). For the whole-cell ELISA assays, HRP-conjugated goat anti-rabbit IgG was used (Sigma).

Table 1
Bacterial strains

E. coli strains	Description	Resistance	Source
BW25113	Wild-type	–	NBRP (NIG,Japan) [14]
Δ*bamC*	BW25113 Δ*bamC::kan*	Kan^r	NBRP (NIG,Japan) [14]
BL21(DE3)Star	Recombinant protein expression strain	–	Novagen

Table 2
Plasmids

Plasmid	Description	Source
pBamC_H_6	Amp^r; cloned in with NcoI–XhoI restriction sites. Contains a C-terminal His_6-tag	[2]

2.4 Immunofluorescence Microscopy

1. Paraformaldehyde (PFA) (4 %): measure 100 mL PBS, add 4 g PFA and in a fume cupboard, use a moderate stir on top of the hotplate/stirrer with a magnetic stirrer heat control on 7 (medium heat). Allow the solution to warm up, it will turn from being cloudy to clear when ready. Inspect regularly to avoid over-heating and consequent spilling. When the PFA has dissolved, switch off the heat but leave to stir: do not handle for safety reasons. Allow to cool. When cooled, aliquot solution into 15 mL falcon tubes and store at −20 °C.
2. PBS (1×) with BSA (2 %): for a 2 % BSA/PBS *w/v* solution, weigh out 2 g bovine serum albumin (BSA) and dissolve in 100 mL 1× PBS. Store at 4 °C.
3. Permeabilization buffer: 0.1 % *v/v* Triton X-100, 10 mM EDTA, 100 μg/mL lysozyme in PBS.
4. Slides/Coverslips: we use 76×26 mm glass microscope slides and 13 mm round glass coverslips that are coated in poly-l-lysine (Menzel-Glaser).
5. DAPI: 400 ng/μL DAPI (4′,6-diamidino-2-phenylindole) in PBS.
6. Mowiol solution (Adapted from NIH-NIAID protocols): To make a 100 mg/mL solution, combine Mowiol®4-88 (Aldrich) with 6 g glycerol in a 50 mL falcon tube. Add in a magnetic stirrer and place on a hotplate/stirrer. While stirring add 6 mL water and leave for 2–3 h at room temperature. Add 12 mL 0.2 M Tris pH 8.5 and continue stirring in a beaker of hot water (50–60 °C) on the hot plate/stirrer for 10 min intervals until the Mowiol dissolves (this can take up to an hour). Centrifuge at 5000 ×*g* for 15 min to remove undissolved solids. Store in 1 mL aliquots at −20 °C and keep one tube at 4 °C for use (will last for over a month).

2.5 Whole-Cell ELISA Stock Solutions

1. PBS (1×) with FCS (1 %): dilute 1 mL fetal calf serum (FCS; Gibco) in 100 mL 1× PBS. Store at 4 °C.
2. Phosphate–citrate buffer (50 mM, pH 5.0): combine 25.7 mL 0.2 M Na_2HPO_4 with 24.3 mL 0.1 M citric acid and make up to 100 mL with water.
3. (NH_4)2(2,2′-Azino-bis-3-ethylbenzothiazoline-6-sulfonicacid) (ABTS) (4 mg/mL): prepare in water. Store aliquots in foil at −20 °C.
4. ABTS Substrate Solution: combine 9 mL of 50 mM phosphate–citrate buffer, pH 5.0, with 3 mL of ABTS. Add 10 μL hydrogen peroxide to the 12 mL of solution JUST BEFORE use.
5. Plates: Nunc-Immuno Plates 96 well “F” Maxisorp from Thermofisher cat#456537 but any 96-well plates with a flat bottom can be used.

2.6 Whole-Cell Co-Immunoprecipitation

1. Spheroplast buffer: 0.75 M sucrose, 50 mM Tris–HCl, pH 7.8, 6 mM EDTA, 0.6 mg/mL lysozyme. (Store buffer at 4 °C without lysozyme in it. Add when about to use.)
2. IP buffer: 25 mM Tris–HCl, pH 7.5, 150 mM NaCl, 1 mM EDTA, 1 % *v/v* NP-40, 5 % *v/v* glycerol (*see* **Note 1**).
3. Protein A/G Agarose beads (Pierce).
4. Triton X-100 (10 %).
5. SDS-loading dye (6×): to make 30 mL, combine 9 mL of 1 M Tris–HCl, pH 6.8, with 3 mL water and add 9 mg bromophenol blue and 2.7 g dithiothreitol (DTT). Once dissolved, add 18 mL glycerol and mix. Lastly add 1.8 g SDS and stir until dissolved (*see* **Note 2**). Store 1.5 mL aliquots at −20 °C. To make a 1× solution dilute 1 mL 6× SDS-loading dye with 5 mL water.

2.7 Protease Shaving Stock Solutions

1. IPTG (1 M): dissolve 2.38 g of isopropyl β-D-1-thiogalactopyranoside (IPTG) in 10 mL water and aliquot 1 mL into tubes. Store aliquots at −20 °C.
2. Trypsin (10 mg/mL): dissolve 5 mg of trypsin in 500 μL of water and aliquot 10 μL into aliquots. Store aliquots at −20 °C. Use at a final concentration of 100 μg/mL.
3. Polymixin B (20 mg/mL): dissolve 10 mg of polymixin B in 500 μL of water and store at 4 °C. Use at a final concentration of 100 μg/mL.
4. Soybean trypsin inhibitor (50 mg/mL): dissolve 25 mg of soybean trypsin inhibitor in 500 μL of water and aliquot into 10 μL aliquots. Store aliquots at 20 °C.

3 Methods

3.1 Immunofluorescence Microscopy

To demonstrate that BamC was surface-exposed, we used immunofluorescence microscopy (*see* **Note 3**).

1. Inoculate a 5 mL LB culture with a fresh colony of wild-type (BW25113) or Δ*bamC* (with 34 μg/mL kanamycin) strains and grow overnight, 37 °C at 220 rpm.
2. The next day, dilute the overnight culture 1:100 into 20 mL LB (containing the appropriate antibiotics) to an $OD_{600} = 0.1$ and incubate at 37 °C, 220 rpm until the $OD_{600} = 0.6–0.8$.
3. Place 1 mL of culture in an Eppendorf tube (for each sample prepare two tubes: for whole-cell and permeabilized cell samples). Centrifuge at 5000 × *g* for 5 min at 4°C to pellet cells and remove supernatant. Resuspend cells in 900 μL PBS.
4. Centrifuge cells at 5000 × *g* for 5 min at 4°C and remove supernatant. Resuspend pellet in 50 μL PBS.

5. Add 400 μL 4 % PFA to each tube, vortex and leave at room temperature for 30 min spinning gently.
6. Centrifuge cells at 5000 × *g* for 5 min at 4°C and remove supernatant. Resuspend cells in 500 μL PBS. Repeat wash steps two more times to remove excess fixative.
7. To prepare permeabilized cell samples, resuspend the pellet in 500 μL Permeabilization buffer and gently rock at room temperature for 25 min.
8. Centrifuge cells at 5000 × *g* for 5 min at 4°C and remove supernatant. Resuspend cells in 500 μL PBS. Repeat wash steps two more times. Resuspend cells in 300 μL.
9. Place a poly-L-lysine-coated cover slip in each well of a flat-bottom 24-well tray. Add 20 μL of either the fixed or permeabilized cells on each cover slip.
10. Centrifuge tray at 4000 × *g* at room temperature for 5 min to evenly distribute the bacteria over the coverslip. Then place the tray at 42 °C for 10 min to adhere the cells to the coverslip.
11. Add 500 μL PBS/2 % BSA to each well and incubate at RT for 20 min (or overnight at 4 °C). Remove solution.
12. Add the primary antibody(s) diluted in PBS/2 % BSA to each well and gently rock for 1 h at 4 °C (*see* **Note 4**). To simultaneously probe BamC and the periplasmic protein, SurA, we add both anti-BamC (mouse) and anti-SurA (rabbit) at a 1:2000 dilution to each well (*see* **Note 5**).
13. Remove the primary antibody solution and wash each well with 500 μL PBS. Repeat wash step two more times.
14. Add AlexaFluor 488 goat anti-mouse IgG and AlexaFluor 594 goat anti-rabbit IgG antibodies at 1:300 dilution in 300 μL PBS/2 % BSA to each well. Wrap the tray in foil and incubate for 1 h at 4 °C (*see* **Note 6**).
15. Remove the secondary antibody solution and add 500 μL PBS to each well. With minimum exposure to light, wash wells two more times with PBS.
16. Add 150 μL DAPI solution to each well and leave at room temperature for 15 min keeping tray covered with foil.
17. Rinse wells with 500 μL PBS followed by 500 μL water.
18. Carefully remove the coverslip from the well and blot to remove excess water. Add 2 μL Mowiol solution onto the cell-side of the coverslip and place the cover-slip facedown onto a microscope slide (*see* **Note 7**).
19. Cover with foil and place slides at 37 °C for ~1 h to allow the Mowiol solution to set. Store slides at 4 °C (*see* **Note 8**).
20. Visualize cells using a fluorescence microscope with the appropriate filters (we use an Olympus IX81 microscope with the 100× objective). Analyze images using a program such as ImageJ.

3.2 Indirect Whole-Cell ELISA

As an alternative method to whole-cell immunofluorescence detection, indirect whole-cell enzyme-linked immunosorbent assay (ELISA) relies on antibody detection of surface-exposed epitopes using horseradish peroxidase (HRP)-conjugated secondary antibodies. Catalysis of the HRP chemiluminescent substrate ABTS is measured at 405 nm to determine the presence of the protein of interest (this protocol is adapted from [15]).

1. Inoculate 5 mL LB cultures with a fresh colony of wild-type (BW25113) or Δ*bamC* (with 34 μg/mL kanamycin) strains and grow overnight, 37 °C at 220 rpm.
2. Dilute the overnight culture 1:100 into 20 mL LB (with the appropriate antibiotics), to an $OD_{600} = 0.1$ and incubate at 37 °C, 220 rpm until the $OD_{600} = 0.5$.
3. Centrifuge cells at 3000 × *g* for 5 min at 4°C and remove supernatant. Resuspend cells in 500 μL PBS (*see* **Note 9**).
4. Centrifuge cells again at 3000 × *g* for 5 min at 4°C and remove supernatant. Resuspend in PBS to obtain an $OD_{600} = 0.2$.
5. Using a multi-channel pipette, aliquot 50 μL of cells into each well of a 96-well tray (leaving some wells empty as a control) and air dry at 65 °C for ~2 h to adhere the cells to the plate surface (*see* **Note 10**).
6. Add 200 μL of PBS/1 % FCS to each well for 1 h at 37 °C. Remove blocking solution and carefully wash the wells with PBS twice.
7. Incubate wells with PBS/0.1 % FCS containing the primary antibody at an appropriate dilution (anti-BamC at 1:500 dilution), for 1 h at 37 °C. Carefully wash three times with PBS.
8. Incubate wells with PBS/0.1 % FCS containing the secondary HRP-conjugated goat anti-rabbit secondary antibody at 1:2000 dilution for 1 h at 37 °C. Carefully wash three times with PBS.
9. To each well, add 100 μL of the ABTS substrate solution. Cover plates with foil immediately (*see* **Note 11**).
10. Using a plate reader, take absorbance measurements at $OD_{405\ nm}$ after 15 and 30 min. Analyze results accordingly.

3.3 Whole-Cell Co-Immuno-precipitation of Surface-Exposed Antigens

To confirm the surface-exposed BamC is integrated into the BAM complex, whole-cell co-immunoprecipitation (co-IP) experiments were performed [2, 16]. Again, for a negative control we used the BamC deletion strain in parallel (*see* **Note 12**). For a positive control, a more typical co-IP protocol is performed where the antibodies are added after cell permeabilization.

1. Inoculate a 50 mL LB culture of wild-type (BW25113) and Δ*bamC* (with 34 μg/mL kanamycin) strains and grow overnight, shaking, 37 °C at 220 rpm.

2. Harvest cells at 4000×g for 5 min at RT. Resuspend cells in 10 mL PBS. Centrifuge again at 4000×g for 5 min at RT.
3. Resuspend cell pellets in 2 mL PBS/2 % BSA and add 5 μL BamC primary antibody, 1:400 dilution. Gently spin for 1 h at RT.
4. Pellet cells at 1500×g for 5 min at room temperature and resuspend the cells in 2 mL PBS/2 % BSA. Wash two more times.
5. Resuspend cells in 10 mL Spheroplast buffer and gently roll for 90 min at room temperature or until spheroplasting has occurred.
6. Add Triton X-100 (1 % final concentration) and gently roll for 15 min at 4 °C.
7. Add 15 mM $MgCl_2$ and incubate for 5 min at 4 °C. Remove unlysed cells by centrifugation at 6000×g for 20 min at 4 °C. Place the cleared lysates into a new tube.
8. For the positive control co-IP, add the primary antibody here for 1 h at room temperature.
9. Aliquot out 40–60 μL Agarose A/G Plus beads per tube into new Eppendorf tubes and add 0.5 mL IP buffer to each, tap to mix.
10. Pellet beads at 3000×g for 1 min at RT. Remove supernatant and add the cleared lysates containing bound antibodies mix to the tubes of washed beads. Incubate for 1 h spinning gently at room temperature.
11. Pellet the bound beads at 3000×g for 1 min at room temperature. Wash with 0.5 mL IP buffer and pellet beads again at 3000×g for 1 min at room temperature. Repeat washes two more times followed by a last wash with 1 mL PBS.
12. Elute proteins by adding 50 μL 1× SDS-loading buffer to beads and boil for 5 min. Centrifuge samples at 3000×g for 1 min at room temperature to pellet beads and place the supernatant in new tubes (*see* **Note 13**).
13. Load ~10 μL of each sample on SDS-PAGE minigel for Western blot analysis. We probed the membrane with anti-BamC as well as other components of the BAM complex and anti-SurA as a negative control.

3.4 Protease Shaving

BamC is resistant to any form of protease shaving in wild-type cells [2]. We reasoned that this was because it is in a stable protease-resistant form when integrated into the BAM complex. To test this we overexpressed BamC to create a population of BamC that would not be incorporated into the BAM complex, and repeated the protease shaving experiments (*see* **Note 14**). In this way we were able to capture stable fragments of BamC that could be cleaved from the cell surface.

1. Inoculate a 5 mL LB culture (with 100 μg/mL ampicillin) with a fresh colony of pBamC_H_6 transformed into BL21(DE3) Star cells and grow overnight, 37 °C at 220 rpm.
2. The next day, dilute the overnight culture 1:100 into 40 mL LB (with 100 μg/mL ampicillin) to an OD_{600} = 0.1. Incubate at 37 °C, 220 rpm until the OD_{600} = 0.5 and then induce with 0.2 mM IPTG. Grow for a further 3 h.
3. Centrifuge cells 3000 × *g* for 5 min at 4°C. Remove the supernatant and resuspend the pellet in 10 mL PBS.
4. Centrifuge cells 3000 × *g* for 5 min at 4°C. Resuspend cells in 900 μL PBS and keep on ice.
5. For each strain, aliquot 90 μL of cells into four Eppendorf tubes (on ice) labeled A, B, C, and D.
6. Add 10 μL polymixin B (20 mg/mL) to tubes B and D and 10 μL PBS to tubes A and C. Incubate for 10 min on ice.
7. Then add 1 μL trypsin (10 mg/mL, *see* **Note 15**) to tubes C and D and incubate for 20 min on ice (*see* **Note 16**).
8. To stop the reaction, add trypsin inhibitor (1 μL of 50 mg/mL stock) to tubes C and D.
9. Take 10 μL of each sample and add 90 μL 1× SDS-loading dye. Heat for 3 min at >95 °C.
10. Load samples onto an SDS-PAGE gel (10–20 μL) for Coomassie staining and/or Western transfer (*see* **Notes 17** and **18**).
11. If possible, isolate the bands of interest for N-terminal sequencing to identify each fragment.

4 Notes

1. You can also purchase the IP buffer directly from Pierce premade.
2. By adding the SDS last it should dissolve in a timely fashion. Also remember to wear a mask when weighing out SDS.
3. A BamC deletion strain was used as a negative control and permeabilizing the cells as a positive control.
4. The amount of antibody used will need to be optimized for your experiments. Typically, we take the dilution used for a western and increase it a 1000-fold. For example the anti-BamC antibody is used at 1:20,000 for a western so we use it at 1:2000 for IF experiments.
5. If you add more than one antibody, make sure they are from different hosts to ensure the secondary antibodies can recognize the individual proteins. For example, do not use two antibodies made in a rabbit.

6. The secondary antibodies lose fluorescence over time in the presence of light so make sure all samples are not exposed by wrapping samples in foil.
7. The amount of Mowiol solution required is just enough to fill the entire coverslip surface area.
8. Fluorescence should last for a week or so but for best results image them as soon as possible.
9. The Δ*bamC* cells tend not to pellet as well as wild type and we therefore used an angled rotor to get the best pelleting result.
10. When planning your experiment, allow at least three or four wells per experiment including four wells that do not contain any cells as a negative control.
11. You will see the wells go bright green over time if you have surface exposure of your antigen.
12. An additional negative control is to use the preimmune serum from the primary antibody bleeds if you have them. If not, you can check for non-specific binding of proteins to the Protein A/G beads by setting up a second falcon tube of your sample but without any antibody added. An additional positive control could be cells containing your protein overexpressed.
13. At this stage you can freeze the samples at −20 °C if necessary.
14. We found trypsin to be the most useful but it is wise to try a range of proteases to see which ones can create stable fragments of your protein of interest.
15. Always use a new tube to ensure efficiency of the protease as freeze/thaw cycles eventually reduced the enzyme's activity.
16. The length of time, temperature, and amount of protease may need to be optimized depending on the protein of interest.
17. Depending on whether the protein of interest is at native levels or is overexpressed will determine how much sample to load onto the SDS-PAGE gel.
18. Always run high percentage gels to make sure you see all the small fragments after proteolysis.

Acknowledgements

We acknowledge support from the National Health and Medical Research Council (NHMRC) for research funding through a NHMRC Program Grant 606788 and the Australian Research Council FL130100038. (to T.L.). C.T.W. is an ARC Laureate Fellowship Postdoctoral Research Associate and T.L. is an ARC Australian Laureate Fellow.

References

1. Rigel NW, Ricci DP, Silhavy TJ (2013) Conformation-specific labeling of BamA and suppressor analysis suggest a cyclic mechanism for beta-barrel assembly in *Escherichia coli*. Proc Natl Acad Sci U S A 110(13):5151–5156
2. Webb CT, Selkrig J, Perry AJ et al (2012) Dynamic association of BAM complex modules includes surface exposure of the lipoprotein BamC. J Mol Biol 422(4):545–555
3. Kim KH, Aulakh S, Paetzel M (2011) Crystal structure of beta-barrel assembly machinery BamCD protein complex. J Biol Chem 286(45):39116–39121
4. Knowles TJ, McClelland DM, Rajesh S et al (2009) Secondary structure and (1)H, (13)C and (15)N backbone resonance assignments of BamC, a component of the outer membrane protein assembly machinery in *Escherichia coli*. Biomol NMR Assign 3(2):203–206
5. Kim KH, Aulakh S, Tan W et al (2011) Crystallographic analysis of the C-terminal domain of the *Escherichia coli* lipoprotein BamC. Acta Crystallogr Sect F: Struct Biol Cryst Commun 67(Pt 11):1350–1358
6. Warner LR, Varga K, Lange OF et al (2011) Structure of the BamC two-domain protein obtained by Rosetta with a limited NMR data set. J Mol Biol 411(1):83–95
7. Pajon R, Yero D, Niebla O et al (2009) Identification of new meningococcal serogroup B surface antigens through a systematic analysis of *Neisserial* genomes. Vaccine 28(2):532–541
8. Cowles CE, Li Y, Semmelhack MF et al (2011) The free and bound forms of Lpp occupy distinct subcellular locations in *Escherichia coli*. Mol Microbiol 79(5):1168–1181
9. Dong C, Beis K, Nesper J et al (2006) Wza the translocon for *E. coli* capsular polysaccharides defines a new class of membrane protein. Nature 444(7116):226–229
10. Epstein EA, Reizian MA, Chapman MR (2009) Spatial clustering of the curlin secretion lipoprotein requires curli fiber assembly. J Bacteriol 191(2):608–615
11. Manning PA, Beutin L, Achtman M (1980) Outer membrane of *Escherichia coli*: properties of the F sex factor traT protein which is involved in surface exclusion. J Bacteriol 142(1):285–294
12. Battisti JM, Hicks LD, Minnick MF (2011) A unique *Coxiella burnetii* lipoprotein involved in metal binding (LimB). Microbiology 157(Pt 4): 966–976
13. Sambrook J, Russell DW (2001) Molecular cloning: a laboratory manual, vol v. 1. Cold Spring Harbor Laboratory Press, New York, NY
14. Baba T, Ara T, Hasegawa M et al (2006) Construction of *Escherichia coli* K-12 in-frame, single-gene knockout mutants: the Keio collection. Mol Syst Biol 2(2006):0008
15. Konieczny MP, Suhr M, Noll A et al (2000) Cell surface presentation of recombinant (poly-) peptides including functional T-cell epitopes by the AIDA autotransporter system. FEMS Immunol Med Microbiol 27(4): 321–332
16. Leyton DL, Sevastsyanovich YR, Browning DF et al (2011) Size and conformation limits to secretion of disulfide-bonded loops in autotransporter proteins. J Biol Chem 286(49): 42283–42291

Chapter 18

Construction and Characterization of an *E. coli bamD* Depletion Strain

Dante P. Ricci

Abstract

The central Bam components BamA and BamD are both essential genes in *E. coli*, a fact that often confounds genetic analysis using classical methods. The isolation of "depletion strains" in which these genes can be conditionally expressed removes this obstacle and facilitates the in vivo characterization of Bam function. This chapter describes an efficient two-step recombineering method for the construction of such a depletion strain, which contains an arabinose-inducible allele of *bamD*, using the λ Red system. Additionally, a simple protocol is presented for the depletion of *bamD* expression in live cells, which is particularly useful for the characterization of mutant alleles of *bamD* (complementation analysis). In principle, the procedures described can be adapted to produce and characterize depletion strains for any essential gene in *E. coli* or any other bacterium that is similarly amenable to genome engineering.

Key words Beta-barrel membrane proteins, BamD, BAM complex, Complementation analysis, *E. coli*, BamA

1 Introduction

Genetic analysis of essential biological processes is complicated by the fact that loss-of-function mutations in essential genes necessarily restrict cell growth and/or viability. The isolation of conditional alleles, which permit the experimental manipulation of the synthesis, stability, and/or function of a gene product, is therefore imperative for the geneticist and a matter of tremendous practical importance for any experimentalist wishing to study an essential process in a living cell. Prior to the development of modern genetic techniques that enable custom engineering of microbial genes and genomes, conditional mutants most often manifested as temperature-sensitive lethality or as nonsense mutants of phages that could be conditionally propagated in host cells bearing amber suppressor mutations [1]. The advent of gene fusion technology and powerful genome modification methods have enabled the construction of conditional alleles for any bacterial gene through the coupling of an essential gene of

Susan K. Buchanan and Nicholas Noinaj (eds.), *The BAM Complex: Methods and Protocols*, Methods in Molecular Biology, vol. 1329, DOI 10.1007/978-1-4939-2871-2_18, © Springer Science+Business Media New York 2015

interest to a tightly regulated inducible promoter whose activity can be readily modified in the laboratory.

Because the components of the BAM complex that are essential for β-barrel assembly (BamA and BamD) are also essential for viability in *E. coli*, much genetic characterization of Bam function over the past decade has made use of conditional "depletion" alleles of *bamA* and *bamD* (constructed in the Silhavy laboratory) that allow the experimentalist to inhibit expression of either gene through the removal of a chemical inducer from the growth medium [2–6]. Here, I describe an efficient process for building an *E. coli bamD* depletion strain and provide a generic protocol for experimental depletion of BamD activity in cell culture. While this procedure is specific to *bamD*, the general approach could be adapted to depletion strain construction for *bamA*, or, in principle, any other essential gene in *E. coli*.

The original *bamA* and *bamD* depletion strains were constructed using the λInCh ("lambda inch") method developed in the Beckwith laboratory [2, 3]. This technique enables the stable integration of genes carried on certain pBR322- or pUC-based vectors at the chromosomal attachment site for the genome of bacteriophage lambda (*attB*, situated between the *gal* and *bio* loci) through sequential site-specific integration and homologous recombination events using specialized lambda lysogens [7]. While this method is exceedingly clever and relatively simple, it has largely been supplanted by another technique (also based on phage-mediated recombination) that is not only faster but also far more versatile than λInCh as it places fewer constraints on the site of chromosomal integration or the sequence of the DNA to be integrated. This system, typically referred to as λ Red, co-opts the functions of three phage genes—*exo*, *bet*, and *gam*—which serve as effectors of homologous recombination between horizontally acquired DNA and the bacterial genome [8]. Using engineered *E. coli* strains in which the λ Red genes can be experimentally induced, linear dsDNA sequences containing short (~50 bp) stretches of homology to a chromosomal locus at each end can be introduced by transformation and efficiently recombined into the genome, thereby enabling the one-step generation of insertions, deletions, and targeted mutations of various sizes at any position in the bacterial genome [8–10]. In light of the numerous advantages afforded by the λ Red system, I will discuss the recombineering-based construction of a *bamD* depletion strain.

I define a depletion strain as a mutant that contains a single copy of the gene of interest, the expression of which can be modulated—minimally between "ON" and "OFF" states—using an inducer. Several independent strain construction pathways leading to such a strain can be imagined, some of which are briefly considered in Subheading 4. Here, I describe a detailed method for creating a genetically stable strain containing a single, arabinose-regulated *bamD* allele at the native *araBAD* promoter in two

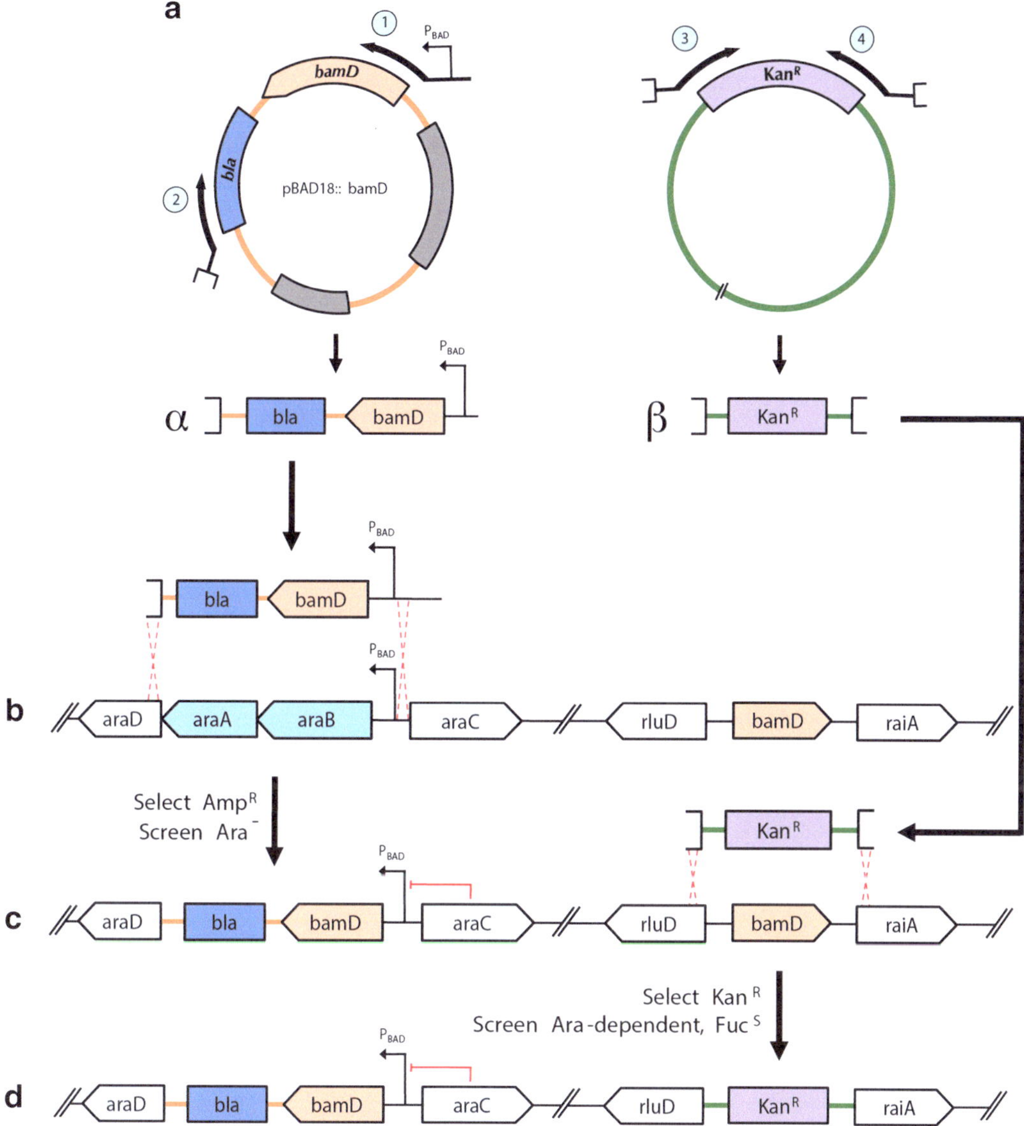

Fig. 1 General scheme for two-step construction of a *bamD* depletion strain. (**a**) The dsDNA constructs that serve as substrates for recombineering (labeled α and β) are, respectively, amplified from pBAD18::*bamD* (*left*) and from any vector or strain that serves as a source of the kanamycin cassette (*right*) using primers (numbered as in Table 1) that add ~50 bp of sequence homology to each end of the PCR product. (**b**) In the first step, two of the three *araBAD* genes are replaced with *bamD-bla* to generate a chromosomal transcriptional fusion of *bamD* to the *araBAD* promoter (P_{BAD}). The resulting strain then serves as the host for a second round of recombineering (**c**) in which the native *bamD* allele is replaced with a drug resistance cassette (in this case, Kan^R). The product of the second reaction (**d**) is the final *bamD* depletion strain, which contains a single copy of *bamD* that is regulated by the arabinose-inducible P_{BAD} promoter

successive recombination steps using recombineering (Fig. 1) (*see* **Note 1**). In the first step, a copy of *bamD* is placed under control of P*araBAD* at the *ara* locus, producing a strain with two copies of *bamD* (Subheading 3.2). The native *bamD* allele in this merodiploid intermediate is then replaced with a selectable marker in the second step (Subheading 3.3), yielding a haploid, conditional *bamD* strain. Note that the two recombination steps must necessarily proceed in the order in which they are presented in Subheading 3; attempting to replace the native *bamD* allele without first installing an inducible module elsewhere in the chromosome will produce an inviable Δ*bamD* mutant.

A stable *bamD* depletion strain can be a remarkably useful tool, allowing overexpression and partial repression of *bamD* in addition to tight inhibition, each of which can be informative. In addition, since the depletion allele is situated within the *araBAD* operon, mutant alleles of *bamD* can be easily transferred to the depletion strain for complementation and further characterization. To complement the strain construction protocol presented in Subheadings 3.1–3.3, I present in Subheading 3.4 a simple generic depletion and complementation protocol that can be used to study the physiological consequences of the complete block in the β-barrel assembly pathway that results from a loss of BamD activity.

2 Materials

2.1 Preparation of Linear dsDNA Recombineering Substrates

1. Phusion High-Fidelity PCR Kit (Thermo Scientific) including Phusion DNA polymerase, 10 mM dNTP mix, 5× Phusion buffer, 50 mM $MgCl_2$ solution, and dimethylsulfoxide (DMSO) (*see* **Note 2**).
2. Betaine (dissolved in dH_2O to final concentration of 5 M).
3. Oligonucleotide primers (see Table 1).
4. Agarose gel (1 %): Dissolve gel-strength, DNase-free agarose in microwave in 1× Tris–borate–EDTA (or preferred buffer)

Table 1
PCR primers and suggested templates for amplification of dsDNA recombineering substrates

PCR product	PCR template(s)	Oligo sequences[a]
Δ*araBA*::*bamD*-*bla* (ampicillin resistance)	**pBAD18**::*bamD* [3]	#1:**ACCTGACGCTTTTTATCGCAACTCTCTACTGTTTC TCCATACCCGTTTTT**TTGAGGAAAGTCAAAACGTCATG
		#2:**TAATACCTGGCGTTTGAGATCTTCTAACATGTTG ACTCCTTCGTGCCGGA**AGAGTTGGTAGCTCTTGATC
Δ*bamD*::*neo* (kanamycin resistance)	**pDUAL** (stratagene) **Tn5**	#3:**GTGTGCTATTGTAGCTGGTCTTAACCGGGAGCAG GAACAGAGAATCTCCC**TATGGACAGCAAGCGAACCG
		#4:**TGTTATGTATTGCTGCTGTTTGCGGCGATGATTT TCGCTACTTTTTCAGC**TCAGAAGAACTCGTCAAGAAG

Depending on the specific construct desired, constructs are described that allow replacement of the *bamD* gene with markers encoding resistance to kanamycin, chloramphenicol, or tetracycline
[a]Sequences necessary for homologous recombination are indicated in boldface

at a final concentration of 1 % (wt/vol) and add ethidium bromide to a final concentration of 0.5 μg/mL. Cast while molten into a gel tray and let polymerize with well comb(s) in place.

5. GeneJET Gel Extraction Kit (Thermo Scientific).

2.2 λ-Red-Mediated Recombineering

1. *E. coli* K-12 *ara*[+] strain containing a plasmid or defective prophage expressing the λ Red recombination system (*see* **Note 3**).
2. M63 liquid minimal medium base (5× stock): Dissolve 10 g $(NH_4)_2SO_4$, 68 g KH_2PO_4, and 2.5 mg $FeSO_4 \bullet 7H_2O$ in 1 L dH_2O. Adjust pH to 7.0 with KOH and autoclave to sterilize.
3. M63 agar plates (1.5 %): Combine 200 mL of 5× M63 stock with 15 g Bacto-agar, bring up to 1 L with dH_2O, and autoclave. Once cooled, add 1 mL 1 M $MgSO_4 \cdot 7H_2O$, 0.1 mL 0.5 % vitamin B1 (thiamine), and 10 mL 20 % carbon source (either glucose or arabinose; final concentration = 0.2 %).
4. Luria-Bertani (LB) broth: 10 g NaCl, 10 g tryptone, 5 g yeast extract, 1 mL of 1 N NaOH per liter of dH_2O, autoclave.
5. LB agar (1.5 %): Add 15 g Bacto-agar to 1 L of LB broth and autoclave. Pour into individual 100 × 15 mm petri plates (30 mL molten agar per plate) and allow to cool.
6. LB broth and agar plates containing antibiotic (depending on the drug cassette used): Prepare broth and agar as described above, but add antibiotic to the final concentrations appropriate for the antibiotic used (ampicillin, 25 μg/mL; chloramphenicol, 20 μg/mL; tetracycline, 12.5 μg/mL; kanamycin, 25 μg/mL). For LB agar, add antibiotic only after molten agar has cooled to 50 °C.
7. Incubator, 30–32 °C.
8. Incubator (37 °C) for incubation of agar plates and rotation/shaking of liquid cultures.
9. Shaking water bath (42 °C).
10. Erlenmeyer flasks, 125 and 250 mL.
11. Glass metal-capped culture tubes.
12. Refrigerated low-speed centrifuge with Sorvall SA-600 rotor (or equivalent).
13. Plastic centrifuge tubes (35–50 mL), chilled on ice.
14. Electroporation cuvettes (0.1 cm) (Bio-Rad), chilled on ice.
15. Electroporator (e.g., Bio-Rad GenePulser).

2.3 Depletion of bamD Expression

1. Inducer (carbohydrate) solutions: L-Arabinose (20 % vol/vol in distilled water), D-fucose (15 % vol/vol in distilled water), D-glucose (20 % vol/vol in distilled water).

2. LB agar plates and prewarmed liquid medium containing 0.2 % L-arabinose (and appropriate antibiotic as needed).
3. Prewarmed LB liquid medium containing 0.15 % D-fucose (*see* **Note 4**).

3 Methods

3.1 Preparation of Linear dsDNA Recombineering Substrates

This protocol describes the use of the pBAD18::*bamD* vector [3] as a template for PCR amplification of *bamD* (together with a β-lactamase-encoding *bla* cassette), the product of which is incorporated into the chromosome using the recombineering method introduced above. Linking a drug resistance gene to *bamD* not only eliminates the need for exhaustive screening of transformants, but also provides a convenient marker for the selection of phage transductants or exconjugants carrying the *bamD* depletion construct, thereby enabling facile and efficient horizontal transfer of this allele once the final strain has been constructed (*see* **Notes 5** and **6**).

1. Using oligonucleotide primers #1 and #2 (Table 1) and the pBAD18::*bamD* plasmid as template, amplify the region containing the *bamD* Shine-Dalgarno sequence and ORF along the *bla* cassette by PCR (*see* **Notes 7** and **8**).
2. Using primers #3 and #4 (which add homology to the sequences immediately flanking the *bamD* ORF) and either a plasmid conferring kanamycin resistance (e.g., pDUAL) or a Tn5-containing strain (Table 1), amplify the kanamycin resistance cassette by PCR (*see* **Note 9**).
3. Visualize both PCR products on a 1 % agarose gel and purify by gel extraction (*see* **Note 10**).

3.2 Replacement of araBA with bamD (First Recombineering Step)

The first recombination step involves the complete replacement of *araA* and *araB* (the products of which catalyze the first and second steps in the arabinose metabolism pathway, respectively) with the ribosome-binding site (RBS) and ORF of *bamD*. The deletion of *araBA* is intentional—a depletion strain that cannot "eat the inducer" (i.e., metabolize arabinose) has several useful properties, including stable, gratuitous induction over long growth periods, a higher induction ratio, and linear response kinetics over a broad range of inducer concentrations [11]. Additionally, the conversion to Ara^- that results from the integration of *bamD* downstream of P*araBAD* begets a convenient phenotype for use in a secondary screen for correct recombinants (*see* **step 14**).

1. Inoculate 5 mL of LB medium with a single colony of the λ Red recombineering strain and grow overnight with shaking at 30 °C (*see* **Note 11**).

2. Add 250 μL of the overnight culture to 25 mL of LB medium in a 250-mL Erlenmeyer flask (preferably baffled to increase aeration) and incubate in a 30 °C shaking water bath to an optical density (OD_{600}) of 0.4–0.6 (*see* **Note 12**).
3. When the culture has reached the appropriate OD_{600}, transfer half (~12.5 mL) to a 125-mL Erlenmeyer flask. Incubate the 125-mL flask in a 42 °C shaking water bath for 15 min to induce the λ Red genes (*see* **Note 13**) and continue to incubate the remaining half of the culture at 30 °C (uninduced control).
4. When the 15-min induction period has ended, immediately cool the induced culture and the uninduced control by swirling both flasks continuously for 10 min in an ice-water slurry (*see* **Note 14**).
5. Transfer the induced and uninduced cultures to pre-chilled centrifuge tubes that are compatible with the rotor used and large enough to accommodate each culture (35–50 mL). Spin for 10 min at 5,000 × *g* with refrigeration (4 °C). Discard the supernatant.
6. Gently resuspend each pellet in 25 mL ice-cold distilled water (dH_2O) by pipetting up and down (avoid more vigorous techniques such as vortexing) and repeat the spin described in **step 5**. Carefully decant and discard the supernatant (*see* **Note 15**).
7. Gently resuspend each pellet in 1 mL ice-cold dH_2O and transfer to pre-chilled microcentrifuge tubes. Spin the tubes in a microcentrifuge in the cold at maximum speed for 1 min. Carefully decant or aspirate the supernatant.
8. Gently resuspend each pellet in 200 μL ice-cold dH_2O (*see* **Note 16**) and immediately place on ice.
9. Chill three 0.1-cm electroporation cuvettes and three microcentrifuge tubes on ice.
10. Mix purified PCR product obtained in Subheading 3.1 with 50–100 μL of the suspension of induced and uninduced cells to prepare three reactions as follows:

    Reaction 1: Induced recombineering strain + dsDNA. Mix 100–1000 ng purified PCR product (*see* **Note 17**) with 50–100 μL of electrocompetent cells prepared from the induced culture in a chilled microcentrifuge tube. This is the experimental condition and, if successful, should yield ampicillin-resistant transformants.

    Reaction 2: Uninduced recombineering strain + dsDNA. Mix 100–1000 ng purified PCR product with 50–100 μL of electrocompetent cells prepared from the uninduced culture in a chilled microcentrifuge tube. This control is useful for estimating background transformation to ampicillin resistance (i.e., false positives).

Reaction 3: Induced recombineering strain only (no DNA). Add 50–100 μL of electrocompetent cells prepared from the induced culture to a chilled microcentrifuge tube. No ampicillin-resistant colonies should arise from this reaction; the appearance of ampicillin-resistant transformants implies the presence of contaminating DNA or bacteria in the competent cell preparation.

11. Transfer each reaction mixture to a pre-chilled electroporation cuvette from **step 9** and electroporate each with the pulse controller set to 1.8 kV (*see* **Notes 18** and **19**).
12. After pulsing each mixture in the electroporator, add 1 mL pre-warmed LB medium (without antibiotic) directly to the cuvette and mix by pipetting up and down. Transfer the cell suspension to a microcentrifuge tube and incubate with shaking at 30 °C for 1–2 h (recovery period) to allow sufficient expression of the *bla* gene.
13. Prepare two tenfold dilutions of each cell suspension by combining 100 μL of the next highest dilution with 900 μL of LB medium (the end result will be three concentrations per transformation: 10^0, 10^{-1}, and 10^{-2}). Spread 200 μL from each dilution for each transformation (nine samples altogether) onto LB agar plates containing 25 μg/mL ampicillin using a sterile spreader or autoclaved glass beads. Incubate the plates overnight, inverted, at 30 °C.
14. When colonies arise (2–4 days), screen for correct recombinants by streaking individual transformants for single colonies on LB agar plates containing ampicillin and M63 agar plates containing ampicillin and 0.2 % arabinose (*see* **Note 20**). Confirm proper integration of *bamD-bla* in Amp^R, Ara^- transformants by PCR (*see* **Note 21**) and DNA sequencing.
15. Prepare a liquid culture of a successful, sequence-verified recombinant in LB medium with ampicillin (and additional antibiotic if the λ Red genes are present on a plasmid in this strain) and grow overnight at 30 °C (*see* **Note 22**). Use the culture to prepare a glycerol stock of this Δ*araB*::*bamD-bla* strain for long-term storage and proceed to Subheading 3.3.

3.3 Replacement of bamD with a Drug Resistance Cassette (Second Recombineering Step)

The procedure by which the native *bamD* allele is replaced with a drug resistance marker is in many ways identical to that used to replace *araBA* with *bamD-bla* as described in Subheading 3.2. Indeed, all steps listed in the preceding section can be essentially repeated (with the modifications below) to complete construction of the *bamD* depletion strain.

1. Prepare electrocompetent cells as described in **steps 1–8** of Subheading 3.2 using the Δ*araB*::*bamD-bla* derivative of the λ Red recombineering strain constructed in Subheading 3.2.

2. Transform the Δ*araB*::*bamD-bla* λ Red strain as described in **steps 9–12** of Subheading 3.2 using the selectable marker amplified in Subheading 3.1.
3. Plate dilutions of the cell suspensions as described in **step 13** of Subheading 3.2, but perform the selection on LB agar plates containing 0.2 % arabinose and the antibiotic appropriate for the resistance cassette to be integrated at the *bamD* locus (*see* **Note 23**).
4. When colonies arise, screen for correct recombinants by streaking individual transformants for single colonies on LB agar plates containing either 0.2 % arabinose (and the appropriate antibiotic) or 0.15 % fucose (*see* **Note 4**). Confirm deletion of the endogenous *bamD* allele in fucose-sensitive (Fuc^S), arabinose-dependent transformants by PCR-amplifying the *bamD* locus (*see* **Note 18**).
5. Successful recombinants carry the genotype of a *bamD* depletion strain and can be cured of λ Red function prior to further experimentation (*see* **Note 24**).

3.4 In Vivo Depletion of BamD and Complementation Analysis

The ability to experimentally manipulate BamD levels in vivo allows direct observation of the myriad physiological effects associated with the wholesale arrest of β-barrel assembly. In addition, the *bamD* depletion strain is a suitable background for the characterization of mutant *bamD* alleles provided in *trans* or transferred to the native chromosomal locus. This property is particularly useful for studying partial or complete loss-of-function alleles of *bamD*, since the ability to conditionally express wild-type *bamD* ensures that a depletion strain carrying a defective or nonfunctional variant of BamD can be stably maintained.

The protocol described below enables the complete inhibition of *bamD* expression in the depletion strain during growth in liquid culture. This can be performed in the *bamD* haploid depletion strain constructed in Subheading 3.3 (in which case the terminal phenotype will be cell death) or in the presence of a complementing allele, in which case the phenotype(s) associated with this additional copy of *bamD* can be determined following inhibition of P*araBAD*-dependent *bamD* expression (*see* **Notes 25** and **26**).

1. In separate tubes, inoculate 5 mL of LB medium containing 0.2 % arabinose (and antibiotic(s), where appropriate) with a single colony of (a) the *bamD* depletion strain constructed in Subheading 3.3, (b) the *bamD* merodiploid strain constructed in Subheading 3.2, which serves as a useful control, and, if desired, (c) a derivative of the depletion strain expressing a second copy of *bamD* in *trans*. Grow to saturation overnight at 37 °C.

2. Pellet 1 mL of the overnight culture(s) in 1.7-mL microcentrifuge tubes by spinning in a benchtop centrifuge at 10,000 × *g* for 1 min.
3. Gently resuspend the pellet(s) in 1 mL LB by pipetting up and down.
4. Repeat **steps 2** and **3** once.
5. For each strain, add 250 μL of the washed and resuspended cell culture to each of the two flasks with 25 mL of prewarmed LB, one containing 0.2 % arabinose and the other containing 0.15 % fucose (*see* **Note 27**). Incubate these back-diluted samples with shaking at 37 °C and record the optical density (OD_{600}) at least once every 30 min (*see* **Notes 28–30**). Collect samples at each time point for downstream analysis (*see* **Note 31**).
6. When the optical density of a culture reaches approximately 0.5, transfer 250 μL of that culture to a new flask containing 25 mL of the same medium (i.e., LB + arabinose or LB + fucose) and incubate the diluted culture with shaking at 37 °C.
7. Repeat **step 6** for each culture for the desired length of time, or until growth of the culture in the repressed (+fucose) condition arrests (*see* **Note 32**).

4 Notes

1. It is possible in principle to construct a *bamD* depletion strain in a single step by using recombineering to replace the native *bamD* promoter with a copy of the *araBAD* promoter that has been artificially linked to an antibiotic resistance cassette. However, because wild-type repression of the *araBAD* promoter requires distant *cis*-acting upstream sequences [12], and because the efficiency of the recombineering method drops dramatically as the length of the transformed dsDNA product increases [13], the dsDNA product required to introduce both a selectable marker and the complete *ara* control region at the *bamD* locus would potentially be too long to allow efficient recombineering using the λ Red method. I encourage the interested reader to explore alternative approaches that permit the chromosomal integration of large constructs [13] if one-step replacement of the native *bamD* promoter is desired.
2. Although many commercially available DNA polymerases may suffice for the PCR amplification step, the use of a high-fidelity polymerase is recommended in order to avoid unintentional mutagenesis of the *bamD* coding region.
3. The method by which *bamD* depletion strains are constructed relies on the use of published strains that express the λ-Red-based homologous recombination system. Suitable strains (and their

Table 2
Common λ Red recombineering strains

Strain	Genotype
W3110	Wild-type *E. coli* K-12; F-, λ-
DH10B	F- *endA1 recA1 galE15 galK16 nupG rpsL* Δ*lacX74* Φ80*lacZ*ΔM15 *araD139* Δ(*ara,leu*)7697 *mcrA* Δ(*mrr-hsdRMS-mcrBC*)
DY330	W3110 Δ*lacU169 gal490 pgl*Δ8 λ*c*I857Δ(*cro-bioA*)
DY331[a]	W3110 Δ*lacU169 srlA::Tn10*Δ*recA gal490 pgl*Δ8 λ*c*I857Δ(*cro-bioA*) (Tet^R)
DY378	W3110 λ*c*I857 Δ(*cro-bioA*)
DY380[b]	DH10B λ(*c*I857*ind1*) Δ{(*cro-bioA*)::*tetRA*} (Tet^R)

[a]The *nadA*::Tn*10* insertion is genetically linked to the lambda prophage and can serve as a marker for the convenient transfer of λ Red functions by generalized transduction

[b]As with DY331, the tetracycline resistance cassette that lies adjacent to the lambda prophage in this strain can be used for efficient P1 transduction of the prophage itself, which is 100 % linked to *tetRA* in this case

sources) are listed in Table 2. If desired, it is also possible to endow a strain of interest with λ Red gene function by introducing a plasmid or defective prophage that carries the λ Red system (*see* footnotes to Table 2), so long as the strain minimally contains *araC*, *araB*, and the intergenic region between them (which contains the *araBAD* promoter), and so long as the drug resistance cassette found on the plasmid is orthogonal to the cassette used for selection of recombinants. The genotypes of the common laboratory strains W3110 and MG1655 are appropriate acceptors for this purpose. Strains that are Ara$^-$ by virtue of a loss-of-function mutation in *araD* (e.g., TOP10, DH10B, MC1061, or MC4100) should not be used in cases where the λ Red genes are controlled by the P*araBAD* promoter (e.g., pKD46-based), as arabinose is toxic to such mutants [14]. However, strains of this genotype can be easily converted to arabinose resistance by selecting Ara$^+$ revertants on minimal medium that contains arabinose as the sole carbon source.

4. D-fucose competitively binds the AraC transcriptional regulator of the *araBAD* operon, leading to tight repression of P*araBAD* [15]. Early *bamD* depletion studies exploited glucose-mediated catabolite repression to minimize leaky *araBAD* expression [3]; while glucose is a viable alternative, the use of fucose, either alone or in combination with glucose [16], is recommended for tight, rapid inhibition of *bamD* expression.
5. There are alternative methods by which a linear dsDNA construct containing both *bamD* and a drug resistance cassette can be generated without the need for pBAD18::*bamD*.

For example, two separate PCR products encoding *bamD* and a drug resistance gene can be joined by Gibson isothermal assembly [17], overlap-extension PCR [18], BioBrick assembly [19], circular polymerase extension cloning [20], traditional restriction/ligation cloning using engineered restriction sites, or any number of other DNA assembly methods. Alternatively, the *bamD* ORF and Shine-Dalgarno (SD) sequence can first be introduced into the multiple cloning site (MCS) of a vector containing a selectable marker downstream of the MCS, and then the marker can be amplified together with *bamD* with flanking homology to the *ara* locus as described for pBAD18::*bamD* in Subheading 3.1.

6. Schemes for marker-less integration of *bamD* downstream of the native *araBAD* promoter by recombineering can be envisioned (example two-step method: replace *araBA* with a counterselectable marker such as *sacB*, which converts sucrose to a toxic compound, then target *sacB* for replacement with *bamD* dsDNA, and select sucrose-resistant transformants; example one-step method: transform a λ Red strain with dsDNA to replace *araBA* with the *bamD* ORF, plate transformants without selection on tetrazolium arabinose [TA] plates, and screen for red colonies). The reader may choose to pursue such a course depending on the desired properties of the final depletion strain.
7. The primers listed add sequences on either end of the PCR product that contain homology to P*araBAD* (immediately upstream of *araB*) and to the 5′ end of *araD* (immediately downstream of *araA*).
8. The addition of DMSO and betaine to the PCR reaction (final concentrations = 5 % DMSO and 1 M betaine) is recommended, as this often increases product yield and decreases nonspecific amplification.
9. The Δ*bamD* insertion-deletion described here leads to the replacement of chromosomal *bamD* with the kanamycin resistance cassette (*neo*), but it is important to note that many other selectable markers can be used instead. Primers #3 and #4 (Table 1) can be modified to permit amplification of several unique antibiotic resistance cassettes from common cloning vectors or mobile elements by appending the homologous sequences required for integration at the *bamD* locus (shown in **boldface**) to the 5′ end of any resistance-cassette-specific oligonucleotides. Note that the *tetRA* cassette should not be used if the recombineering strain selected is already resistant to tetracycline (e.g., DY331 or DY380; *see* Table 2).
10. Gel extraction (as opposed to simple PCR cleanup) is strongly recommended in order to avoid contamination of the purified product with the pBAD18::*bamD* vector that was used as

template DNA in the PCR reaction. Since residual plasmid in the purified product can efficiently transform *E. coli* to ampicillin resistance, it is essential to purify the linear dsDNA product away from the plasmid template in order to limit the false-positive rate. To further decrease unwanted background, digesting the unpurified PCR product with DpnI prior to gel extraction is advised, as this promotes the selective degradation of methylated template DNA without affecting the unmethylated PCR product. The background transformation frequency, which reflects the presence of intact template plasmid, can be estimated by performing a control experiment wherein the purified PCR product is used to transform cells in which λ Red functions have not been induced.

11. For recombineering strains in which the λ Red genes are controlled by a temperature-sensitive allele of the gene encoding the lambda repressor (i.e., *cI857*), it is necessary to maintain growth at temperatures below 32 °C to limit expression of the lambda genes found within the defective prophage, which are toxic when continuously induced [9]. If the recombineering strain used carries the λ Red genes on a plasmid, the appropriate antibiotic should be added to the culture to maintain the plasmid during overnight growth.
12. The incubation time required for the culture to reach mid-exponential phase will vary depending on the precise strain background and dilution factor. It is critical to harvest cells during mid-exponential growth, as recombination functions are poorly expressed in stationary phase and the recombination efficiency will decrease sharply. If the strain used carries the λ Red genes under the control of an inducible promoter (e.g., P*araBAD* in the case of pKD46; *see* Ref. [21]), the inducer should be included in the growth medium and a separate culture lacking the inducer should be inoculated as an uninduced control. Additionally, the appropriate antibiotic(s) should also be included to maintain plasmids contained within the strain, as noted above.
13. **Step 3** of Subheading 3.2 can be omitted if λ Red gene expression is controlled by a chemical inducer (e.g., arabinose) rather than by a temperature shift.
14. Rapid cooling of a thermally induced recombineering strain is necessary in order to prevent continued synthesis of lambda genes, which can cause toxicity with constitutive induction.
15. The pellet may be relatively loose after this centrifugation step; take care to remove the supernatant without agitating the pellet and inadvertently discarding cells.
16. Although optimal recombination efficiencies are achieved using freshly prepared competent cells, these aliquots can be resuspended in 200 μL ice-cold 20 % glycerol (instead of

dH_2O), snap-frozen in liquid nitrogen, and stored at −80 °C for future use.

17. There is generally a positive relationship between DNA concentration and recombinant yield. Although the highest yields are observed with amounts of DNA greater than ~300 ng (for a 1 kb PCR product), acceptable recovery of ampicillin-resistant recombinants can be expected when using at least 100 ng of dsDNA [9].

18. 0.2-cm electroporation cuvettes may be used in lieu of 0.1-cm cuvettes, but the voltage setting should be increased to 2.5–3.0 kV in that case.

19. The time constant is a useful indicator of the pulse length and, consequently, the transformation efficiency. Optimal results are achieved at a time constant greater than or equal to 5 ms; a lower time constant may indicate problems that compromise electroporation efficiency such as high salt concentrations in the cell suspension.

20. Because the target strain contains *bamD* in place of *araBA*, which are essential for arabinose utilization, correct recombinants should be not only ampicillin resistant (Amp^R) but should also be unable to metabolize arabinose (Ara^-). The Ara^- character can be unambiguously determined by growing candidates on minimal medium containing arabinose as a sole carbon source (M63 + arabinose); the expected phenotype is an inability to grow on M63 + arabinose (note that minimal medium must be supplemented with biotin if a lysogenic recombineering strain such as DY378 is used). As an alternative, it is also acceptable to screen transformants for white colony color on rich MacConkey agar with arabinose or for red colony color on TA plates (*see* Subheading 1).

21. Although the transformed dsDNA product is designed to integrate specifically at the *ara* locus, improper targeting may be observed. For this reason, it is important to screen transformants for the correct genotype. This is accomplished by PCR-amplifying the *araBA* locus using primers that anneal just outside of the substituted region; the product of this reaction from a wild-type ($araBA^+$) strain will be approximately the size of *araBA* (~3.0 kb), whereas the product from successful recombinants will correspond to the complete *bamD*-*bla* construct (~2.2 kb). The appropriately sized PCR product can be further analyzed by Sanger sequencing of the ectopic *bamD* allele using sequencing primers that anneal to the *araBAD* promoter (oriented in the direction of transcription), the pBAD18 backbone (downstream of the MCS), or within the *bamD* ORF itself.

22. It is essential that products of the first recombination step be grown at 30 °C to prevent expression of λ Red genes

(or, in the case of plasmids such as pKD46, to prevent plasmid loss due to growth at the non-permissive temperature for plasmid replication). The λ Red functions must be retained, as a second recombineering step is required to complete construction of the depletion strain (Subheading 3.3).

23. The strain resulting from the intended recombination event contains a single allele of the *bamD* gene under control of the *araBAD* promoter and consequently requires arabinose for growth. It is essential, therefore, that arabinose be included in the selective medium to allow expression of *bamD*. Since fucose is a competitive inhibitor of *araBAD* expression, this strain should also be inviable in the presence of fucose, a phenotype that can be used to screen for correct recombinants.

24. There are several methods by which λ Red genes can be cured from a recombineering strain after the desired genotype is achieved. Certain plasmids containing λ Red genes (such as pKD46; *see* Ref. 21) can be easily cured by simply growing the strain on LB agar plates with 0.2 % arabinose at the non-permissive temperature for replication of these vectors (37 °C or higher). If the λ Red genes are carried chromosomally on a lambda prophage, the prophage itself can be deleted by λ-Red-mediated homologous recombination. Alternatively, since the lysogenic strains commonly employed in recombineering are biotin auxotrophs (*bio*$^-$) by virtue of a partial deletion of the *attB*-proximal *bio* locus, the prophage can be easily replaced with the "empty" λ$^-$ attachment site (*attB*$^+$) by raising a P1 lysate from a wild-type strain and transducing the *bamD* depletion strain to *bio*$^+$ by selecting for growth on minimal media (which lacks biotin) following transduction. It is also possible to separately transfer the Δ*araB*::*bamD-bla* and Δ*bamD*::Ω alleles horizontally via conjugation or P1 transduction, although this requires two sequential transduction steps.

25. A simple, qualitative assay to determine whether a mutant *bamD* allele complements *bamD* depletion can be performed by transforming or transducing the *bamD* depletion strain with a *bamD* mutant of interest, selecting transformants/transductants on medium containing arabinose, and streaking the resulting colonies on medium containing arabinose or fucose. BamD variants that permit arabinose-independent growth are therefore likely to be at least partially functional.

26. While complementation tests can be performed in the depletion strain described here, it may be desirable to first convert the depletion strain to *recA*$^-$, particularly when characterizing hypomorphic *bamD* alleles. This is because RecA-mediated homologous recombination between the two *bamD* alleles present in the merodiploid can produce *bamD*$^+$ revertants,

thereby restoring constitutive expression of the wild-type allele and confounding further analysis.

27. The culture grown in the permissive condition (0.2 % arabinose) serves as a useful control and should be conducted in parallel with all experiments performed using cells grown in the non-permissive condition (0.15 % fucose).

28. It is often useful to collect samples over the course of the experiment for downstream analysis (e.g., Western blotting). This can be done by collecting a 1 mL sample of the culture at each time point, pelleting the cells in microcentrifuge tubes by spinning in a benchtop centrifuge at 10,000 × *g* for 1 min, discarding the supernatant, and storing the pellets at −80 °C. The frozen pellets can be thawed and resuspended at a later point, normalized by OD_{600}, and processed for the desired application (e.g., adding an appropriate volume of SDS-PAGE sample buffer and boiling for 10 min to prepare the sample for SDS-PAGE).

29. The interval time can be shortened if higher resolution data is desired.

30. Raising the dilution factor may eliminate the need for multiple back dilutions over the course of the experiment; if the dilution factor is too high, however, growth of the cell population may arrest before sufficient material can be recovered for downstream applications (e.g., Western blotting or affinity purification of the BAM complex).

31. In addition to the cessation of growth, several other measurable effects that are characteristic of strains with limiting Bam activity can be observed following BamD depletion. These include the accumulation of unfolded outer membrane proteins in the periplasmic space, increased expression of the periplasmic protease DegP, and the inhibition of porin synthesis [22]; the latter two effects are stimulated by the former through activation of an alternate sigma factor (σ^E) that can be monitored using transcriptional fusions of reporter genes to promoters representing primary regulatory targets of σ^E (e.g., *rpoHP3*, *ycbK*, *ybfG*, or *degP* itself; [23]).

32. Depletion times can vary depending on the strain background and specific conditions used; growth of the original *bamD* depletion strain ceases within ~5–7 h following subculturing into medium lacking arabinose, but the addition of fucose/glucose as discussed here likely leads to faster, tighter inhibition. It has been shown previously that fucose/glucose can reduce ParaBAD expression to <1 % of induced levels within 10 min of addition, even in the presence of arabinose [16]. This may indeed be particularly important in an Ara^- background, where the inducer is not destroyed.

Acknowledgements

I wish to express my indebtedness to Marcelo Sousa (University of Colorado Boulder) for guidance and insightful comments on this manuscript, and to Tom Silhavy (Princeton University) and Natacha Ruiz (Ohio State University) for support and encouragement.

References

1. Edgar RS (1966) Conditional lethals. In Phage and the Origin of Molecular Biology, edited by J. Cairns, G. S. Stent and J. B. Watson. Cold Spring Harbor Laboratory Press, Cold Spring Harbor, NY pp. 166–170
2. Wu T, Malinverni J, Ruiz N et al (2005) Identification of a multicomponent complex required for outer membrane biogenesis in *Escherichia coli*. Cell 121:235–245
3. Malinverni JC, Werner J, Kim S et al (2006) YfiO stabilizes the YaeT complex and is essential for outer membrane protein assembly in *Escherichia coli*. Mol Microbiol 61:151–164
4. Misra R, Stikeleather R, Gabriele R (2014) *In vivo* roles of BamA, BamB and BamD in the biogenesis of BamA, a core protein of the β-barrel assembly machine of *Escherichia coli*. J Mol Biol. (in press)
5. Rossiter AE, Leyton DL, Tveen-Jensen K et al (2011) The essential BAM complex components BamD and BamA are required for autotransporter biogenesis. J Bacteriol 193(16): 4250–4253
6. Kim S, Malinverni JC, Sliz P et al (2007) Structure and function of an essential component of the outer membrane protein assembly machine. Science 317:961–964
7. Boyd D, Weiss DS, Chen JC et al (2000) Towards single-copy gene expression systems making gene cloning physiologically relevant: lambda InCh, a simple *Escherichia coli* plasmid-chromosome shuttle system. J Bacteriol 182:842–847
8. Court DL, Sawitzke JA, Thomason LC (2002) Genetic engineering using homologous recombination. Annu Rev Genet 36:361–388
9. Yu D, Ellis HM, Lee EC et al (2000) An efficient recombination system for chromosome engineering in Escherichia coli. Proc Natl Acad Sci USA 97:5978–5983
10. Thomason L, Court DL, Bubunenko M, et al. (2007) Recombineering: genetic engineering in bacteria using homologous recombination. Curr Protoc Mol Biol Chapter 1:Unit 1.16 (Editors: Frederick M, Ausubel Al)
11. Guzman LM, Belin D, Carson MJ et al (1995) Tight regulation, modulation, and high-level expression by vectors containing the arabinose PBAD promoter. J Bacteriol 177:4121–4130
12. Schleif R (2010) AraC protein, regulation of the l-arabinose operon in *Escherichia coli*, and the light switch mechanism of AraC action. FEMS Microbiol Rev 34:779–796
13. Kuhlman TE, Cox EC (2010) Site-specific chromosomal integration of large synthetic constructs. Nucleic Acids Res 38:e92
14. Sheppard DE, Englesberg E (1967) Further evidence for positive control of the L-arabinose system by gene araC. J Mol Biol 25:443–454
15. Wilcox G (1974) The interaction of L-arabinose and D-fucose with AraC protein. J Biol Chem 249:6892–6894
16. Brockmann B, Koop Genannt Hoppmann KD, Strahl H et al (2013) Time-delayed *in vivo* assembly of subunit a into preformed *Escherichia coli* FoF1 ATP synthase. J Bacteriol 195:4074–4084
17. Gibson DG, Young L, Chuang R-Y et al (2009) Enzymatic assembly of DNA molecules up to several hundred kilobases. Nat Methods 6: 343–345
18. Horton RM, Hunt HD, Ho SN et al (1989) Engineering hybrid genes without the use of restriction enzymes: gene splicing by overlap extension. Gene 77:61–68
19. Shetty RP, Endy D, Knight TF (2008) Engineering BioBrick vectors from BioBrick parts. J Biol Eng 2:5
20. Quan J, Tian J (2009) Circular polymerase extension cloning of complex gene libraries and pathways. PLoS One 4:e6441
21. Datsenko KA, Wanner BL (2000) One-step inactivation of chromosomal genes in *Escherichia coli* K-12 using PCR products. Proc Natl Acad Sci U S A 97:6640–6645
22. Rhodius VA, Suh WC, Nonaka G et al (2006) Conserved and variable functions of the sigmaE stress response in related genomes. PLoS Biol 4:e2
23. Bury-Moné S, Nomane Y, Reymond N et al (2009) Global analysis of extracytoplasmic stress signaling in *Escherichia coli*. PloS Genet 5:e1000651

Chapter 19

Expression, Purification, and Screening of BamE, a Component of the BAM Complex, for Structural Characterization

Mark Jeeves, Pooja Sridhar, and Timothy J. Knowles

Abstract

In Gram-negative bacteria, integral outer membrane β-barrel proteins (OMP) are assembled by the β-barrel assembly machine complex, or BAM complex. This complex includes the essential components BamA, an OMP composed of a carboxyl terminal β-barrel domain and five polypeptide transport-associated domains (POTRA), and the lipoprotein BamD. In *Escherichia* coli, the complex contains an additional three lipoproteins, BamB, C and E required for efficient delivery of OMPs to the outer membrane. Here we provide methods for production, isotope labeling, purification, and functional screening of BamE for research purposes. Purification strategies of both the soluble and wild-type membrane-tethered forms of BamE are described using techniques including osmotic shock, Ni-NTA purification, and size-exclusion chromatography. Functional screening using a simple plate assay is also described which allows screening for defects in outer membrane permeability.

Key words Beta barrel membrane proteins, BamE, BAM complex, POTRA, Protein labeling, Structural biology

1 Introduction

The outer membrane of Gram-negative bacteria is a selective barrier essential for protection against environmental factors whilst permitting the flow of nutrients and metabolites into and out of the cell. This membrane is asymmetric, composed of an inner leaflet of phospholipids and an outer leaflet of lipopolysaccharide, and harbors both lipoproteins and integral outer membrane proteins (OMPs). Many of these proteins are associated with basic physiological functions, virulence and multidrug resistance, and therefore play fundamental roles in maintaining cellular viability [1]. A single outer membrane complex known as the β-barrel assembly machine complex, or BAM complex, has been recognized as essential for the efficient assembly of almost all OMPs into the outer membrane [1, 2]. In *E. coli* this complex consists of five Bam proteins,

Susan K. Buchanan and Nicholas Noinaj (eds.), *The BAM Complex: Methods and Protocols*, Methods in Molecular Biology, vol. 1329, DOI 10.1007/978-1-4939-2871-2_19, © Springer Science+Business Media New York 2015

designated BamA through E. BamA is an essential integral OMP whilst BamB–E are peripheral lipoproteins of which only BamD is essential [3–5].

BamE is a 10 kDa lipoprotein with an α-α-β-β-β topology consisting of an antiparallel β-sheet packed against a pair of α-helices [6] and is found in all proteobacteria of the α, β, and γ classes. Although not essential, null mutants exhibit OMP folding and membrane barrier defects that are manifest as hypersensitivity to various toxic molecules including vancomycin, rifampin, SDS, and EDTA [3, 7]. Recently BamE has been shown to modulate the structure of BamA, and together with BamD is believed to be responsible for regulating BamA's conformation during the OMP assembly reaction [8].

This chapter describes methods for the over-expression, isotope labeling, and purification of BamE from *E. coli* both as a soluble protein (lacking its N-terminal lipoprotein acylation) and in its native membrane-tethered form using techniques including osmotic shock, Ni-NTA chelating resins, and size-exclusion chromatography (SEC). Furthermore we describe a simple plate assay that allows screening for defects in outer membrane permeability and can be used to probe for functionally important residues in BamE.

2 Materials

2.1 Media, Buffers, and Solutions

Prepare all media, buffers, and solutions using ultrapure water, which is deionized water which has been further purified through a purification system to attain a sensitivity of 18.2 MΩ-cm at 25 °C.

1. LB medium: Dissolve 10 g tryptone, 5 g yeast extract, and 10 g NaCl in 1 L of water, autoclave, and let cool. Ampicillin can be added prior to use at a final concentration of 100 μg/mL.
2. M9 medium stock metal solution: Dissolve 1.15 g $ZnSO_4 \cdot 7H_2O$, 0.169 g $MnSO_4 \cdot 5H_2O$, 0.229 g of H_3BO_3, and 0.175 g $CuSO_4 \cdot 5H_2O$ in 1 L of water and filter sterilize using a 0.4 μm filter.
3. M9 medium nutrient mix: Mix 1.0 mL thiamine (20 mg/ml), 0.4 mL $FeCl_3$ (3 mM), 1.0 mL M9 medium stock metal solution (*see* **item 2**), 2.0 mL $MgSO_4$ (1 M), 2.0 mL $CaCl_3$ (50 mM), 2 g glucose (or ^{13}C-glucose depending on labeling strategy), and 1 g NH_4Cl ($^{15}NH_4Cl$ depending on labeling strategy) to 8 mL of water, adjust the volume to 20 mL, and filter sterilize using a 0.4 μm filter.
4. M9 salts: Dissolve 6 g Na_2HPO_4, 3 g KH_2PO_4, and 0.5 g NaCl in 800 mL of water, adjust the pH to 7.3 with 10 N NaOH or 5 N HCl, adjust the volume to 1 L, autoclave, and let cool.

5. M9 medium: Immediately prior to use, aseptically add 20 mL of M9 medium nutrient mix (*see* **item 3**) to 1 L of M9 salts (*see* **item 4**), and mix.
6. Ni-NTA binding buffer: 50 mM HEPES-NaOH, 500 mM NaCl, pH 7.5. Dissolve 11.92 g HEPES and 29.22 g of NaCl in 800 mL of water, adjust the pH to 7.5 using 10 N NaOH or 5 N HCl, adjust the volume to 1 L, and filter sterilize using a 0.4 μm filter.
7. Ni-NTA wash buffer: 50 mM HEPES, 500 mM NaCl, 50 mM imidazole, pH 7.5. Dissolve 11.92 g HEPES, 29.22 g of NaCl, and 3.40 g imidazole in 800 mL of water, adjust the pH to 7.5 using 10 N NaOH or 5 N HCl, adjust the volume to 1 L, and filter sterilize using a 0.4 μm filter.
8. Ni-NTA elution buffer: 50 mM HEPES, 500 mM NaCl, 500 mM imidazole, pH 7.5. Dissolve 11.92 g HEPES, 29.22 g of NaCl, and 34.04 g of imidazole in 800 mL of water, adjust the pH to 7.5 using 10 N NaOH or 5 N HCl, adjust the volume to 1 L, and filter sterilize using a 0.4 μm filter.
9. Membrane Ni-NTA binding buffer: 50 mM HEPES, 500 mM NaCl, 50 mM imidazole, 0.5 % (w/v) *n*-dodecyl β-D-maltoside (DDM), pH 7.5. Dissolve 11.92 g HEPES, 29.22 g of NaCl, 3.40 g imidazole, and 20 g DDM in 800 mL of water, adjust the pH to 7.5 using 10 N NaOH or 5 N HCl, adjust the volume to 1 L, and filter sterilize using a 0.4 μm filter.
10. Resuspension buffer: 50 mM HEPES and 500 mM NaCl pH 7.5. Dissolve 11.92 g HEPES and 29.22 g of NaCl in 800 mL water, adjust the pH to 7.5 using 10 N NaOH or 5 N HCl, adjust the volume to 1 L, and filter sterilize using a 0.4 μm filter.
11. Membrane Ni-NTA wash buffer: 50 mM HEPES, 500 mM NaCl, 50 mM imidazole, 0.1 % (w/v) DDM, pH 7.5. Dissolve 11.92 g HEPES, 29.22 g of NaCl, 3.40 g imidazole, and 1.0 g DDM in 800 mL of water, adjust the pH to 7.5 using 10 N NaOH or 5 N HCl, adjust the volume to 1 L, and filter sterilize using a 0.4 μm filter.
12. Membrane Ni-NTA elution buffer: 50 mM HEPES, 500 mM NaCl, 500 mM imidazole, 0.1 % (w/v) DDM, pH 7.5. Dissolve 11.92 g HEPES, 29.22 g of NaCl, 34.04 g imidazole, and 1.0 g DDM in 800 mL of water, adjust the pH to 7.5 using 10 N NaOH or 5 N HCl, adjust the volume to 1 L, and filter sterilize using a 0.4 μm filter.
13. SEC buffer: 50 mM HEPES, 200 mM NaCl, pH 7.5. Dissolve 11.92 g HEPES and 11.69 g of NaCl in 800 mL water, adjust the pH to 7.5 using 10 N NaOH or 5 N HCl, adjust the volume to 1 L, and filter sterilize using a 0.4 μm filter.

14. Membrane SEC buffer: 50 mM HEPES, 200 mM NaCl, 0.1 % (w/v) DDM, pH 7.5. Dissolve 11.92 g HEPES, 11.69 g of NaCl, and 1.0 g DDM in 800 mL of water, adjust the pH to 7.5 using 10 N NaOH or 5 N HCl, adjust volume to 1 L, and filter sterilize using a 0.4 μm filter.
15. EDTA solution (500 mM): Add 146.12 g of EDTA to 800 mL of water, dissolve by addition of 10 N NaOH until all EDTA has been dissolved, adjust pH to 8.0, adjust the volume to 1 L, and filter sterilize using 0.4 μm filter.
16. Osmotic shock solution: 20 % (w/v) sucrose, 50 mM Tris–HCl (pH 8), and 5 mM EDTA. Dissolve 6.057 g of Trizma base and 200 g of sucrose in 800 mL of water, add 10 mL of 500 mM EDTA solution (*see* **item 15**), adjust the pH to 8.0 using 10 N NaOH or 5 N HCl, adjust the volume to 1 L, and filter sterilize using a 0.4 μm filter.
17. TBS Tween: 50 mM Tris, 150 mM NaCl, and 0.1 % Tween 20, pH 7.4. Dissolve 6.06 g of Trizma base, 8.77 g of NaCl, and 1 mL of Tween 20 in 800 mL of water, adjust the pH to 7.4 using 10 N NaOH or 5 N HCl, adjust the volume to 1 L, and filter sterilize using 0.4 μm filter.

2.2 Other Reagents and Equipment

1. Culture tubes (15 mL) and large-volume baffled flasks.
2. Shaking incubator: for small- and large-scale growth cultures.
3. UV-vis spectrometer: for optical density (OD) measurements.
4. Isopropyl β-D-1-thiogalactopyranoside (IPTG) (1 M).
5. Benchtop microcentrifuge and an ultracentrifuge.
6. $CaCl_2$ solution (5 mM).
7. Sephadex S75 gel filtration column or similar.
8. Automated protein purification system such as an ÄKTA Purifier (GE Healthcare).
9. Amicon Ultra-15 centrifugal filter unit (3000 NMWL) (Merck Millipore) or similar.
10. BL21(DE3) cells: for protein expression.
11. Electrocompetent *bamE::Kan* BW25113 cells: for functional assays.
12. French press or high-pressure homogenization: for lysing cells.
13. Dounce homogenizer: for resuspending membranes.
14. Ni-NTA agarose beads or similar immobilized metal affinity chromatography resin.
15. Gravity flow columns (empty).
16. Quikchange Site-Directed Mutagenesis kit (Agilent).
17. DH5α or XL1-Blue cells (included with some kits): for making plasmids.

18. Mini-prep kit such as the QIAprep Spin Miniprep Kit (Qiagen) or similar.
19. pET-upstream and T7 terminator primers (Merck Millipore) for sequencing pET system plasmids.
20. Vancomycin: to be used at a final concentration of 150 μg/mL in functional assays.
21. SDS-PAGE loading buffer (2×).
22. Microtip sonicator: to disrupt membranes and shear genomic DNA.
23. Polyvinylidene difluoride (PVDF) membrane: for Western blot analysis.
24. Standard SDS-PAGE and Western blotting methods: for monitoring protein expression and purification and for functional assays. We will not describe this in detail since it is described sufficiently in other chapters and since it is beyond the scope of the methods we present here.
25. Milk powder: for use as a blocking agent during Western blot analysis.
26. Anti-BamE polyclonal antibody (use at 1:4000 dilution).
27. Rabbit anti-mouse IgG-HRP conjugate secondary antibody (use at 1:10,000 dilution).
28. Chemiluminescent detection reagent, such as ECL reagent (GE Healthcare).
29. Detection film for developing Western blots. HyperFilm ECL (GE Healthcare) or similar. Alternatively this can be done using digital imaging systems as well.

3 Methods

3.1 Expression of Soluble BamE

BamE within the cell is expressed with an N-terminal 19 residue signal peptide that targets it for transport through the Sec machinery and lipidation at position Cys20; it is then transported via the Lol pathway to its final location at the inner leaflet of the outer membrane. However, a soluble form of BamE can be produced by replacing the signal peptide with a pelB leader sequence and mutating Cys20 to a serine yielding a protein lacking acylation but maintaining expression into the periplasm where correct folding takes place. Expression of the protein at low temperature followed by extraction from the periplasmic space via osmotic shock is required as mis-folding artifacts yielding dimers and aggregates are observed via excessive over-expression into the cytoplasm. The native membrane-tethered form of BamE can be produced using the native sequence of BamE codon optimized for optimal expression. Like soluble BamE expression, low temperature is required to minimize aggregation and inclusion body formation. As native BamE is tethered to the membrane via lipidation of Cys20, purification is not trivial and requires the use of detergents to extract the protein from the membrane and maintain it in its monomeric form.

E. coli BL21 (DE3) pBamE-S strain carries the *E. coli bamE gene*, codon optimized for optimum expression in *E. coli*, lacking its signal peptide and harboring the C20S mutation to remove any possible site for lipidation, under the control of the T7 promoter expression vector in pET22b. The gene is coded between restriction nuclease sites MscI and XhoI giving a construct that utilizes the pET22b pelB leader sequence and ends with a C-terminal 6xHis tag.

3.1.1 Cell Growth

1. Aseptically pick a single *E. coli* BL21 (DE3) pBamE-S colony from a freshly streaked LB agar plate into 10 mL of LB medium supplemented with 100 μg/mL ampicillin. Grow culture overnight at 37 °C shaking at 250 rpm.
2. Use the 10 mL overnight culture to inoculate 1 L of LB medium (*see* **Note 1**) in a 5 L baffled flask. Culture at 37 °C shaking at 200 rpm until an OD_{600} of 0.4 is reached.
3. Reduce the temperature to 18 °C and continue to grow until an OD_{600} of 0.6 is reached.
4. Induce the expression of soluble BamE by sterile addition of IPTG to a final concentration of 1 mM. Continue to grow at 18 °C for an additional 16 h.
5. Harvest cells by centrifugation at 8000 × *g* for 15 min at 4 °C. Discard supernatant (*see* **Note 2**).

3.1.2 Osmotic Shock to Extract Periplasmic Proteins

Osmotic shock procedure used based on ref. 9.

1. Wash freshly harvested cells by resuspension in water followed by centrifugation. Repeat twice. Discard supernatants.
2. Resuspend cells in a solution of 5 mM $CaCl_2$ for 5 min at 4 °C (100 mL/L growth pellet) followed by centrifugation at 8000 × *g* for 10 min at 4 °C. Discard supernatant.
3. Resuspend cell pellet in osmotic shock solution (80 mL/L growth pellet, 4 °C). Incubate at 4 °C for 10-min shaking at 200 rpm.
4. Centrifuge at 8000 × *g* for 10 min at 4 °C and discard supernatant.
5. Resuspend cells in water (80 mL/L growth pellet 4 °C). Incubate at 4 °C for 10-min shaking at 200 rpm.
6. Centrifuge at 8000 × *g* for 10 min at 4 °C, recover supernatant containing soluble BamE, and discard pellet.

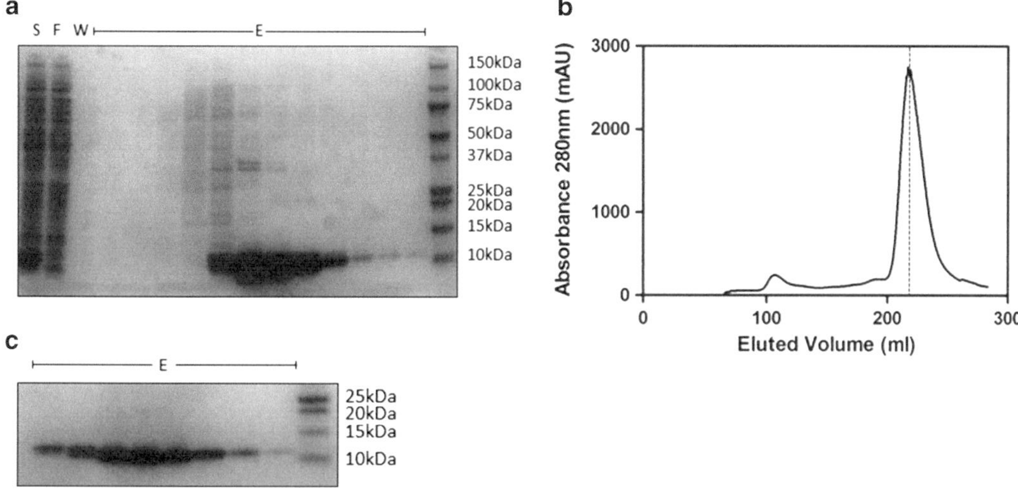

Fig. 1 Representative purification of soluble BamE. (**a**) 4–12 % SDS-PAGE showing a representative Ni-NTA purification of soluble BamE. S = Supernatant following cell disruption and centrifugation. F = Flow-through following supernatant binding. W = 50 mM imidazole wash and E = 500 mM imidazole elutions. (**b**) Representative S75 26/60PG size-exclusion profile at 280 nm of soluble BamE post-Ni-NTA purification. (**c**) 4–12 % SDS-PAGE showing the purity of soluble BamE following size-exclusion purification shown in (**b**)

3.1.3 Ni-NTA Purification

1. Equilibrate a Ni-NTA capture column (i.e., HisTrap 5 mL column (GE Healthcare)) with 5 column volumes of Ni-NTA binding buffer at 4 °C.
2. Pass the supernatant slowly (1–2 mL/min at 4 °C) through the Ni-NTA capture column and collect all flow-through.
3. Wash the column to remove weakly associated proteins by passing 10 column volumes of Ni-NTA wash buffer through the column at a rate of 1–2 mL/min at 4 °C and collect the flow-through.
4. Elute the protein from the column by passing at least 3 column volumes of Ni-NTA elution buffer through the column and collecting 1 mL elution fractions.
5. Detect fractions containing soluble BamE by performing standard SDS-PAGE on all fractions (Fig. 1a).
6. Pool the fractions containing soluble BamE.

3.1.4 Size-Exclusion Chromatography

1. Second-stage purification is performed by SEC utilizing a Sephadex S75 gel filtration column or similar. Recommended column for large-scale purification is a S75 26/60 PG column attached to a suitable purification system such as ÄKTA (GE Healthcare).
2. Wash column with 1.2 column volumes of water (*see* **Note 3**) followed by 1.2 column volumes of SEC buffer at 4 °C. Use recommended flow rates stipulated by column manufacturer.

3. Concentrate the pooled soluble BamE sample approximately five- to tenfold to a final volume of ~2.5 mL using a 3000 NMWL centrifugal filter such as an Amicon Ultra-15 centrifugal filter unit (Merck Millipore) at 4000 × *g* using a swing-out rotor (*see* **Note 4**). More dilute samples can be used if precipitation is observed.
4. Load a maximum of 2.5 mL of sample onto the column at a flow rate of 1.0 mL/min to ensure maximum resolution.
5. Elute protein using 1 column volume of SEC buffer at a rate of 1 mL/min and 4 °C and collect protein in 4 mL fractions. The elution of the protein can be visualized by monitoring the absorbance at 280 nm (Fig. 1b).
6. Identify which fractions contain the protein and the purity by analyzing each sample using standard SDS-PAGE methods (Fig. 1c).
7. Pool the fractions containing pure protein and concentrate to desired concentration using a 3000 NMWL centrifugal filter.
8. Repeat as necessary until the entire sample has been purified.

3.2 Expression and Purification of Membrane-Associated BamE

E. coli BL21 (DE3) pBamE-FL strain carries the *E. coli bamE* gene, codon optimized for optimum expression in *E. coli*, but harboring the WT sequence including signal peptide, under the control of the T7 promoter expression vector in pET20b encoded between restriction nuclease sites NdeI and Xho1 giving a construct that targets the protein for lipidation and outer membrane localization but with a C-terminal 6xHis tag.

3.2.1 Cell Growth

1. Aseptically pick a single *E. coli* BL21 (DE3) pBamE-FL colony from a freshly streaked LB agar plate into 10 mL of LB medium supplemented with 100 μg/mL ampicillin. Culture overnight at 37 °C shaking at 250 rpm.
2. Use the overnight culture to inoculate 1 L of LB medium (*see* **Note 1**) in a 5 L baffled flask. Culture at 37 °C shaking at 200 rpm until an OD_{600} of 0.4.
3. Reduce the temperature to 18 °C and continue growing until an OD_{600} of 0.6.
4. Induce the expression of full-length BamE by sterile addition of IPTG to a final concentration of 0.1 mM. Continue to grow at 18 °C with shaking at 200 rpm for 16 h.
5. Harvest the cells by centrifugation at 8000 × *g* for 15 min at 4 °C. Discard the supernatant (*see* **Note 2**).

3.2.2 Membrane Isolation

1. Resuspend the cell pellet in ice-cold resuspension buffer (20 mL per L growth pellet).
2. Lyse the cells by high-pressure disruption (i.e., French press or high-pressure homogenization). A minimum of three passages is required for optimal lysis. Do not use sonication (*see* **Note 5**).

3. Remove cell debris by centrifugation at 10,000 × *g* for 20 min at 4 °C. Discard pellet.
4. Centrifuge the supernatant at 100,000 × *g* for 1 h to pellet the membranes. Discard supernatant (*see* **Note 6**).
5. Store the pellet at −80 °C until required.

3.2.3 Ni-NTA Purification

1. Solubilize the pellet in ice-cold membrane Ni-NTA binding buffer (10 mL/L growth membrane pellet) (*see* **Note 7**). An optimal method is homogenization using a Dounce (hand) homogenizer to ensure that all membranes are uniformly resuspended. Allow to mix for 1 h at 4 °C using a rotary tube mixer.
2. Remove the insoluble material by centrifugation at 100,000 × *g* for 30 min at 4 °C and discard the pellet.
3. Wash 0.5 mL Ni-NTA agarose beads per L of growth of cell membrane with membrane Ni-NTA binding buffer.
4. Add the supernatant to the pre-washed Ni-NTA agarose beads and mix gently overnight on a rotary mixer at 4 °C.
5. Load the sample into an empty gravity flow column and collect the flow-through. The Ni-NTA beads will settle to the bottom of column.
6. Wash the beads by gravity using 10 column volumes of membrane Ni-NTA wash buffer at 4 °C. Keep the flow-through.
7. Elute the protein by gravity using membrane elution buffer (10 mL per mL of beads at 4 °C).
8. Identify which fractions contain FL-BamE and the purity by analyzing each sample using standard SDS-PAGE methods (Fig. 2a).
9. Pool the elution fractions containing FL-BamE.

3.2.4 Size-Exclusion Chromatography

1. Second-stage purification is performed by SEC utilizing a Sephadex S75 gel filtration column or similar. Recommended column for large-scale purification is a S75 26/60 PG column attached to a suitable purification system such as ÄKTA (GE Healthcare).
2. Wash column with 1.2 column volumes of water (*see* **Note 2**) followed by 1.5 column volumes of membrane SEC buffer at 4 °C. Use the recommended flow rates stipulated by the column manufacturer.
3. Concentrate the pooled soluble FL-BamE sample to a final volume of ~2.5 mL using a 3000 NMWL centrifugal filter such as an Amicon Ultra-15 centrifugal filter unit (Merck Millipore) at 4000 × *g* using a swing-out rotor (*see* **Note 4**).
4. Load a maximum of 2.5 mL sample onto column at a flow rate of 1 mL/min to ensure maximum resolution.

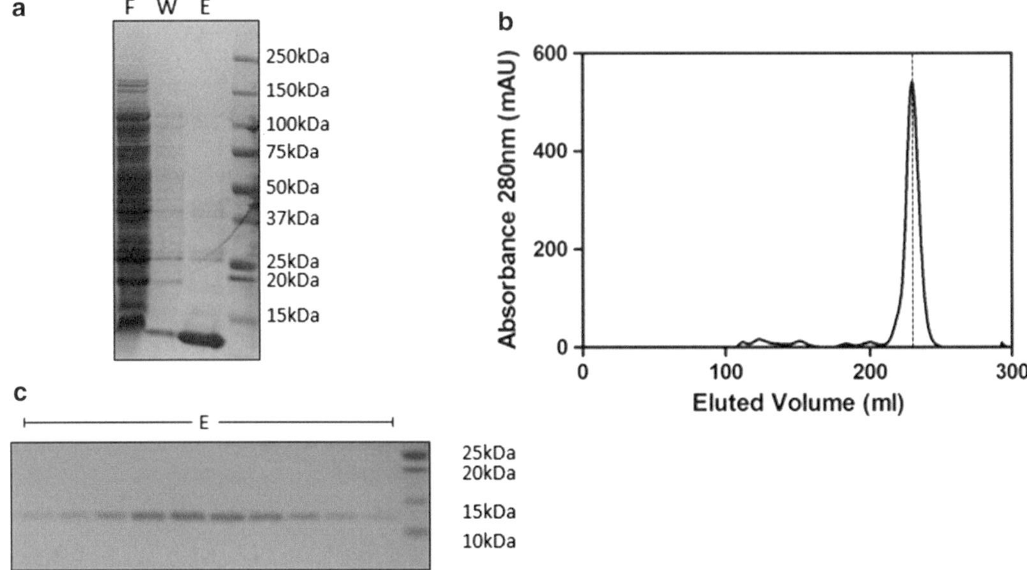

Fig. 2 Representative purification of membrane-associated BamE. (**a**) 4–12 % SDS-PAGE showing a representative Ni-NTA purification of solubilized membrane-associated BamE. F = Flow-through following solubilization of membrane and binding to Ni-NTA agarose. W = 50 mM imidazole wash and E = 500 mM imidazole elution. (**b**) Representative S75 26/60PG size-exclusion profile at 280 nm of solubilized membrane-associated BamE post-Ni-NTA agarose purification. (**c**) 4–12 % SDS-PAGE showing the purity of solubilized membrane-associated BamE following size-exclusion purification shown in (**b**)

5. Elute protein using 1 column volume of membrane SEC buffer at a rate of 1 mL/min and 4 °C and collect protein in 4 mL fractions. The elution of the protein can be visualized by monitoring the absorbance at 280 nm (Fig. 2b).
6. Identify which fractions contain the protein and the purity by analyzing each sample using standard SDS-PAGE methods (Fig. 1c).
7. Pool the fractions containing pure protein and concentrate to desired concentration using a 3000 NMWL centrifugal filter (*see* **Note 8**).
8. Repeat as necessary until the entire sample has been purified.

3.3 Screening for Functionally Relevant Mutations

Plasmid pBamE-WT contains the wild-type *bamE* coding sequence amplified from *E. coli* K12 and cloned between sites NdeI and EcoR1. The BamE knockout strain *bamE::Kan* is derived from *E. coli* BW25113 cells and is taken from the Keio collection [10]. The pBamE-WT plasmid is sufficiently leaky that when present within strain *bamE::Kan* BW25113 cells enough BamE protein is produced to rescue the *bamE::Kan* phenotype.

Single-point mutations can be introduced throughout the protein using a suitable single-point mutagenesis kit, such as

Quikchange (Agilent). Suitable primers can be designed by using available software (i.e., Quikchange Primer Design Tool from Agilent) and chemically synthesized by commercial companies.

3.3.1 Construction of a Cys-Gly Library

1. With plasmid pBamE-WT as a template, design and perform single-point mutations throughout the DNA sequence.
2. Transform mutant plasmids into a high-plasmid-copy-number cell line such as DH5α or XL1-Blue as described in the kit.
3. Aseptically pick single colonies produced from mutagenesis kit and grow overnight in 55 mL LB medium supplemented with appropriate antibiotic (for pBamE-WT—100 μg/mL ampicillin) at 37 °C shaking at 250 rpm.
4. Pellet cells at 6000 × *g* for 15 min and discard supernatant.
5. Harvest plasmid by using a suitable plasmid mini-prep kit such as Qiagen. Elute the plasmid into 50 μL water.
6. Sequence plasmid stocks for desired mutation, with correct primers for plasmid, for example pET-upstream primer (Merck Millipore) and T7 terminator primer (Merck Millipore) for pET system.

3.3.2 Selection of Functionally Important Mutants

1. Transform chemically competent *bamE::Kan* BW25113 cells using 1 μL (approx. 0.2 μg DNA) of desired mutant pBamE-WT DNA using established methods (*see* pET System Manual for additional information if needed, Merck Millipore).
2. Plate out onto LB agar plates supplemented with 100 μg/mL ampicillin and grow overnight at 37 °C inverted.
3. Aseptically pick single colonies of *bamE::Kan* pBamE WT mutant cells and streak onto both LB amp plates supplemented with 150 μg/mL vancomycin and plates supplemented with 0.5 % SDS and 0.5 mM EDTA.
4. Grow overnight inverted at 37 °C.
5. Compare complementation by assessing growth on SDS/EDTA plates and vancomycin plates in comparison to *bamE:Kan* BW25113 cells carrying either pBamE-WT or empty pET20b vector. Mutants which fail to grow on LB agar plates supplemented with either SDS/EDTA or vancomycin have a compromised outer membrane barrier and suggest that the mutation may play a role in the function of BamE.

3.3.3 Testing Mutant Protein Expression Levels

Not all mutations solely affect the function of the protein; many affect the folding. Testing of protein folding can be achieved by measuring protein expression levels in the cell. For those proteins whose fold is compromised, levels of protein will be reduced or nonexistent as the protein is broken down within the cell.

1. Aseptically pick a single colony of desired pET20b-BamE mutant in *bamE::Kan* cells and grow overnight in 10 mL of LB medium supplemented with 100 μg/mL ampicillin at 37 °C shaking at 250 rpm.
2. Use 1 mL of overnight growth culture to inoculate a fresh 10 mL of LB medium supplemented with 100 μg/mL ampicillin.
3. Grow at 37 °C shaking at 250 rpm until an $OD_{600} = 0.6$. Record OD_{600} precisely for each sample.
4. Harvest the cells by centrifugation at 6000 × *g* for 15 min at 4 °C.
5. Resuspend cells in water to a final $OD_{600} = 6$.
6. Take 20 μL of cells and add 20 μL of 2× SDS-PAGE loading buffer.
7. Sonicate sample using microtip sonicator to disrupt membrane and break DNA (5 micron amplitude for 30 s).
8. Run 10 μL each sample on an SDS-PAGE gel.
9. Transfer to PVDF membrane by Western blotting, using the manufacturer's suggested method for transfer.
10. Block membrane for 1 h using 5 % (w/v) milk powder in TBS Tween at room temperature with gentle agitation.
11. Stain with mouse anti-BamE polyclonal antibody (1:4000 dilution) in 5 % (w/v) milk powder in TBS Tween overnight at 4 °C with gentle agitation.
12. Wash membrane three times for 5 min each with 50 mL TBS Tween at room temperature with gentle agitation.
13. Stain with rabbit anti-mouse IgG-HRP conjugate secondary antibody (1:10,000 dilution) in 5 % milk powder in TBS Tween for 1 h at room temperature with gentle agitation.
14. Wash membrane three times for 5 min with 50 mL TBS Tween at room temperature with gentle agitation.
15. Stain membrane with a suitable chemiluminescent detection reagent, such as ECL reagent (GE Healthcare). Follow the manufacturer's recommended method.
16. Detect using suitable detection film, such as HyperFilm ECL (GE Healthcare) or by digital methods.
17. Compare levels between mutants and WT pET20b-BamE in *bamE::Kan* cells.

4 Notes

1. If BamE is to be studied by nuclear magnetic resonance, isotope labelling is required. Simply substitute LB media for M9 media supplemented with ^{15}N-ammonium chloride

and/or ^{13}C-glucose depending on isotope labeling strategy wanted.

2. For those cells not requiring osmotic shock, cells can be frozen at −80 °C until required. For osmotic shock, cells must be used immediately.
3. Washing a column with water prior to your buffer ensures that no salts are present that could interfere with your buffer; for example calcium chloride in a column prior to addition of sodium phosphate will result in calcium phosphate formation leading to column blockage.
4. When using a centrifugal concentrator ensure that the sample is mixed thoroughly every 10 min during the concentration process to minimize the development of a concentration gradient and precipitation of the sample due to overconcentration.
5. Sonication is not recommended as the power of ultrasound has been found to be damaging especially to membranes and membrane proteins.
6. The pellet should be a deep brown/red color that is slightly translucent at the edges. If the majority of the pellet is white this suggests that inclusion bodies have been formed. If this is the case the pellet is useless for further purification.
7. For membrane protein solubilization, the initial concentration of detergent is high to enable solubilization of all components within the system. Following binding to the resin, all remaining buffers only require a detergent concentration just greater than the critical micelle concentration (cmc) of the detergent. For example the cmc of DDM is ~0.009 % (w/v); thus in this study a concentration of 0.1 % (w/v) was used.
8. For FL-BamE concentrating the sample results in concentrating the detergent also.

Acknowledgements

We would like to thank Michael Overduin for discussions and BBSRC Grants BB/G022054/1 and BB/F000472/1 for funding this research.

References

1. Bos MP, Robert V, Tommassen J (2007) Biogenesis of the gram-negative bacterial outer membrane. Annu Rev Microbiol 61:191–214
2. Knowles TJ, Scott-Tucker A, Overduin M et al (2009) Membrane protein architects: the role of the BAM complex in outer membrane protein assembly. Nat Rev Microbiol 7(3): 206–214
3. Sklar JG, Wu T, Gronenberg LS et al (2007) Lipoprotein SmpA is a component of the YaeT complex that assembles outer membrane proteins in *Escherichia coli*. Proc Natl Acad Sci U S A 104(15):6400–6405

4. Voulhoux R, Bos MP, Geurtsen J et al (2003) Role of a highly conserved bacterial protein in outer membrane protein assembly. Science 299(5604):262–265

5. Wu T, Malinverni J, Ruiz N et al (2005) Identification of a multicomponent complex required for outer membrane biogenesis in *Escherichia coli*. Cell 121(2):235–245

6. Knowles TJ, Browning DF, Jeeves M et al (2011) Structure and function of BamE within the outer membrane and the beta-barrel assembly machine. EMBO Rep 12(2):123–128

7. Ruiz N, Falcone B, Kahne D et al (2005) Chemical conditionality: a genetic strategy to probe organelle assembly. Cell 121(2): 307–317

8. Rigel NW, Schwalm J, Ricci DP et al (2012) BamE modulates the *Escherichia coli* beta-barrel assembly machine component BamA. J Bacteriol 194(5):1002–1008

9. Chen YC, Chen LA, Chen SJ et al (2004) Modified osmotic shock for periplasmic release of a recombinant creatinase from *Escherichia coli*. Biochem Eng J 19(3):211–215

10. Baba T, Ara T, Hasegawa M et al (2006) Construction of *Escherichia coli* K-12 in-frame, single-gene knockout mutants: the Keio collection. Mol Syst Biol 2:2006.0008

Chapter 20

Purification and Bicelle Crystallization for Structure Determination of the *E. coli* Outer Membrane Protein TamA

Fabian Gruss, Sebastian Hiller, and Timm Maier

Abstract

TamA is an Omp85 protein involved in autotransporter assembly in the outer membrane of *Escherichia coli*. It comprises a C-terminal 16-stranded transmembrane β-barrel as well as three periplasmic POTRA domains, and is a challenging target for structure determination. Here, we present a method for crystal structure determination of TamA, including recombinant expression in *E. coli*, detergent extraction, chromatographic purification, and bicelle crystallization in combination with seeding. As a result, crystals in space group $P2_12_12$ are obtained, which diffract to 2.3 Å resolution. This protocol also serves as a template for structure determination of other outer membrane proteins, in particular of the Omp85 family.

Key words Outer membrane protein, Membrane protein purification, Bicelle crystallization, X-ray crystallography, β-Barrel, TamA, Autotransporter, Omp85, Bacterial outer membrane

1 Introduction

The insertase proteins of the Omp85 family are responsible for the assembly of β-barrel proteins in the outer membrane of Gram-negative bacteria, mitochondria, and chloroplasts [1–3]. Omp85 insertases comprise a 16-stranded C-terminal transmembrane β-barrel and between one and five attached periplasmic POTRA domains [4–6], which are involved in initial substrate interaction via β-strand augmentation [7]. The β-barrel domain is characterized by an unusually weak lateral connection between the N-terminal and C-terminal strand. The latter strand is not fully zipped up but kinks to the inside of the barrel, thus creating a gate to the lipid phase of the membrane [5, 6].

TamA and BamA are the only Omp85 proteins in *E. coli* K12. BamA is the core subunit of the general β-barrel assembly machinery (BAM) [1]. TamA is more specialized and involved in the biogenesis of a subset of autotransporters [8, 9], which typically consist of a 12-stranded membrane-embedded β-barrel linked to a large extracellular passenger domain [10, 11].

Susan K. Buchanan and Nicholas Noinaj (eds.), *The BAM Complex: Methods and Protocols*, Methods in Molecular Biology, vol. 1329, DOI 10.1007/978-1-4939-2871-2_20, © Springer Science+Business Media New York 2015

For structure determination, TamA is recombinantly expressed into the outer membrane of its native host *E. coli*, using a pET vector that provides an N-terminal *malE* signal sequence for periplasmic targeting, followed by a hexahistidine purification (HIS-) tag and a TEV protease cleavage site preceding the insertion site for the coding sequence of mature TamA without its native export signal sequence. TamA is overexpressed in a partially OMP-deficient *E. coli* strain and extracted from membrane pellets using the mild detergent β-octylglucoside (β-OG). Immobilized metal affinity chromatography (IMAC) via the HIS-tag is employed for initial purification, followed by cation-exchange chromatography (CIEX) for buffer exchange for subsequent TEV cleavage of the HIS-tag. A second IMAC step removes TEV protease and uncleaved TamA. Finally, the cleaved protein is subjected to size-exclusion chromatography (SEC), yielding monodisperse monomeric TamA of high purity as determined by SDS-polyacrylamide gel electrophoresis (SDS-PAGE). After buffer exchange and detergent concentration equilibration via dialysis, the protein is integrated into bicelles by mixing it with an aqueous solution of preformed DMPC/CHAPSO bicelles. The mixture is used directly for crystallization screening and optimization. Crystals diffracting to high resolution are obtained by adding seeds prepared from lower quality crystals grown in similar conditions to the protein-bicelle mixture prior to crystallization setup. This seeding strategy is generally applicable to bicelle crystallization of other membrane proteins.

2 Materials

2.1 Plasmid Construction

1. Vector: pET22b (Novagen).
2. Cloning strain and genomic DNA source: *E. coli* DH10B.
3. Oligonucleotides:

   sigseq_down: 5′-ATAGGAATTCCATATGAAAATAAAAACAGGTGCACGCATCCTCGCATTATCCGCATTAACGACG-3′

   sigseq_up: 5′-GAGCGCATGCCATGGCGAGAGCCGAGGCGGAAAACATCATCGTCGTTAATGCGGATAATGCGAGG-3′

   histev_down: 5′-CATGGCCCACCACCACCACCACCACGAGAATCTGTATTTCCAGGG-3′

   histev_up: 5′-AATTCCCTGGAAATACAGATTCTCGTGGTGGTGGTGGTGGTGGGC-3′

   tama_down: 5′-GGAATTCAAGCGAACGTCCGTCTACAGGTCG-3′

   tama_up: 5′-CCGCTCGAGTCATAATTCTGGCCCCAGACCGATG-3′

4. DNA polymerase for PCR amplification.
5. Restriction endonucleases: NdeI, NcoI, EcoRI, XhoI (New England BioLabs).
6. Ligation enzyme: T4 DNA ligase (New England BioLabs).

2.2 Expression

1. Expression strain: *E. coli* BL21(DE3)omp3 [12].
2. LB medium (400 mL) for pre-cultures.
3. LB medium (18 L) for main cultures.
4. Flasks (4×500 mL) for pre-cultures.
5. Baffled flasks (12×5 L) for main cultures.
6. Ampicillin.
7. Isopropyl β-D-thiogalactopyranoside (IPTG).

2.3 Purification

1. Stock solutions for buffer preparations: (1) 500 mM Na_2HPO_4, (2) 500 mM NaH_2PO_4, (3) 5 M NaCl.
2. Lysis buffer (500 mL): 25 mM NaPi, pH 7.5 (86.6 % Na_2HPO_4, 13.4 % NaH_2PO_4), 150 mM NaCl (*see* **Note 1**).
3. High-pressure homogenizer (e.g., Microfluidizer by Microfluidics).
4. Extraction buffer (250 mL): 25 mM NaPi, pH 7.5, 150 mM NaCl, 3 % β-OG (*n*-octyl-β-D-glucopyranoside, >99 % chemical purity and >98 % β-anomer purity, Anatrace) (*see* **Note 1**).
5. High-performance liquid chromatography system (e.g., Äkta by GE Healthcare, NGC by Bio-Rad).
6. HisTrap FF column (5 mL) (GE Healthcare).
7. Immobilized metal affinity chromatography (IMAC) buffers (*see* **Note 1**):
   (a) A (250 mL): 25 mM NaPi, pH 7.5, 150 mM NaCl, 0.8 % β-OG.
   (b) B (150 mL): 25 mM NaPi, pH 7.5, 150 mM NaCl, 0.8 % β-OG, 500 mM imidazole.
8. Cation exchange (CIEX) column: 6 mL column with MacroPrep High S resin (Bio-Rad).
9. CIEX buffers (*see* **Note 1**):
   (a) Dilution buffer (150 mL): 25 mM NaPi, pH 6.9 (53.3 % Na_2HPO_4, 46.7 % NaH_2PO_4), 0.8 % β-OG.
   (b) A (300 mL): 25 mM NaPi, pH 7.0, 50 mM NaCl, 0.8 % β-OG.
   (c) B (200 mL): 25 mM NaPi, pH 7.0, 1 M NaCl, 0.8 % β-OG.
10. TEV protease (HIS-tagged) (Addgene plasmid 8827 [13]).
11. β-Mercaptoethanol.
12. Syringe filters (0.45 μm) (PTFE based for particle removal).

13. Superdex 75 16/60 column (GE Healthcare).
14. Size-exclusion chromatography (SEC) buffer (400 mL) (*see* **Note 1**): 25 mM NaPi, pH 7.5, 150 mM NaCl, 0.8 % β-OG.
15. Concentrator: Amicon Ultra unit 30 kDa c/o (Millipore).
16. Slide-A-Lyzer cassette, 3.5 kDa c/o (Pierce).
17. Dialysis buffer (200 mL): 20 mM Tris–HCl, pH 7.5 (adjusted at 10 °C), 150 mM NaCl, 0.8 % β-OG.

2.4 Crystallization

1. 1,2-Dimyristoyl-*sn*-glycero-3-phosphocholine (DMPC, Anatrace).
2. 3-[(3-Cholamidopropyl)dimethylammonio]-2-hydroxy-1-propanesulfonate (CHAPSO, Anatrace).
3. Vortex mixer.
4. Crystallization screens: MembFac (Hampton Research), MemGold (Molecular Dimensions), MemStart & MemSys (Molecular Dimensions).
5. Crystallization solutions: Initial crystals: 0.1 M imidazole pH 6, 0.6 M Na-acetate; crystals with seeding: 0.1 M imidazole pH 6, 1.2 M Na-acetate.
6. Crystallization plates: MRC 2 Well (Swissci, Hampton Research).
7. Crystallization robot (e.g., OryxNano by Douglas Instruments, optional but recommended).
8. Cryoprotectant: Perfluoropolyether oil (dried, Hampton Research).

2.5 Data Processing and Model Building Software

1. XDS [14].
2. CCP4 [15].
3. PHENIX [16].
4. COOT [17].

3 Methods

3.1 Plasmid Construction

1. Anneal the oligonucleotides sigseq_down and sigseq_up to obtain the *malE* signal sequence DNA.
2. Digest individually the *malE* signal sequence DNA and the pET22b vector using NdeI and NcoI restriction enzymes to remove the *pelB* signal sequence from the vector and purify digested DNA via agarose gel electrophoresis (*see* **Note 2**).
3. Ligate the purified DNA into the pET22b vector using T4 DNA ligase.

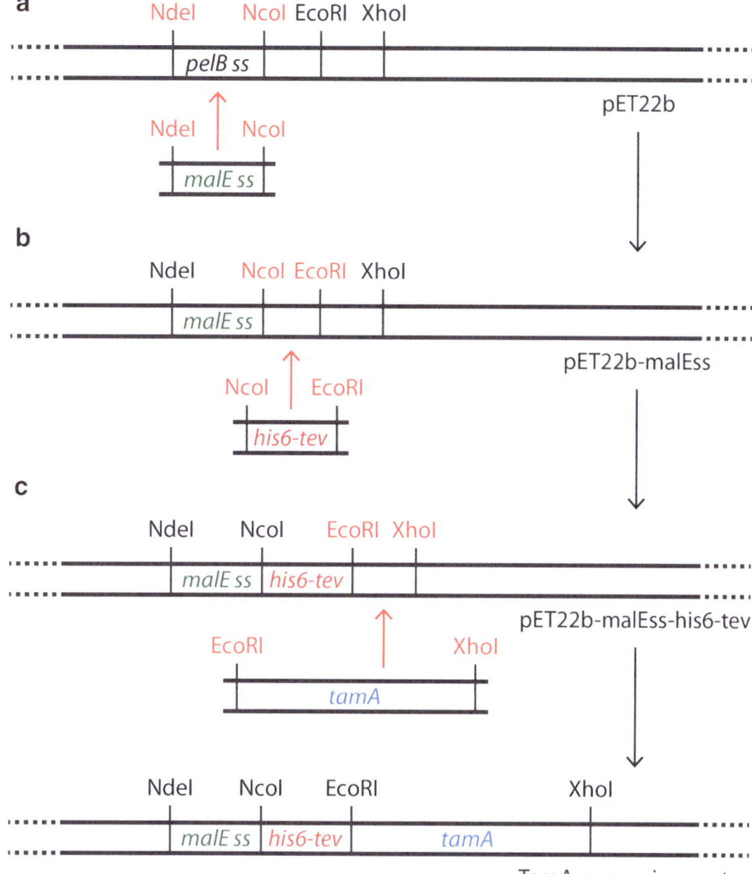

Fig. 1 Cloning strategy for the TamA expression vector. (**a**) In the first step, the *pelB* signal sequence of the pET22b vector is replaced by the *malE* signal sequence. (**b**) In the second step, DNA coding for a HIS-tag followed by a cleavage site for TEV protease is put 3′ of the signal sequence. (**c**) Last, the *tamA* DNA without its own signal sequence is inserted 3′ of the TEV site

4. Transform the ligation product into electrocompetent *E. coli* DH10B, and plate out and select positive colonies for plasmid purification to obtain the pET22b-malEss vector (Fig. 1a).
5. Repeat **steps 1–4** using the oligonucleotides histev_down and histev_up, the pET22b-malEss vector obtained from **step 4** and NcoI and EcoRI restriction enzymes to obtain the pET22b-malEss-his6-tev vector containing the *malE* signal sequence followed by a HIS-tag and a cleavage site for TEV protease (Fig. 1b).
6. Amplify genomic *tamA* DNA without the *tamA* signal sequence from an *E. coli* K12 strain (e.g., DH10B) using oligonucleotides tama_down and tama_up and purify the PCR product via agarose gel electrophoresis.

7. Repeat **steps 2–4** using the *tamA* PCR product from **step 6**, the pET22b-malEss-his6-tev vector obtained from **step 5**, and EcoRI and XhoI restriction enzymes to obtain the final TamA expression vector containing the *malE* signal sequence followed by a HIS-tag and a cleavage site for TEV protease followed by *tamA* (Fig. 1c).

3.2 Expression

1. Transform chemically competent *E. coli* BL21(DE3)omp3 cells [12] with the TamA expression vector obtained from Subheading 3.1. Inoculate 4×100 mL LB medium containing 100 mg/L ampicillin with the transformed cells and grow overnight by shaking at 37 °C.
2. Inoculate 12×1.5 L LB medium in baffled 5 L flasks containing 100 mg/L ampicillin with the overnight cultures from **step 1** to an OD_{600} (optical density at 600 nm) of around 0.06 and grow by shaking at 100 rpm and 37 °C.
3. When the cultures reach $OD_{600}=0.8$ set temperature to 20 °C and continue shaking the cells for 2 h before inducing protein expression by addition of IPTG to a final concentration of 0.1 mM.
4. Continue shaking at 20 °C for 18 h before harvesting. The obtained bacterial pellet is used directly or stored at −80 °C (*see* **Note 3**).

3.3 Purification

1. Resuspend the bacterial pellet obtained from Subheading 3.2 in 450 mL lysis buffer and lyse the cells by processing the suspension twice with the microfluidizer (*see* **Note 4**). Centrifuge for 10 min at 10,000×*g* and 4 °C to remove inclusion bodies and cell debris. Centrifuge the supernatant containing soluble protein and membrane vesicles for 1 h at 100,000×*g* and 4 °C to pellet membrane vesicles containing TamA. Remove the supernatant, and gently wash the pellet three times using lysis buffer.
2. Add 250 mL extraction buffer to the membrane pellet and stir overnight at 10 °C to extract TamA. Centrifuge the resuspended membranes for 1 h at 100,000×*g* and 10 °C to pellet the detergent-insoluble components. Harvest the supernatant containing solubilized TamA.

   All the following steps should be performed at temperatures between 10 and 15 °C, because the detergent-solubilized protein solution tends to precipitate reversibly at low temperatures. Chromatographic purification steps are carried out at room temperature (*see* **Note 5**).
3. Add 4 % v/v IMAC buffer B to the protein solution to adjust the imidazole concentration to approx. 20 mM. Load this solution onto a 5 mL or larger HisTrap FF column equilibrated

with IMAC buffer A (*see* **Note 6**). Wash the column with a mixture of 96 % IMAC buffer A/4 % IMAC buffer B until the OD_{280} reaches a stable baseline (typically 5–6 column volumes) and collect the flow-through. Elute the bound protein with a linear gradient from 4 to 100 % IMAC buffer B in 5 column volumes. When using a 5 mL column, repeat this step from column loading to elution using the flow-through of the first loading step to increase the yield by capturing remaining unbound protein.

4. Pool the fractions of the elution peak (resp. peaks if using a 5 mL HisTrap FF column) to obtain about 30–40 mL total volume of elution fractions containing TamA. Slowly add three times the volume of dilution buffer to adjust ionic strength for CIEX (cation exchange).
5. Load the protein solution onto a 6 mL MacroPrep High S column, equilibrated with CIEX buffer A. Wash the column with CIEX buffer A until the OD_{280} reaches a stable baseline. Elute the protein with a linear gradient from 0 to 100 % CIEX buffer B in 100 mL volume. Pool peak fractions to obtain a total volume of 30–50 mL (Fig. 2a).
6. Add 5 mM β-mercaptoethanol to the solution before adding about 1 mg of HIS-tagged TEV protease. Digest the protein overnight. The digested protein solution may appear slightly cloudy due to precipitation. Centrifuge or filtrate to remove the precipitate.
7. Load the solution onto a HisTrap FF 5 mL column, equilibrated with IMAC buffer A without imidazole. Collect the flow-through containing cleaved TamA and concentrate it with 30 kDa c/o centrifugal concentrator units to a volume of 1 mL. At this step, TamA should be essentially pure (Fig. 2b).
8. Load the protein onto a Superdex 75 16/60 column, equilibrated with SEC buffer, collect the peak fractions to obtain about 10 mL total volume, and concentrate it with 30 kDa c/o centrifugal units to a concentration of 15–20 mg/mL and about 0.5 mL volume (Fig. 3).
9. Dialyze the protein solution overnight against 80 mL dialysis buffer in a 3.5 kDa c/o Slide-A-Lyzer cassette. Dialyze it again against 80 mL fresh dialysis buffer for 5 h. Adjust the protein concentration to 12.5 mg/mL by adding the appropriate volume of dialysis buffer to obtain the final protein sample for crystallization. Typical yields of purified TamA are in the range of 5–10 mg from an 18 L LB culture.

3.4 Crystallization

1. To prepare 500 μL 40 % w/w bicelle stock solution composed of DMPC and CHAPSO in a molar ratio of 2.43:1, weigh out 144.6 mg DMPC in a 1.5 mL vial, and add 55.4 mg CHAPSO

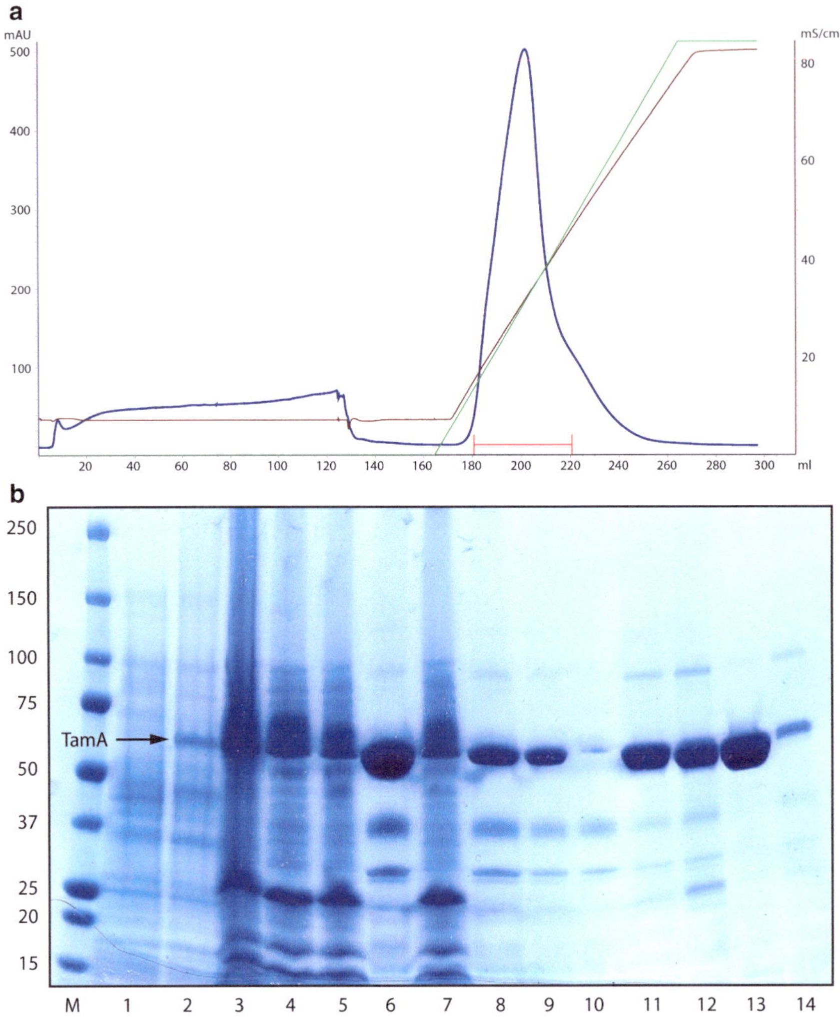

Fig. 2 CIEX chromatogram and SDS-PAGE analysis of TamA purification. (**a**) Chromatogram of CIEX chromatography. *Blue curve*: absorption at 280 nm; *green curve*: buffer B concentration; *brown curve*: conductivity; the *red line* indicates pooled fractions. (**b**) SDS-PAGE analysis of TamA expression and purification. Lane M shows a molecular weight standard and molecular weights in kDa for the bands are indicated on the *left-hand side*. Lane 1: *E. coli* expression cells before induction with IPTG; lane 2: *E. coli* expression cells after induction and growth overnight; lane 3: membrane pellet containing TamA suspended with extraction buffer; lane 4: supernatant after ultracentrifugation of the suspension; lane 5: first load onto HisTrap FF 5 mL column, flow-through; lane 6: pooled peak fractions of first HisTrap elution; lane 7: second load onto HisTrap FF 5 mL column, flow-through; lane 8: pooled peak fractions of second HisTrap elution; lane 9: Macroprep High S column load; lane 10: Macroprep High S flow-through; lane 11: MacroPrep High S pooled peak fractions; lane 12: TamA after TEV digest; lane 13: flow-through of HisTrap FF 5 mL column after TEV digest, containing TamA without HIS-tag; lane 14: elution from the HisTrap FF 5 mL column with 500 mM imidazole, containing TamA with non-cleaved HIS-tag and HIS-tagged TEV protease

on top. Cover it with 300 μL H_2O. Mount the 1.5 mL vial on a vortex mixer and vortex overnight at 4 °C to dissolve the bicelle mixture. If it has not dissolved completely, apply a few cycles of heating to 50 °C, cooling on ice, vortexing, and

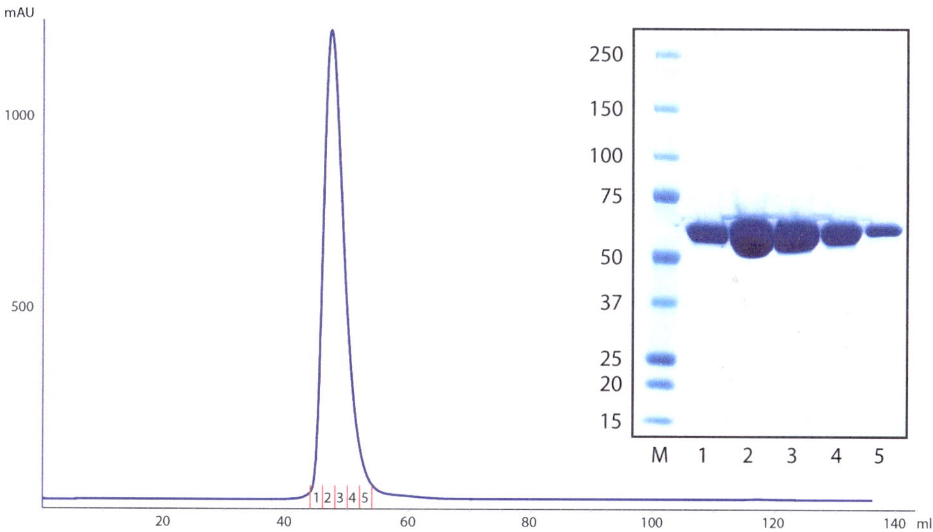

Fig. 3 Size-exclusion chromatogram and SDS-PAGE analysis of peak fractions. *Blue curve*: absorption at 280 nm; *red lines* indicate pooled fractions. Lane M shows a molecular weight standard and molecular weights in kDa for the bands are indicated on the *left-hand side*. Lanes 1–5: SEC fractions 1–5

freezing [18] until the solution is homogenous and clear. You can store the bicelle solution at −20 °C or keep it at 4 °C for immediate use.

2. To prepare a solution of TamA in 8 % bicelles (*see* **Note 7**), add one-fourth the volume bicelle stock solution to TamA solution (e.g., add 50 μl bicelle stock to 200 μl TamA), mix by pipetting up and down, and incubate for at least 30 min at 4 °C. Do not store this solution on ice as a phase transition may occur resulting in a cloudy inhomogeneous sample (*see* **Note 8**).
3. Setting up crystallization plates must be performed at 4 °C either using a robot (e.g., OryxNano) or manually. Pipette the crystallization solution on top of the protein-bicelle mixture or add both at the same time when using an OryxNano robot. After pipetting, immediately seal the crystallization plates and keep at 4 °C for 10 min before shifting to 20 °C. Initial crystallization trials may be set up using the sitting-drop vapor diffusion method with commercial membrane protein crystallization screens (e.g., MembFac, MemStart & MemSys, MemGold). Initial platelike crystals for seeding can be obtained in 0.1 M imidazole pH 6 and 0.6 M sodium acetate with a ratio of protein to reservoir solution in the drop of 1:2 after about 1 week.
4. Seed preparation: Prepare a solution resembling the mother liquor of the crystals: Mix dialysis buffer in a 4:1 ratio with bicelle stock solution and mix the resulting solution in a 1:2 ratio with 0.1 M imidazole pH 6 and 0.6 M sodium acetate.

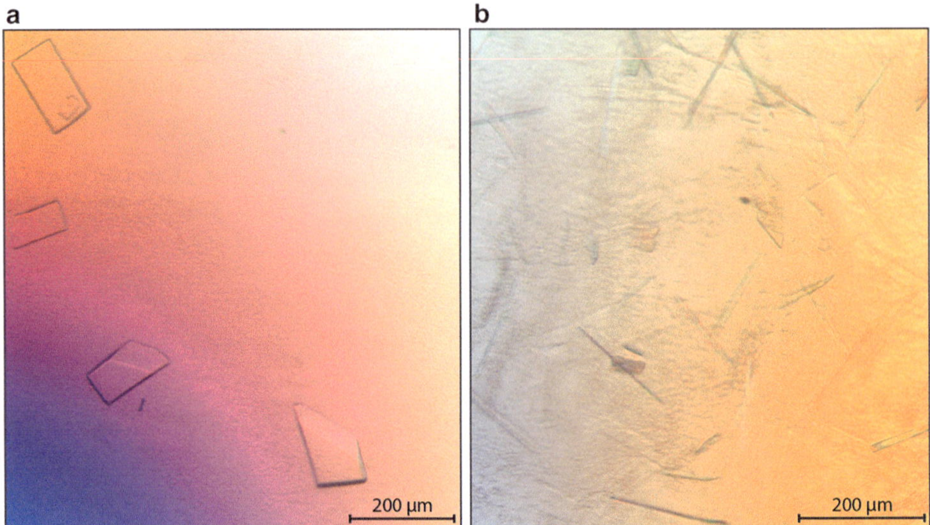

Fig. 4 TamA bicelle crystallization with seeding. (**a**) Crystals grown using TamA in a concentration of 7 mg/mL and a 1:1000 diluted seed stock, which diffracted to a resolution of 2.25 Å. (**b**) Lower quality crystals grown using TamA in a concentration of 8 mg/mL and a 1:100 diluted seed stock

Transfer crystals into a 1.5 mL vial containing 50–100 μL mother liquor solution using a loop or a pipette. Pipette up and down to crush crystals. Finally, vortex for 1 min. Prepare a dilution of the seed stock of 1:10 by mixing part of the seed solution 1:9 with mother liquor solution. Repeat to obtain a dilution series of 1:1, 1:10, 1:100, 1:1000, 1:10,000, and 1:100,000 (*see* **Note 9**).

5. To grow high-quality crystals, prepare TamA-bicelle solutions of different protein concentrations (e.g., 5–10 mg/mL in 1 mg/mL steps) by diluting the TamA solution with dialysis buffer before mixing with bicelle stock solution. Split each TamA-bicelle mixture into 6 parts and add one-ninth the volume of each of the seed stocks (e.g., add 1 μl 1:1000 seed stock to 9 μl TamA in bicelles with a protein concentration of 6 mg/mL). Set up a 2D-grid screen of protein concentration versus seed dilution using 0.1 M imidazole pH 6 and 1.2 M sodium acetate as reservoir solution and set up drops as before, mixing the TamA-bicelle/seed solutions in a 1:2 ratio with reservoir solution. The seed stocks can be frozen in liquid nitrogen and stored at −80 °C to later reproduce conditions producing optimal crystals (*see* **Note 9**) (Fig. 4).
6. To freeze crystals, cover the respective drop with perfluoropolyether oil for cryo-protection before fishing crystals with a loop, moving them through the oil layer and immediately vitrifying them in liquid nitrogen.

3.5 Data Processing and Model Building

1. Diffraction data can be processed using standard programs, e.g., XDS [14]. The space group should be $P2_12_12$ with unit cell axis dimensions of $a = 77.5$ Å, $b = 261.1$ Å, $c = 57.8$ Å, and $\alpha = \beta = \gamma = 90°$.
2. Initially, phases could be solved by molecular replacement with the CCP4 program Phaser [15, 19] using the separately solved structure of the POTRA domains of TamA (PDB entry 4BZA [5]) and the barrel of FhaC (PDB entry 2QDZ [4]).
3. Model building and refinement can be done using the programs COOT [17] and PHENIX [16], respectively. The TamA crystal structure is provided under the PDB entry 4C00 [5].

4 Notes

1. The pH of NaPi buffers can be adjusted at room temperature by mixing Na_2HPO_4 and NaH_2PO_4 stock solutions in appropriate ratios to obtain near-final values and fine-tuning by addition of HCl or NaOH to the buffer solutions containing all ingredients.
2. It is possible that the *pelB* signal in the pET22b vector can also be used for expression and does not need to be replaced by the *malE* signal sequence.
3. After induction of expression and overnight growth, the bacterial pellet appears partially lysed, which has no detectable effect on protein quality.
4. French pressure cell and sonication may be suitable alternative lysis methods.
5. When keeping TamA below 4 °C it may aggregate. This aggregation is reversible upon shifting to 4 °C again.
6. The choice of the IMAC resin material may affect yield and purity.
7. The β-OG concentration of final TamA samples before mixing with bicelles is measured by NMR spectroscopy and may be adjusted to 2.0 %.
8. It is recommended to keep the TamA-bicelle mixture at 4 °C. Neither put it on ice nor warm it up as phase transitions occur, which may affect sample quality. Phase transitions are usually reversible but for a warmed-up sample it may take hours or even days until the solution turns clear again at 4 °C.
9. Seed dilution series and crystal growth testing have to be repeated each time a new seed stock is prepared as the concentration of seeds differs between preparations. For these 2D-grid screens manual setup may be necessary as different protein solutions are used.

References

1. Kim S, Malinverni JC, Sliz P et al (2007) Structure and function of an essential component of the outer membrane protein assembly machine. Science 317(5840):961–964
2. Chacinska A, Koehler CM, Milenkovic D et al (2009) Importing mitochondrial proteins: machineries and mechanisms. Cell 138(4): 628–644
3. Walther DM, Rapaport D, Tommassen J (2009) Biogenesis of beta-barrel membrane proteins in bacteria and eukaryotes: evolutionary conservation and divergence. Cell Mol Life Sci 66(17):2789–2804
4. Clantin B, Delattre AS, Rucktooa P et al (2007) Structure of the membrane protein FhaC: a member of the Omp85-TpsB transporter superfamily. Science 317(5840):957–961
5. Gruss F, Zahringer F, Jakob RP et al (2013) The structural basis of autotransporter translocation by TamA. Nat Struct Mol Biol 20(11): 1318–1320
6. Noinaj N, Kuszak AJ, Gumbart JC et al (2013) Structural insight into the biogenesis of beta-barrel membrane proteins. Nature 501(7467):385–390
7. Delattre AS, Saint N, Clantin B et al (2011) Substrate recognition by the POTRA domains of TpsB transporter FhaC. Mol Microbiol 81(1):99–112
8. Stegmeier JF, Gluck A, Sukumaran S et al (2007) Characterisation of YtfM, a second member of the Omp85 family in *Escherichia coli*. Biol Chem 388(1):37–46
9. Selkrig J, Mosbahi K, Webb CT et al (2012) Discovery of an archetypal protein transport system in bacterial outer membranes. Nat Struct Mol Biol 19(5):506–510
10. Dautin N, Bernstein HD (2007) Protein secretion in gram-negative bacteria via the autotransporter pathway. Annu Rev Microbiol 61:89–112
11. van den Berg B (2010) Crystal structure of a full-length autotransporter. J Mol Biol 396(3): 627–633
12. Prilipov A, Phale PS, Van Gelder P et al (1998) Coupling site-directed mutagenesis with high-level expression: large scale production of mutant porins from *E-coli*. FEMS Microbiol Lett 163(1):65–72
13. Kapust RB, Tozser J, Fox JD et al (2001) Tobacco etch virus protease: mechanism of autolysis and rational design of stable mutants with wild-type catalytic proficiency. Protein Eng 14(12):993–1000
14. Kabsch W (2010) Xds. Acta Crystallogr D 66:125–132
15. Winn MD, Ballard CC, Cowtan KD et al (2011) Overview of the CCP4 suite and current developments. Acta Crystallogr D 67: 235–242
16. Adams PD, Afonine PV, Bunkoczi G et al (2010) PHENIX: a comprehensive Python-based system for macromolecular structure solution. Acta Crystallogr D 66:213–221
17. Emsley P, Cowtan K (2004) Coot: model-building tools for molecular graphics. Acta Crystallogr D 60:2126–2132
18. Ujwal R, Bowie JU (2011) Crystallizing membrane proteins using lipidic bicelles. Methods 55(4):337–341
19. Mccoy AJ, Grosse-Kunstleve RW, Adams PD et al (2007) Phaser crystallographic software. J Appl Crystallogr 40:658–674

Chapter 21

Strategies for the Analysis of Bam Recognition Motifs in Outer Membrane Proteins

Nagarajan Paramasivam and Dirk Linke

Abstract

Well-structured proteins interact with other proteins through surface–surface interactions. In such cases, the residues that form the interacting surface are not necessarily neighboring residues on the level of protein sequence. In contrast, unfolded or partially unfolded proteins can interact with other proteins through defined linear motifs. In the case of the β-barrel assembly machinery (BAM) in the outer membrane of Gram-negative bacteria, unfolded β-barrel proteins are recognized through a C-terminal linear motif, and are inserted into the membrane. While the exact mechanism of recognition is still under investigation, it has been shown that mutations in the recognition motif can partially or completely abolish membrane insertion. In this chapter, we demonstrate the workflow for motif discovery, motif extraction, and motif visualization on the example of the C-terminal motifs in transmembrane β-barrel proteins.

Key words Linear motif, C-terminal motif, Protein–protein interaction, Motif extraction, BAM recognition

1 Introduction

In Gram-negative bacteria, two membranes separate the cytoplasm from the environment: the inner membrane (IM) and outer membrane (OM), with the space between them called the periplasm. Different specialized machineries help to target membrane proteins, but also periplasmic and extracellular proteins to their final destination. In the IM, the major machineries are the Sec translocon, the YidC insertase, and the Tat system [1]. OM proteins (OMPs) typically cross the IM with help of the signal-dependent Sec pathway. OMPs are almost exclusively transmembrane β-barrels, which are inserted into the OM by the β-barrel assembly machinery (BAM) complex [2]. The BAM complex operates without an external energy source such as ATP or ion gradients [3], and consists of an essential transmembrane core component, BamA, and multiple lipoproteins, called BamB to BamE in *E. coli*. While the exact mechanism of OMP membrane insertion and

Susan K. Buchanan and Nicholas Noinaj (eds.), *The BAM Complex: Methods and Protocols*, Methods in Molecular Biology, vol. 1329, DOI 10.1007/978-1-4939-2871-2_21, © Springer Science+Business Media New York 2015

folding is still under investigation, it is clear that recognition of OMPs by the BAM complex depends on a linear motif at the C-terminus of the β-barrel sequence [4]. The C-terminal β-strand of an OMP β-barrel domain typically contains an aromatic residue at its C-terminus. It has been reported that deletion or substitution of this C-terminal residue alone already negatively affects the biogenesis of OMPs [5]. We have recently explored the sequence variation of the C-terminal recognition motifs in β-barrel proteins, and found that there is a significant conservation of sequence features across all Gram-negative species [6], again emphasizing the common evolutionary origin of all transmembrane β-barrels [7, 8]. In this chapter, we review the workflow of motif discovery, analysis, and visualization for short protein motifs on the example of the Bam recognition motif.

2 Materials

This section is a representative collection of useful online resources and tools. In many cases, alternative tools for a single purpose are presented. It is advisable to try different tools in parallel for all individual steps and to compare their results, and where applicable, their error rates.

2.1 General Protein Databases

1. Uniprot [9]: http://www.uniprot.org/, a database of curated protein sequence information and functional information.
2. NCBI protein: http://www.ncbi.nlm.nih.gov/protein, a large collection of protein sequences from different sources.
3. PFAM [10]: http://pfam.sanger.ac.uk, a database of protein families.

2.2 Specific Databases for Membrane Proteins, Including Transmembrane β-Barrel Proteins

1. ClubSub-P [11]: http://toolkit.tuebingen.mpg.de/clubsubp, a clustering-based subcellular localization database of proteins from Gram-negative bacteria and Archaea.
2. PDBTM [12]: http://pdbtm.enzim.hu/, a protein data bank of transmembrane proteins.
3. TMBB-DB [13]: http://beta-barrel.tulane.edu/, a transmembrane β-barrel proteome database.

To extract a motif, it is essential to have a database of protein sequences that are known to contain the motif. For the purpose of C-terminal Bam recognition motifs, we can start with a database of sequences with known subcellular localization (SCL) annotation, ClubSub-P, a SCL annotation database for Gram-negative bacteria and Archaea. Alternatively, more general databases, or manually curated databases as listed above, can be used as a starting point for motif extraction.

In the ClubSub-P database, sequences are clustered based on sequence similarity and a consensus SCL is annotated to the cluster, based on the assumption that highly similar sequences will be sorted to the same subcellular localization. As the cluster's SCL is based on a majority vote, this database heavily reduces the false-positive and false-negative errors which can arise in SCL predictions of single sequences. The subcellular localization at protein level is a consensus prediction from different tools available to predict each step in the biological protein sorting. This includes signal peptide prediction, which is the initial step in protein sorting mechanism. Note that the proteins targeted for the outer membrane in Gram-negative bacteria *must* possess a signal peptide to cross the inner membrane.

In ClubSub-P, OMPs are identified by the presence of the general signal peptide (see below) and by a predicted transmembrane β-barrel structure, based internally on individual predictions with tools such as HHomp, CELLO, and PSORTb.

2.3 General Tools for Secondary Structure Prediction and Other Sequence Parameters

1. Secondary structure prediction tools:
   (a) HMMTOP [14]: http://www.enzim.hu/hmmtop/, a tool to find transmembrane segments.
   (b) Phobius [15]: http://phobius.sbc.su.se/, a combined transmembrane and signal peptide prediction tool.
   (c) TMHMM [16]: http://www.cbs.dtu.dk/services/TMHMM/, a tool to predict transmembrane helices in proteins.
   (d) PSIPRED [17]: http://bioinf.cs.ucl.ac.uk/psipred/, a secondary structure prediction program based on PSI-BLAST.
2. Signal peptide prediction:
   (a) SignalP [18]: http://www.cbs.dtu.dk/services/SignalP/, a general signal peptide prediction tool.
   (b) TatP [19]: http://www.cbs.dtu.dk/services/TatP/, a twin arginine translocation signal peptide prediction tool.
   (c) LipoP [20]: http://www.cbs.dtu.dk/services/LipoP/, a lipoprotein signal peptide prediction tool.

2.4 Specific Tools for Transmembrane β-Barrel Protein Detection and Analysis

1. HHomp [21], http://toolkit.tuebingen.mpg.de/hhomp, a tool for homology-based detection of OMPs using HMM–HMM comparison.
2. PRED-TMBB [22]: http://bioinformatics.biol.uoa.gr/PRED-TMBB/, a HMM-based method to predict transmembrane beta-strands in outer membrane proteins.
3. BOCTOPUS [23]: http://boctopus.cbr.su.se/, a tool to predict transmembrane β-barrel topology.
4. ProfTMB [24]: https://rostlab.org/owiki/index.php/Proftmb, a transmembrane β-barrel secondary structure prediction tool.

2.5 Tools for Linear Motif Detection

1. MEME/MAST [25]: http://meme.nbcr.net/meme/cgi-bin/meme.cgi, a program suite for online bioinformatics motif extraction, detection, and alignment.
2. GLAM2 [26]: http://meme.nbcr.net/meme/cgi-bin/glam2.cgi, a gapped sequence motif discovery tool.

2.6 Tools for Linear Motif Visualization

1. WebLogo [27]: http://weblogo.berkeley.edu/, an online tool to generate motif logos from the aligned motif sequences.

3 Methods

3.1 Getting Started: Obtaining a Set of OMP Sequences

1. The C-terminal motif or the β-barrel domain is not always located at the C-terminus of the protein—this can interfere with automated motif extraction (*see* **Note 1**).
2. To show the procedure with a simple test set of sequences, we will start with an alignment of β-barrel proteins based on a seed sequence that has the β-barrel domain at the C-terminus, so that the motif is also at the C-terminal end of the protein and easier to extract.
3. We can use difference resources like HHomp, PFAM to extract seed sequences or complete pre-computed alignments. As an example, we use the *E. coli* β-barrel protein OmpW, a small protein of known structure [28, 29] which has the recognition motif at its C-terminal end.
   (a) Go to the HHomp database at http://toolkit.tuebingen.mpg.de/hhomp.
      - Click "browse database" and "view table overview of clusters," for OmpW, select the cluster OMP.8.2.1 in the drop-down menu.
      - Notice the secondary structure prediction by PsiPred [17] and β-barrel configuration by ProfTMB [24] in the Clustal alignment tracks as ss_pred and bb_conf respectively.
      - Select any RefSeq sequence identifier from the cluster, which will be our seed sequence (e.g. 110834783).

   (b) Alternatively, go to PFAM at http://pfam.sanger.ac.uk/.
      - In the keyword search tab, search for "OmpW."
      - Select the sequence family with OmpW ID or PF03922.
      - Under the alignment tab, select view seed sequence via HTML.
      - Select one of the UniProt sequence identifiers and proceed to the next "obtaining OMP cluster" step or select all the seed sequences and skip the next step.

- Readers can also notice the secondary structure annotation in the seed alignment.

3.2 Obtaining an OMP Cluster

1. Use the above seed sequence identifier to obtain the sequence in FASTA format from databases like NCBI-Protein or UniProt.
2. Now BLAST the seed sequence against different databases like ClubSub-P, UniProt or NCBI-BLAST to obtain the OMP sequence cluster.
   (a) Go to UniProt at http://www.uniprot.org/.
      - Select the BLAST tab and enter the seed sequence, use the default settings for BLAST.
      - Once the BLAST is done, use the "customize" tab to increase the rows to maximum 250 and randomly select a minimum of 15 sequences with varying similarities to the "Query." Retrieve the above selected sequence in FASTA format.

   (b) Alternatively, go to ClubSub-P at http://toolkit.tuebingen.mpg.de/clubsubp.
      - Enter the seed sequence and BLAST it against gram-negative-607 database.
      - Once the BLAST is done, select the "sequence cluster" to which the seed sequence belongs and retrieve a minimum of 15 sequences randomly.

3.3 Discovering the Motif

1. Copy the sequences files to notepad or other text editor.
2. Since we are interested in the C-terminal motif, splice the sequences by keeping only the last line in each of the FASTA sequences. Make sure we have more than ten amino acids left in each sequence.
   (a) When dealing with a large-scale dataset, users have to automate this procedure using their favorite scripting language to extract the sub-sequences around the last predicted beta-sheet of the OMPs (GLAM2 only allows a limited number of sequences to check at the same time).
3. We will now use GLAM2 for extracting the motifs from the above sequence cluster. Go to the GLAM2 service in the MEME/MAST suite, at http://meme.nbcr.net/meme/cgi-bin/glam2.cgi.
   (a) Enter the sequences and provide your e-mail address.
   (b) Since the C-terminal motifs are approx. 10 AAs in length, enter the minimum and maximum number of aligned columns and also the initial number of aligned columns as 10.
   (c) Once the GLAM2 run is finished, the output html or text file will display the motif discovered, and will provide a list

based on the obtained scores. The motif discovered at the C-terminal end of the sub-sequences is the C-terminal motif.

(d) Download the GLAM2.txt output to the local system.

3.4 Generating the Motif or Sequence Logo

1. The GLAM2 html output file already contains a motif logo, but it does not provide the flexibility to manipulate the final logos.
2. Instead, we can use WebLogo at http://weblogo.berkeley.edu/logo.cgi to generate a new logo using the motif discovered in the previous step. Open the GLAM2.txt with excel and using "space" as delimiter, separate the motif results into different columns. This will extract the motif sequences into a separate column and can be easily used for further analysis.
3. Enter the motif sequences in the WebLogo service and manipulate the final image with advanced and color options. This will help you to generate publication quality motif logos. The size and length of the amino acids in the sequence logo represent the enrichment of these residues in the sequence motif. We can use different colors to highlight the physical and chemical groups to which each of the residues belongs.

4 Notes

1. Since the β-barrel may be found at/near the N-terminus or C-terminus of the protein, you cannot always just use the last ten amino acids from the protein.

Acknowledgements

The basis of this work was funded by a Grand Challenges Explorations grant from the Bill & Melinda Gates Foundation. Additional funding by the Max Planck Society and by the Collaborative Research Center SFB766 is gratefully acknowledged.

References

1. Kudva R, Denks K, Kuhn P et al (2013) Protein translocation across the inner membrane of Gram-negative bacteria: the Sec and Tat dependent protein transport pathways. Res Microbiol 164(6):505–534
2. Rigel NW, Silhavy TJ (2012) Making a beta-barrel: assembly of outer membrane proteins in Gram-negative bacteria. Curr Opin Microbiol 15(2):189–193
3. Hagan CL, Kim S, Kahne D (2010) Reconstitution of outer membrane protein assembly from purified components. Science 328(5980):890–892
4. Robert V, Volokhina EB, Senf F et al (2006) Assembly factor Omp85 recognizes its outer membrane protein substrates by a species-specific C-terminal motif. PLoS Biol 4(11): e377

5. Struyvé M, Moons M, Tommassen J (1991) Carboxy-terminal phenylalanine is essential for the correct assembly of a bacterial outer membrane protein. J Mol Biol 218(1):141–148
6. Paramasivam N, Habeck M, Linke D (2012) Is the C-terminal insertional signal in Gram-negative bacterial outer membrane proteins species-specific or not? BMC Genomics 13:510
7. Remmert M, Biegert A, Linke D et al (2010) Evolution of outer membrane beta-barrels from an ancestral beta beta hairpin. Mol Biol Evol 27:1348–1358
8. Arnold T, Poynor M, Nussberger S et al (2007) Gene duplication of the eight-stranded beta-barrel OmpX produces a functional pore: a scenario for the evolution of transmembrane beta-barrels. J Mol Biol 366(4):1174–1184
9. UniProt-Consortium (2010) The Universal Protein Resource (UniProt) in 2010. Nucleic Acids Res 38:D142–D148
10. Finn RD, Bateman A, Clements J et al (2014) Pfam: the protein families database. Nucleic Acids Res 42(Database issue):D222–D230
11. Paramasivam N, Linke D (2011) ClubSub-P: cluster-based subcellular localization prediction for Gram-negative bacteria and Archaea. Front Microbiol 2:1–14
12. Kozma D, Simon I, Tusnady GE (2013) PDBTM: protein data bank of transmembrane proteins after 8 years. Nucleic Acids Res 41(Database issue):D524–D529
13. Freeman TC Jr, Wimley WC (2012) TMBB-DB: a transmembrane beta-barrel proteome database. Bioinformatics 28(19):2425–2430
14. Tusnady GE, Simon I (2001) The HMMTOP transmembrane topology prediction server. Bioinformatics 17(9):849–850
15. Kall L, Krogh A, Sonnhammer ELL (2007) Advantages of combined transmembrane topology and signal peptide prediction–the Phobius web server. Nucleic Acids Res 35: W429–W432
16. Krogh A, Larsson B, von Heijne G et al (2001) Predicting transmembrane protein topology with a hidden markov model: application to complete genomes. J Mol Biol 305:567–580
17. Buchan DW, Minneci F, Nugent TC et al (2013) Scalable web services for the PSIPRED Protein Analysis Workbench. Nucleic Acids Res 41(Web Server issue):W349–W357
18. Petersen TN, Brunak S, von Heijne G et al (2011) SignalP 4.0: discriminating signal peptides from transmembrane regions. Nat Methods 8(10):785–786
19. Bendtsen JD, Nielsen H, Widdick D et al (2005) Prediction of twin-arginine signal peptides. BMC Bioinformatics 6:167
20. Juncker AS, Willenbrock H, Von Heijne G et al (2003) Prediction of lipoprotein signal peptides in gram-negative bacteria. Protein Sci 12(8):1652–1662
21. Remmert M, Linke D, Lupas AN et al (2009) HHomp–prediction and classification of outer membrane proteins. Nucleic Acids Res 37:W446–W451
22. Bagos PG, Liakopoulos TD, Spyropoulos IC et al (2004) PRED-TMBB: a web server for predicting the topology of beta-barrel outer membrane proteins. Nucleic Acids Res 32(Web Server issue):W400–W404
23. Hayat S, Elofsson A (2012) BOCTOPUS: improved topology prediction of transmembrane beta barrel proteins. Bioinformatics 28(4):516–522
24. Bigelow H, Rost B (2006) PROFtmb: a web server for predicting bacterial transmembrane beta barrel proteins. Nucleic Acids Res 34(Web Server issue):W186–W188
25. Bailey TL, Boden M, Buske FA et al (2009) MEME SUITE: tools for motif discovery and searching. Nucleic Acids Res 37(Web Server issue):W202–W208
26. Frith MC, Saunders NF, Kobe B et al (2008) Discovering sequence motifs with arbitrary insertions and deletions. PLoS Comput Biol 4(4):e1000071
27. Crooks GE, Hon G, Chandonia JM et al (2004) WebLogo: a sequence logo generator. Genome Res 14(6):1188–1190
28. Albrecht R, Zeth K, Soding J et al (2006) Expression, crystallization and preliminary X-ray crystallographic studies of the outer membrane protein OmpW from *Escherichia coli*. Acta Crystallogr Sect F Struct Biol Cryst Commun 62(Pt 4):415–418
29. Hong H, Patel DR, Tamm LK et al (2006) The outer membrane protein OmpW forms an eight-stranded beta-barrel with a hydrophobic channel. J Biol Chem 281(11): 7568–7577

Chapter 22

Summary and Future Directions

Nicholas Noinaj and Susan K. Buchanan

Abstract

β-barrel outer membrane proteins (OMPs) are found in the outer membranes (OMs) of all gram-negative bacteria, yet exactly how they are folded and inserted remains unknown. The last decade has provided a wealth of discovery including the identification of the BAM complex, a multicomponent complex responsible for the biogenesis of all OMPs into the OM. It is anticipated that the next decade will further advance our knowledge of how the BAM complex is able to perform its unique and interesting function.

Key words BamA, OMP, Outer membrane protein, β-barrel membrane protein, BAM complex, Protein folding, Lateral opening

The BAM complex serves an essential role in gram-negative bacteria by orchestrating the biogenesis of all β-barrel membrane proteins found within the outer membrane. The last decade has witnessed not only the discovery of the BAM complex itself, but an experimental tour de force in dissecting the individual components and how those components come together to function as a molecular machine. Functional assays, both in vitro and in vivo, along with clever genetics experiments have given clues to how each Bam component interacts with the others, the role each may play within the full complex, and how each may be regulated with the cell. The structures of all the Bam components have also contributed significantly and together have provided a molecular blueprint to serve as a guide in steering all on-going and future experiments aimed at understanding exactly how the BAM complex folds and inserts new β-barrel membrane proteins into the outer membrane.

Significant work has been done, yet significant work remains to be done in order to fully understand the mechanism for how the BAM complex functions. The last decade saw a flurry of exciting findings that have provided the foundation for what we anticipate will be an even more enlightening next decade of discovery. At the forefront is the importance of determining structures of the fully assembled complex, as well as, structures of subcomplexes, both in

Susan K. Buchanan and Nicholas Noinaj (eds.), *The BAM Complex: Methods and Protocols*, Methods in Molecular Biology, vol. 1329, DOI 10.1007/978-1-4939-2871-2_22, © Springer Science+Business Media New York 2015

apo form and with substrate. The power of structural biology in providing molecular insight and guiding functional experiments remains unparalleled and therefore knowing the structure of the fully assembled complex would be invaluable in unraveling how it works. Further, an improved in vitro functional assay that can yield information about how the BAM complex sorts and handles the various sizes and properties of new β-barrel membrane proteins would be necessary to verify and validate all current and future mechanistic models. And lastly, the therapeutic potential of the BAM complex and its components both as vaccine targets and as targets for antibiotic development, particularly against multidrug-resistant gram-negative bacteria, should be explored in more depth toward efforts to discover new drugs that are more effective and sustainable. Multidrug-resistant bacteria are an immediate threat to the world that requires an immediate response from the research community.

This volume presents 20 chapters of experimental protocols based on methods that have been used to study the BAM complex. It was these methods that provided us with the wealth of knowledge we now know about this system and while new technology will bring new methods that will further assist in studying the BAM complex, it will be these methods and protocols presented here that will continue to be used in making novel discoveries about this fascinating system.

ERRATUM TO

Chapter 5
The Role of a Destabilized Membrane for OMP Insertion

Ashlee M. Plummer, Dennis Gessmann, and Karen G. Fleming

Susan K. Buchanan and Nicholas Noinaj (eds.), *The BAM Complex: Methods and Protocols*, Methods in Molecular Biology, vol. 1329, DOI 10.1007/978-1-4939-2871-2_5, © Springer Science+Business Media New York 2015

DOI 10.1007/978-1-4939-2871-2_23

The publisher regrets that in the print and online versions of this book, Dennis Gessmann should have been listed as contributing equally to chapter 5 along with Ashlee M. Plummer.

The updated original online version for this chapter can be found http://dx.doi.org/10.1007/978-1-4939-2871-2_5

Susan K. Buchanan and Nicholas Noinaj (eds.), *The BAM Complex: Methods and Protocols*, Methods in Molecular Biology, vol. 1329, DOI 10.1007/978-1-4939-2871-2_23, © Springer Science+Business Media New York 2016

Index

Susan K. Buchanan and Nicholas Noinaj (eds.), *The BAM Complex: Methods and Protocols*, Methods in Molecular Biology, vol. 1329, DOI 10.1007/978-1-4939-2871-2, © Springer Science+Business Media New York 2015

I

K

L

M

N

O

P

R

S

T

V

W

X

Y

MIX
Papier aus verantwortungsvollen Quellen
Paper from responsible sources
FSC® C105338

If you have any concerns about our products,
you can contact us on
ProductSafety@springernature.com

In case Publisher is established outside the EU,
the EU authorized representative is:
Springer Nature Customer Service Center GmbH
Europaplatz 3, 69115 Heidelberg, Germany

Printed by Libri Plureos GmbH
in Hamburg, Germany